中国盐文化研究中心重点科研项目（编号：YWHZ21-02）成果

Inheritance and Innovation

Yanjing Salt Culture Protection and Its Overseas Translation

Yang Wei

NEW VISION PRESS

Printed in the United States of America
Cover design by Brent Yan

Published by New Vision Press
Las Vegas, Nevada, U.S.A.
http://www.usnvp.com/

Printed in the United States of America
9 8 7 6 5 4 3 2 1

First Printing: December 30, 2024
Words count: 175,330
ISBN: 978-1-963469-00-4

Preface

Salt, a tiny salty white granular matter, has always been known as the "initiator of all kinds of flavors", "the general of food", and "the great treasure of the country". It is an indispensable and important element for the human body to maintain normal physiological activities, and has always accompanied human development. From the proverb "The normal operation of a family consists of seven things, i.e. firewood, rice, cooking oil, salt, soy sauce, vinegar and tea", it can be seen that it is not only a necessity in people's daily life, but also a substance that is synchronized with the formation and development of traditional national culture and closely linked with human social life. From the perspective of comprehensive analysis of the factors of national social development, historical archaeologists believe that convenient conditions for salt supply must have existed in any national cultural birthplace. Some people say that when human beings walked out of caves and towards civilization, salt was an insurmountable bottom line of life, which shows the irreplaceable position and value of salt in the entire process of human civilization evolution and transformation.

We are used to calling natural salt "bittern" and processed brine "salt". *Principles of Chinese Character Composition* records, "Salt and bittern are one and the same thing, the natural form being white brine, while the processed being white salt". The discovery and initial use of natural salt (bittern) by humans, like the animals' licking halite and drinking saline water, was often out of physiological instincts. Such stories as "white deer drinking spring", "cattle licking the ground for salt", "monkeys licking the ground", and "sheep licking the soil" have been passed down from ancient China, and the Kanawha Licks in Virginia, North America, has always been a place where ancient Indians and bisons and other animals frequented, etc.; all illustrate this point. [1] As time went by, people were no longer satisfied with the natural salt bestowed by nature, and began to explore ways to make salt from sea water, lake water, halite, and saline soil. "The source of salt is either the sea, the lakes, the wells, or the mountains...As a result, sea salt, lake salt, well salt and rock salt are produced."[2]. Human salt production has since opened a new chapter.

With a long history of salt industry, China is one of the first countries in the world to develop and utilize salt resources. As far back as the Paleolithic Age, our ancestors learned to "flavor the food with salt"; as to salt production, it can be traced back to the Neolithic Age about 6000 years ago. China's natural salt industry resources, which are very rich and diverse, spread throughout the territory; there are sea salt along the southeast coast, well salt in the southwest region, rock salt in the northwest region, and lake salt in the central plains. A history of Chinese culture and civilization is one of searching for using salt along the water. The development and utilization of salt resources have resonated in tandem with the pace of Chinese social civilization. The early human settlement and the formation of cities were inseperable from salt and its emotional appeal. For example, early cultural relics in the Yangtze River basin and the North are gathered around natural saline lakes and springs, and Yao, Shun and Yu all built their capitals around Xie Lake (which, located in Yuncheng City in the southwest of Shanxi Province, is the most famous lake salt producing area in China). What is more, the origin of the southwest "Land of Abundance" was also closely related to the exploitation and utilization of salt and iron in Sichuan. Besides, early mankind wars also originated from the fight for and control of the special resource salt, and it was in the fight for lake salt in Shanxi that Emperor Yan and Emperor Huang started the "first war of China". Since the beginning of the class society, salt tax had become one of the three pillars of the state's fiscal revenue, thus emerging a prosperous situation "salt accounts for a half of the taxes throughout the state"; in the feudal society, salt industry and salt profits became important factors directly affecting social stability and the security of the regime, and it was nothing but the exhaustion of salt profits that led to the decline of the powerful Tang Empire. In modern society, salt still ranks among the five basic

1 Li Xingkui's *Zhang Salt--the Earliest Manually Processed Salt in China.*

2 Zeng Yangfeng's *History of Chinese Salt Administration*, Beijing: Commercial Press, 1984: 49.

industrial raw materials (salt, iron, petroleum, limestone, and sulfur), and its output, consumption and usages are one of the indicators reflecting a country's industrial development level. And salt products, of more than 14,000 kinds, is an irreplaceable and valuable resource for people's survival and development. Even today, in the 21st century, the role of salt in the survival and health of human beings and the progress and development of our economic society are not overshadowed by the highly developed science and technology and the unimpeded advancement of information; instead, it is sparkling in all its glory, because its profound influence and important part in many fields such as politics, economy, science and technology, military affairs, law, education, health service, food and drink, tourism, religion, literature and art, ethnic folklore, urban development, ecological environment, etc., as well as its rich and unique cultural characteristics and connotations are being increasingly recognized, understood and explored. [1]

Our ancestors recognized the importance of salt long long ago. In *Guan Zi*, the pathogenic theory of "edema induced by salt-free diet" was originally proposed, and in *Salt Manufacturing of Heavenly Creations* it was recorded "Not eating salt for ten days, a man will be as lethargic and weak as if he were seriously ill, and will not even be able to catch a chicken." Salt is not only edible, but can also "implement the policy of emphasizing agriculture and suppressing commerce, combat local divisions and separations, put an end to extravagance, and eliminate the possibility of any merger". As a very important national strategic material, salt has been under strict control of the supreme ruler throughout the ages. For this reason, sinologists have their own unique perspective on the interpretation of salt. The traditional Chinese character "鹽", which means "salt", is composed of four characters, i.e., "臣" (minister), "人" (manpower), "卤" (bittern), and "皿" (vessel), graphically reflecting the monopoly of salt by the ancient Chinese regime. What's more, salt was also an crucial item for national sacrifices. Until the Qing Dynasty, salt had still been regarded as one of the important ritual contents in national ceremonies. (*Legend of Sunshine and Wind in Yanjing, Tibet*, by Li Zhongdong and Jiang Hai, Resources and Human Settlements, July 2015) People value salt and believe that it plays an

1 *Salt Flavor and Charm: A Study of Well Salt Culture and Salt City Writers*, by Wang Yu, Li Shumin, and Wang Xiaoping.

indispensable role in their lives. They would often say: “I have eaten more salt than you have walked”, or “I have eaten 100 grams more salt than you.” They measure the course of life and describe life experience with salt, which highlights the status of salt in their minds. They even integrated salt flavor into their philosophy of life by saying “Eat salty food and treat everything bland (inessential)”, which is another understanding of salt.

In its long production practice, the breadth and depth of salt’s involvement in many fields of human life, as well as the incredible magic it displayed, has inspired us to examine salt as a human cultural phenomenon with rich connotations. As for salt culture, this concept has not emerged for a long time. It was put forward by some experts and scholars in the 1980s, and its connotation was later perfected and supplemented. According to Professor Zeng Fanying, its research objects include three aspects, i.e., material culture, institutional culture and spiritual culture, and its research content covers more than such ten disciplines as politics, economics, history, culturology and folklore.[1] Salt culture has gradually developed from a single food culture or industrial product culture into a rich and diverse cultural system. In other words, by covering both tangible and intangible content, salt culture has transformed from various phenomena and manifestations into a symbol of profound content and spirit.

In the past half century, scholars at home and abroad have made unremitting explorations on the exploitation and utilization of salt resources, pushing the research on the history of China’s salt industry--especially the economic history of salt industry and the history of science and technology in salt industry--to a new peak, and achieved fruitful results, which has significantly raised the academic status of China’s salt industry research and greatly enhanced its influence. But in the meantime, the study of Chinese salt culture also reflects the problems of unbalanced research scope and content. Many researches on salt culture are concentrated in major salterns in such salt zones as Changlu County, Hedong (South of Yuncheng City), Shandong Province, Lianghuai (Huainan and Huaibei regions), Liangguang (Guangdong and Guangxi provinces), Sichuan Province, Yunnan Province, etc.,

1 Zeng Fanying. *The Connotation and Characteristics of Salt Culture*. Journal of Sichuan University of Science & Engineering (Social Sciences Edition), 2006(1):47-52.

but the salt culture of Tibet Autonomous Region has never been a research hotspot. For example, in the thick book *History of Chinas Salt Industry: the Ancient Part*, which spanned from ancient times to the Qing Dynasty with detailed historical data and quotations, the development of salt industry in such western provinces (or municipality) as Sichuan, Chongqing, Yunnan and Guizhou is mentioned occasionally, whereas the history of salt industry in Tibet is completely ignored.

In fact, Tibet, the mysterious snowy plateau, has always been one of China's important salt-producing regions, too. Tibetan people started to utilize and exploit salt very early (such records could be found throughout the entire Tibet Autonomous Region), and Tibetan salt culture, especially the precious salt cultural heritage created by the people from Yanjing, Markam County, constitutes an important part of China's "salt culture". As early as the Tang Dynasty, poeple already learned to make salt by evaporating brine in the ancient Yanjing salt pans under the the sunshine, and this skill survived to this day after more than 1,500 years. It intensively shows the inheritance and development of local handicraft production and its tools, salt making methods and sites. In particular, as the only primitive salt making method remaining in the world now, the technique of salt making here is of great historical and cultural value, occupying a glorious page in the history of well salt culture in Tibet. In December 2012, Yanjing of Markam County in Tibet was officially included in the "Preliminary list of China's World Cultural Heritage sites", ending the lack of nominations for Chinese salt culture world heritage. The most unique and highly technical method of evaporating brine under the sunshine, which has lasted for thousands of years, is engraved with the ancestors' legends of wisdom, and serves as an invaluable historical evidence for the study of the economic and cultural exchanges of various ethnic groups along the "The Ancient Tea-Horse Road" for more than a thousand years. Its development and primitive salt making technique are known as a great wonder in the history of human salt civilization, which not only has profound regional cultural significance, but also plays an important role in the process of Chinese cultural history. However, due to the lack and fragmentation of historical data and the inconvenience caused by its geographical location, Yanjing salt culture has been systematically discussed by almost no one at home and abroad except for a few scholars, thus becoming a gap in the study of Tibetology and the history of Chinese salt industry. Therefore, this book attempts to fill the gap by making an exhaustive search of Han, Tibetan and foreign historical data, in addition to conducting a field investigation of Yanjing area in Markam County. And through the introduction of the profound historical origin and unique production technique there, as well as the distinct and exceedingly fascinating "salt culture" phenomena and images bred and derived from the long historical process, it tries to uncover the mystery of Yanjing salt culture.

At the same time, this book will also analyze the dilemma and problems in the development of Yanjing salt culture industry, and put forward some ideas on the protection, inheritance, and development of Yanjing salt culture--especially its translation and dissemination overseas--in the new era. On the one hand, we hope to carry forward the splendid culture of the Chinese nation and, by summing historical experience, promote the reform and development of current salt industry. On the other hand, we also hope that this book would contribute to the study of Chinese salt history and economic history, to the cultivation of people's patriotism, and to the study of Tibetan salt industry culture by Chinese and foreign scholars. Salt culture is a very extensive and profound cultural phenomenon. When writing, the authors consulted a large number of ancient and modern historical literature. Without the research findings of many experts and scholars as a solid foundation, it is inconceivable that we could have finished such a challenging task. Here, I would like to express my special thanks to Professor Li Hechun from Yunnan Minzu University. He is a very authoritative expert on Yanjing salt cultur, with rigorous scholarship and a very modest personality. In the preparation of this book, he generously offered to send his own research monograph, providing me with a great deal of important and reliable fieldwork data. Moreover, his high-quality academic papers have greatly broadened my research horizon and stimulated my way of thinking. Due to the tight schedule of our research project, the harsh field investigation environment, extremely inconvenient transportation, and the impact of the haunting COVID-19 pandemic, we have encountered many unexpected difficulties and challenges throughout the whole process; so, the shortcomings and mistakes in the book are simply inevitable. The author sincerely asks experts and scholars to offer their kind advice and earnestly welcome readers' comments.

Contents

CHAPTER ONE

A Marvelous Place for Salt Industry in the Lantsang River Canyon

Tibetan salt can be divided into three types: lake salt, rock salt and well salt, which are abundant in reserves. Either in terms of geographic products or culture, Markam County in Qamdo Prefecture, Tibet is a very unique and mysterious place. Located at the junction of Tibet, Sichuan and Yunnan provinces (region) to the eastern boundary of Tibet, where Hengduan Mountains stand tall and imposing and the three rivers (Chin-sha River, Lantsang River and Nujiang River) run in parallel from north to south, it boasts breathtaking and secluded scenery, with vast grasslands, fertile fields, rich products, abundant water resources, lush forests, as well as the brilliant culture of the Tibetan people. Yet, Yanjing Township in the southeast part, as an earliest salt production area in China and one of the salt production areas with the highest altitude and the harshest natural environment in the world, is even more well-known.

Yanjing, Tibetan name "tshwa-khron" (meaning "the land of salt making"), earned the name for salt production here. In June, the sixtieth year of the Kangxi period (1721), Du Changding, an aide to the governor of Yungui (Yunnan and Guizhou provinces), went to Yanjing from Adunzi (present-day Deqin County), Yunnan Province. He mentioned "From now on, after traveling northward through Yanjing, it leads to Little Thindu and Great Thindu within a few days; the only route for mass mobilization of Yunnan and Sichuan troops, with Lantsang to the west."[1] This is the earliest record of the Chinese name "Yanjing" so far. Yanjing

1 Wu Fengpei. *A Compilation of Sichuan-Tibet Travel Records: Journey to Tibet*. Chengdu: Sichuan Ethnic Publishing House, 1985, p. 46.

area, once under the control of Batang chieftain, was converted to a county during the bureaucratization of native officers in the late Qing Dynasty. In 1959, Yanjing merged with Ningjing County. In 1965, with the approval of the Central People's Government, the Tibet Autonomous Region renamed Ningjing County as Markam County, under the jurisdiction of which Yanjing became a region and now known as Yanjing Nakhi Ethnic Township.

1. Natural Conditions for Salt Making

Owing to the magic power of the Creator's hands, each place has its own way of supporting its own inhabitants. For Lantsang River, only this section produces bittern. High-class bittern, dry and hot air, and strong sunlight in the Lantsang River canyon, have created a treasure land of Yanjing.

1) Geography, Geology, and Climate

a. Geographical Location

Yanjing Township, the east longitude of 98°34'~98°39' and the north latitude of 28°59'~29°02', is located within the hinterland of Hengduan Mountains in the southeast of Qinghai-Tibet Plateau, with rugged terrain and many high hills and valleys. Adjacent to Batang of Sichuan to the northeast, bordered by Deqin of Yunnan in the south, and joined with Zhayu, Bitu, and Menkong in Zogang County of Tibet in the west, it is 1,336 kilometers away from Lhasa, the capital of the Tibet Autonomous Region, 550 kilometers away from Qamdo, 290 kilometers away from Shangri-La County, Diqing Prefecture of Yunnan Province, only 110 kilometers away from Deqin county town (Shengping Town) of Yunnan Province--about 3 hours by car, and only 10 kilometers away from the junction (the border river) of Yunnan and Tibet.

Yanjing, whose geographical location is relatively special, has always been an important strategic pivot. As the gateway through which the ancient ethnic groups in eastern Tibet migrated and the Tibetans exchanged with other ethnic groups afterwards, it has been not only the first stop from Yunnan into Tibet, but also an important trading spot on The Ancient Tea-Horse Road. Currently, Yanjing is more easily accessible. The National Highway 214 directly passes through the part where the township government is situated, connecting Kunming, Shangri-La, Lijiang, Dali in Yunnan with such important cities and towns as Markam, Zogang, Qamdo, Nyingchi and Lhasa in Tibet.

Yanjing, within a canyon area in the Three Rivers Basin, where the Hengduan Mountains and the Himalayas intersect, presents a geomorphic landscape of "one river between two mountains, and one mountain between two rivers". The average altitude in this region is about 2,500 meters, and the major mountains include Serenity Mountains, Nushan Mountain, etc., the highest peak Mount Pembori being 5,084 meters above sea level. All the mountains and rivers here are north-south oriented, with the rivers divided in between. The Lantsang River canyon, as if split by a giant axe, formed a relative height difference of thousands of meters. The riverbed is the very bottom of the canyon, and there are many rapids and dangerous shoals. At the top of the canyon, the snow-capped mountains look like crowns, and the part beneath them appear to be mahogany brown. The terrace covered with bluish yellow highland barley, dark green fruit trees, and white Tibetan houses, like a jigsaw, are embedded between the mountains and rivers. Down in the canyon, the Lantsang River rushes all the way through the township from north to south, forming a small bend in the shape of an "S". And salt wells and salt pans of all sizes are hidden in this bend which stretches almost 2 kilometers.

Fig. 1-2 the S-shaped canyon in Yanjing (satellite overhead view)

Figure 1-3 the S-shaped canyon in Yanjing (realistic head-up view)

Due to the narrow river canyon and small hirst area, the Nakhi villages in Yanjing are distributed on a stretch of flat hilltops and hillsides, with a drop height of 340 meters from the surface of the Lantsang River. The river surges away from the foot of these surrounding hills, and on the other side Tibetan villages, as part of the upper Yanjing district, are thickly dotted. In terms of topography, both banks are high and precipitous mountains; the east bank is very steep, whereas the west side slopes comparatively gently. Covering an area of 11,654 square kilometers, the township has jurisdiction over four administrative villages: Jiada, Nakhi (also known as Lower Yanjing), Upper Yanjing, and Jiujia, of which Jiada Village is on the west bank, while Nakhi Village, Upper Yanjing Village and Jiujia Village are on the east bank, with a suspension bridge connecting the salt pans on both sides. Although it is a Nakhi Autonomous Township, Tibetans make up the majority of its population. As water sources are abundant here, people channel the mountain streams into numerous canals, which flow around the houses, through the farmland, and finally into the Lantsang River. In Yanjing Township, the earth streets are fairly wide, with low-rise shops lined on both sides, and there is an open square in the central area, where the horse caravans may stop--apparently the relics of ancient stations.

Some scholars believe that the distinction between civilizations in Chinese society is affected by the topography, terrain and landform distribution on Chinese Mainland. According to the terrain, Chinese Mainland can be divided into three levels from west to east. The first level mainly covers most part of the Qinghai-Tibet Plateau in western China, whose prominent feature is the overall high altitude of more than 4,000 meters; the second covers the margin of the Qinghai-Tibet Plateau, which is characterized by a large number of highlands and basins, with an altitude rapidly dropping to about 1,500 meters; the third covers the eastern part of China, which is scattered with hills and plains, with an altitude below 500 meters.[1] According to this three-level classification, Yanjing lies right at the transition zone between the first and second levels. It differs from both the first level with high-altitude mountain ranges and the second level with vast basins. The altitude here is between 2,000 and 3,500 meters. Therefore, either from the perspective of

---

1 See *Overall Scarcity and Cultural Adaptation: Pacuo, Red Sect and Folklore in Sanyan* co-authored by Xu Shaoming and He Guoqiang, Sun Yat-sen University Press, 2013, p. 4.

geographical environment or humanistic history, Yanjing has its own uniqueness.

Surprisingly, it is in such a narrow space that human beings actually use the accumulated bits of land left by the surge of floods to sow seeds, cultivate, live and mulitiply. What's more, they dig salt wells, put up salt pans, and make salt by drawing and carrying bittern bestowed by nature and evaporating it under the sunlight. Thus, the legend of human and nature continues to this day like the inexhaustible Lantsang River.

b. Geology and Climate

The well salt making in Yanjing is closely related to its local geological structure and climatic conditions.

The Lantsang River, when flowing through the Yanjing canyon, deeply cuts into the Triassic red bed, resulting in steep slopes of about 30°. The upper part is a relatively gentle "V" shaped gorge, whereas the lower part is a precipitous narrow gorge. As these slopes are quite unstable, collapses, landslides, and debris flows are common occurrences here. At the confluences of tributary gullies, landslides and debris flows gradually formed valley floors after being washed by the brownish red water of the Lantsang River. These valley floors are fan-shaped, known in geology as alluvial fans or diluvial fans. Villages on the valley floors are divided into Upper Yanjing and Lower Yanjing, and beautiful villages and farmlands are situated on such "fanlike planes". Taken a bird's eye view, such fans are mostly scattered in irregular strip shape, on both sides of the tributary gullies along the Lantsang River. They differ in height from each other by 50 meters, of which the terrace in Lower Yanjing is comparatively larger, with an area of 5 square kilometers, and its front edge is 300 meters above the river surface.

Arable land is scarce in this region, but several rivers run across more than one hundred ore zones, of which salt resource is very rich. Geological surveys show that 200 million years ago, the Qinghai-Tibet Plateau was originally a vast ocean, with the Indian plate pressing against the Eurasian plate at a rate of 45 millimeters per year. This press resulted in the most spectacular orogeny on the Earth. After more than 20 million years, the seabed was uplifted, and the ancient Himalayas were greatly raised, forming a plateau landform, and also causing one major earthquake after another. Owing to Yanjing's location in a zone of intense crustal uplift in the eastern part of the Qinghai-Tibet Plateau, over a very long geological age, nature had created carbonate rocks here, which, subjected to violent press in the east-west direction, formed belts of folded strata and large faults, and resulted in salt ore outcropping. Such an outcrop, mainly Upper Triassic (Jiapila Formation) purple-red sandshale, with a small amount of siltstone, is the lagoon facies sedimentation. Geologically, Yanjing belongs to 'the Indus River--the Brahmaputra River and Pangong Tso--Zigetang Lake--Nujiang River' tectonic belt, which is one of the most intense hydrothermal activity belts within the continental territory of our country. This region, located in the 'Qamdo--Yanjing' landmass, is bounded by the 'Beijianggong--Chali Snow Mountain' fault zone and the Womuzha fault zone in the east and west respectively, adjoining the Mangling junction zone (in the northern section, the 'Chin-sha River--Red River' rift zone is sandwiched in between) to the east, and the Zhayu Bitu junction zone to the west. Faults and folded strata underwent tectonic development here, the fault fold system composed of faults controlling the local distribution of sedimentation and magmation"[1].

As no salt outcrop has been found in the red bed near the salt pans, it is speculated that such saline spring with high concentration might have been formed by deep underground hot water, which, on its way up, dissolved part of the salt bed and gushed out of the earth surface along the fault zone. According to our interview, within the nearly 20-kilometer stretch from Yanjing in the north to the Yunnan-Tibet border area in the south, saline springs unceasingly emerge along the Lantsang River canyon, and salt-sinter, which is called salt frost by local people, also comes out on the ground nearby. This phenomenon is closely linked to the geological background and hydrothermal activity in the region.

The exceptional geological structure here, and its movement, also result in an underground hot water system with deep and long circulation paths. Due to geothermal energy, the groundwater recharge system is characterized with a hot spring effect. In addition, the river water, having washed stones and sediments away from the earth surface covered with carbonate rocks, cut it open and melted part of halite into brine with the help of its dissolution effect , thus providing natural salt making conditions for the Tibetan and Nakhi inhabitants. For geothermal reasons, the

1 "Research on Circulation Mode For Salt Springs by Isotope Tracing in Yanjing, Tibet" by Qi Jihong and others, published in *Earth and Environment*, 2008 (3).

temperature of these saline springs on the bed of Lantsang River is generally pretty high; except for a few whose temperature has been lowered by the immersion of river water, all of them are around 40 degrees Celsius. Standing at the well mouth, one can often see hot steam emitted from below. It was recorded that "in Yanjing Township, within a stretch of about 300 meters in length and 20 meters in width from north to south on both sides of the Lantsang River, there are many springs of different sizes gushing out, warm and salty, with a maximum temperature of 41.4 °C, and a total water yield of 20 L/S."[1] According to the data, its salinity is as high as 30.7 g/L. "As a particular liquid mineral resource, the formation of deep underground brine water is influenced by various factors, such as geological structure, hydrogeology, and climate, etc."[2] Such signs of rock dissolution as karst columns and small corroded caves can be found near the riverside of the Jiada Salt Well, indicating that the brine has a very strong corrosion effect. The most typical spot is near the riverside to the east of Nyima Champhu's and the Dondubs' private brine wells. In Yanjing, saline springs are concentrated in two areas. One is the section on either bank of the Lantsang River where salt is made by evaporating brine under the sunlight, bounded by the bridge leading to Jiada Village in the south and Jiujia Village[3] in the north; within this stretch of three or four kilometers, there are saline springs everywhere. The other, now mainly developed as a hot spring zone, is near the location of Quzika Township Government. According to villagers, in the past there were 108 spring mouths, where eggs could be hard-boiled, but now only about ten remaining. After the development of Quzika Hot Spring, it has been welcomed by people from far and near. Visitors from Markam, Foshan, Yanjing, and Muxu, etc., often come here to take a bath.

According to the preliminary analysis of water quality, the brine contains a certain amount of sodium chloride, with such cation content as sodium ions

---

1 *Records of Hot Springs in Tibet*, by Tong Wei, Liao Zhijie, Liu Shibin, et al., Science Press, 2000 edition, p. 258.

2 "Study on the formation and evolution mechanisms of underground thermal bittern in Yanjing County of Tibet" by Qian Lin, Master's Thesis, School of Environmental and Civil Engineering, Chengdu University of Technology, 2007.

3 Jiujia Village, originally under the jurisdiction of Upper Yanjing Administrative Village, now belongs to Lower Yanjing Administrative Village.

and potassium ions being very high, while such anion content as chloride ions being naturally the highest. The outcrop stratum near salt pans is lagoon facies sedimentation. The red bed in the vicinity of Jiada Village, sloping up to the north with an inclination angle of 45-60 °, is the west wing of the Yanjing anticline. Within a stretch from Yanjing salterns in the north to the border river between Yunnan and Tibet, 20 kilometers away in the south, saline springs often emerge along both river banks.

The salt deposit is a type of evaporite deposits, thus its formation requires arid climatic conditions. The climate in Yanjing belongs to a plateau temperate semi-humid monsoon climate zone. There is plenty of sunlight here, with distinct dry and wet seasons, mild and humid in summer, and cold and dry in winter. The annual average temperature is 10 °C, with an average temperature of -2 °C in January and 16 °C in July. The precipitation is mainly concentrated from June to September, with an average annual precipitation of about 500 millimeters. Possibly, such a climate bears no relation to the age of mineralization, but it meets the conditions for brine evaporation. According to *Local Geographical Records of Yanjing, Batang*[1] during the Xuantong period in the Qing Dynasty: "The climate in Yanjing is mild almost throughout the four seasons, the highest temperature in summer being 70 degrees (Fahrenheit)...And south winds blow constantly all year round, day and night...However, the air is dry here, and whenever it rains or snows, they would be blown away by strong winds in the air...Besides, it seldom rains or snows in Yanjing.[2]Rain and snow are rare, with intense sunlight and strong wind, so it is particularly suitable for brine evaporation in this area.

Meanwhile, it is also influenced by the circulation system of the Qinghai-Tibet Plateau, forming a unique composite climate zone in the transition area, and resulting in a more complex weather condition here, which is characterized by the common climate phenomenon that "the weather varies within ten miles". Due to the vertical nature of the climate, vegetation and cultivation species vary with altitude. Moreover, the soil here is fertile and the water source is abundant, which is conducive to the

---

1 During the Xuantong period, Yanjing came under the jurisdiction of Batang.

2 Duan Pengrui: *Local Geographical Records of Yanjing, Batang*. National Library of China, stereotype edition of the second year of Xuantong (1910), p.4.

growth of plants. Wild plant resources are also quite abundant; in addition to the pine, China fir, cypress and other tree species in the primeval forests, and a variety of medicinal herbs such as fritillary, cordyceps, glossy ganoderma, and codonopsis under the woods, there are also such edible fungi as auricularia auricula, Tricholoma matsutake, and hericium erinaceus, etc. While little importance was attached to agriculture traditionally, and potatoes and turnips were regarded as the staple crops, now villagers have started scientific farming, introducing high-yielding highland barley, winter wheat, and corn. Such fruits as apples, peaches, oranges, grapes, and walnuts, etc. are widely planted, and have become a source of income. As to animal husbandry, it has been a traditional industry of the Nakhi people in Yanjing.

2) The Principle of Evaporative Crystallization

Crystallization refers to the process by which solutes are precipitated in the form of crystals after the cooling of a hot saturated solution, generally in two ways: One is the solution-evaporation method, and the other is the cooling hot saturated solution method. Salt-making by evaporating brine under the sunlight belongs to the former, while salt-making by boiling it off the water belongs to the latter. Their principle is much the same; by volatilizing the water in the salt solution, the solute gradually precipitates, and eventually becomes crystals. The difference lies in the different sources of volatilizing force, the former derives from wind evaporation (combined with the action of light), and the latter from artificial heat. Which of these two methods emerged earlier is currently inconclusive. Hans Ulrich Vogel said, "The solarization method was by no means the chief salt-making method in the ancient Chinese Empire, but it became increasingly important over time and eventually replaced the salt-making method of concentrating brine through boiling"[1], which did not answer the question of which came first, either.

To make salt with the first method, it is necessary to evaporate the brine water to a certain extent first by carrying it into the brine storage tank, forming a saturated

---

1 See Hans Ulrich Vogel's "From Boiling to Solarization: A Further Discussion on Chinese Sea Salt-making Technology in the Imperial Era", translated by Lin Guizhen, and collected in *Archaeology of China's Salt Industry (Episode 2): A Comparative Observation from an International Perspective* edited by Li Shuicheng and Lothar von Falkenhausen, Science Press, 2010 edition, p. 20.

solution of sodium chloride; then pour it into salt pans, and evaporate the water in the brine through continued exposure to the sun and wind, until sodium chloride crystals (crude salt) are precipitated. Regardless of geographical differences, dry monsoons and warm sunlight are essential conditions for brine evaporation. And many salterns at home and abroad conduct salt making in line with the principle of evaporative crystallization. There are salterns in China from Bohai Bay to Beibu Gulf, and from the Qinghai-Tibet Plateau to Hainan Island, among which the weather in Yanjing, Markam County, Tibet, is quite unusual. According to Duan Pengrui, in summer, the maximum temperature in Upper and Lower Yanjing is 70 degrees Fahrenheit (70°F), while in Jiada Village on the west bank it may rise to 73 degrees Fahrenheit (73°F). 70 degrees Fahrenheit (70 °F) equals 21.1 degrees Celsius (21.1 °C) and 73 degrees Fahrenheit (73 °F) equals 22.8 degrees Celsius (22.8 °C). As the Lantsang River flows through this region, the process of heat absorption and release by the water raises the temperature here, with an average daily temperature of 20°C.

As mentioned earlier, Yanjing is located in an "S"-shaped canyon, with towering cliffs on both sides. High-altitude airflow, blocked by several large mountains, spirals downward into the Lantsang River canyon. Whether the wind blows from the direction of Quzika Township in the north or from the direction of Foshan Township in the south, they meet at this section of the canyon, and the airflow, restricted by the terrain, moves in all directions, forming strong winds under the influence of such tortuous and concave topography. In this way, the canyon has a wind tunnel effect, and the "whining" wind sound can be heard everywhere. The wind, blowing all over the river and the slopes, whether it is southward or northward, takes away the water vapor from the saltern, with particularly strong evaporation on sunny days. According to Jin Fei's description, "there's a perennial southerly wind, day and night; if there are crevices in the walls, the wind would be constantly felt everywhere within the house, making one feel dizzy and impeding his breath"[1]. It is thus clear that the intense sunshine, the strong wind, and a high temperature here are all good for natural brine evaporation.

August of the solar calendar every year is the off-season for salk-making, with

1 Duan Pengrui: *Local Geographical Records of Yanjing, Batang*. National Library of China, stereotype edition of the second year of Xuantong (1910), p. 4.

a lot of rain and humidity. Dr. Li Hechun measured the wind speed and temperature at this time in his investigation. When measuring the wind speed, an indigenous method was applied first: When there was wind, a handful of grass was picked up and dropped freely from a height of about 1 meter, and how long it took the grass to fall from the air was recorded; then the horizontal distance between the landing point and the starting point was measured, and the wind speed was calculated with a formula (v=s/t). After that, the result was checked with a portable anemometer and an ordinary thermometer was used to record the temperature. The collected data are shown in Table 1-1.

Table 1-1 wind speeds and temperatures in the salterns in August

measuring time	wind speeds (m/s)	temperatures (°C)	measuring spots
2015.8.15/11:00	1.0~1.5	28.9	the saltern in Jiada Village
2015.8.15/13:00	0.3~0.5	32.0	the saltern in Jiada Village
2015.8.16/12:10	0.8~1.5	29.6	the saltern in Lower Yanjing Village
2015.8.16/16:40	2.0~5.6	29.4	the saltern in Lower Yanjing Village
2015.8.17/14:41	2.3~11.4	30.1	the saltern in Jiada Village

The data show that the outside temperatures were relatively high at the time of measurement, and there was not much difference between the morning and afternoon temperatures of the day (the temperatures remained above 28°C). The wind speed varies greatly in the morning and evening, with a lower one in the morning, ranging from 0.8 to 1.5 m/s. After 4 p.m., the wind speed increased, reaching a maximum of 5.6 m/s. It can be seen that during the day, the wind speed decreases when the temperature is high, and it increases when the temperature is low. This complementarity between the two ensures the crystallization conditions for brine in the morning, noon, and evening.

Of course, it is not for nothing that Yanjing people stick to solarization method in their salt making. The situation in Yanjing is characterized by low vegetation coverage and a lack of wood resources, and even if firewood is available in the high mountains, it is too far away and difficult to be transported due to the precipitous paths. Besides, there's no coal or natural gas in Yanjing. According to *Investigation in Yanjing County* written by Jin Fei in the Qing Dynasty, "there are neither coal mines nor firewood here..." and it is mentioned that "there is no coal nearby, and the firewood is too far away to fetch due to the high mountains, increasing the cost by more than twenty times"[1]. In the literature of the late Qing Dynasty, it was pointed out that in the salt industry "people once intended to switch to the boiling method, and once refined, the color and taste of the salt were excellent." However, it was also repeatedly mentioned that "since there was no coal mine outside the Jiayu Pass, if firewood became the only fuel, the cost would almost increase by more than 20 times; thus, the old method had to be retained temporarily"[2]. It is conceivable that the wood used for building salt shelves and salt pans is extremely scarce, needless to say the application of the boiling method which requires a large amount of firewood. In that case, it is normal for the cost of salt making to exceed 20 times. Obviously, it is not that the salt making technology in Yanjing doesn't need improving, but that it is "extremely difficult to improve". Therefore, only by maintaining this traditional method can they adapt to their natural environment.

Natural evaporation needs no fuel, thus reducing the salt-making cost, and the product, which was of high quality and low price, became most popular in Kangba district near Sichuan border. According to *A Survey of Salt in Different Counties in Xikang Province*: "The salt sources in the bordering counties in Xikang Province are approximately four: The salt consumed in such counties as Ya'an, Rongjing, Hanyuan, and Tianquan is produced in Zhugentan, Leshan, with a price of about 4,000 to 6,000 *wen*[3] per 500 grams. The salt consumed in such counties as Xichang,

---

1 "Investigation in Yanjing County" by Jin Fei, published in Political Affairs of the Border Region, 1931(8).

2 *Archival Historical Data of Sichuan-Yunnan Border Affairs in the Late Qing Dynasty (Middle Volume),* compiled by Sichuan Institute of Ethnic Studies, and the editorial group of *Archival Historical Data of Sichuan-Yunnan Border Affairs in the Late Qing Dynasty*, published by Zhonghua Book Company, 1989 edition, p. 446.

3 1 wen equals about 0.2 yuan RMB at that time.

Yuejun, Mianning, and Zhaojue is produced in Yanyuan County, with a price of over 2,000 to over 4,000 *wen* per 500 grams. And the salt consumed in Yajiang, Lihua, and Ba'an in Xikang, and in the counties near the border in Yunnan, is produced in Yanjing County, Xikang, with a price of over 1,000 to over 2,000 *wen* per 500 grams...Among them, the salt produced in Leshan is the best and its price is the highest, with Qinghai salt being the farthest and Yanjing salt the cheapest." The root cause for the low price of Yanjing salt is that its making process needs no fuel, entirely dependent on wind and sunlight. Therefore, the low cost of technology became an advantage in the competition.

2. A Strategic Post on the Ancient Tea-Horse Road from Yunnan to Tibet

"The Ancient Tea-Horse Road" is an academic concept proposed at the end of the 20th century. It is an ancient transportation corridor formed by the long-term movement of people and exchange of goods between Sichuan, Yunnan, and Tibet. "According to archaeological cultural relics and related historical clues, the history of its coming into existence and use can be traced back at least to the late Neolithic Age, about 4,000-5,000 years ago, or even earlier."[1] Some scholars pointed out: "(The Ancient Tea-Horse Road,) as the fifth channel for China's foreign exchange, has the same historical value and status as the Maritime Route, the Road to Western Regions, the Southern Silk Road, and the 'Musk Silk Road' from the Tang Empire to the Tubo kingdom."[2]

The Ancient Tea-Horse Road consist of two parts: the Sichuan-Tibet line and the Yunnan-Tibet line. The ancient Yunnan-Tibet tea-horse road, which came into being in the late 6th century AD, led from the tea-producing areas of Yunnan through Dali, Lijiang and Shangri-La to Tibet, heading west to Lhasa, and finally to the South Asian subcontinent. It was one of the oldest routes to Tibet. Originating from the

---

1 "Qamdo, A Hub of the Ancient Tea-Horse Roadand and Its Ancient Culture: and the Historical Features of the Ancient Tea-Horse Road" by Shi Shuo, published in *Journal of Tibet University*, 2003(4).

2 *Cultural Exploration of the "Great Triangle" of Yunnan, Tibet, and Sichuan* by Mu Jihong, Chen Baoya, Li Xu, et al., Yunnan University Press, 1992 edition, p. 11.

ancient tea horse trade in the southwestern borderland, it began to develop in the Tang and Song dynasties, and flourished in the Ming and Qing dynasties. From the perspective of the trade route between Yunnan and Tibet, Yanjing is an important courier station, and it is not unreasonable to call it the key link of the ancient Yunnan-Tibet tea-horse road. For thousands of years, this Tibetan village alongside the Ancient Tea-Horse Road has developed as a strategic passage, and thrived by virtue of salt.

1) A Long history of Salt Industry

Yanjing Township has always been a strategic post for salt making in eastern Tibet since ancient times, for local people learned to make pure and tasty salt very long ago by renovating saline spring mouths and digging brine wells. The salt made here became very famous in Tibet, especially in southeast Tibet, northwest Yunnan, and the Kham regions, etc. According to some archaeologists, the traces of human activities 30 kilometers downstream of the Yanjing are of a time "about the early or mid-Spring and Autumn Period, or as early as the late Western Zhou Dynasty"[1]. The salt industry here has a very long history, as is shown in "the Battle of Jiangling", a story of King Gesar. According to historical research, there was already salt making through solarization during the Tang Dynasty; people first enclosed brine water by piling up stones, and then drew it for evaporation under the sunlight. As early as before the Tubo Dynasty (633-842 AD), when each tribe in Tibet seized control of an area, there were salterns here, and legend has it that among the six sgangs[2] of Duokang, Markam sgang was famous for its salt production.

Yanjing, once a stop for ancient Qiang tribes in search of saline springs along the river corridor southward, preserved the cultural spark for the development of this ethnic group. Through an analysis of Tubo salt-making technology, it can be considered that the salt production method in such places as Yanjing during the Tubo period was "charcoal-based salt production method". During the historical period, from the control of Tubo to the arrival of Chieftain Mu in the Ming Dynasty, salt

1 "Nagu Sarcophagus Tombs in Deqin County, Yunnan" by Cultural Relics Working Crew of Yunnan Provincial Museum, published in *Archaeology*, 1983 (3).

2 "sgang" means the highland between two rivers.

making through solarization might not have appeared yet in Yanjing.

During the Jiajing period of the Ming Dynasty (1522-1566), Chieftain Mu, after arriving in Yanjing, seized control of it through military means, and with the saltern construction skills mastered in Lijiang by the accompanying Nakhi people, he had wells dug to draw saline springs on the Lantsang River shore and wooden shelves put up to build salt pans, beginning large-scale salt making through solarization. Some scholars believed that the technique of salt making through solarization here appeared at least in 1526 in the Ming Dynasty.[1] Zhang Yulin reported in his investigation materials of Batang: "The local prefect of Lijiang once immigrated in large numbers to the Batang area. These immigrants, at ordinary times, were Mu's serfs devoted to his farmland reclamation; in wartime, they became soldiers who would be garrisoned in a blockhouse...Nakhi people were adept in dredging ditches and reclaiming farmland, building walls and houses, and cultivating rice. A large stretch of terraced fields in the southeast part of Batang was reclaimed due to the encouragement by Nakhi people's example. "[2] One point is worth emphasizing: Chieftain Mu brought solarization salt-making skill to Yanjing, but he was not the one that first discovered saline springs and started the salt industry here. Judging from the fact that there were already human activities in Yanjing before the Ming Dynasty, the time when natural brine was discovered and small-scale salt making was conducted in Yanjing might be pushed forward to the Tubo period, or even earlier.

In the sixtieth year of the Kangxi period (1721), Du Changding, an aide to the Governor of Yungui (Yunnan and Guizhou provinces), noticed the brine wells in Yanjing when cruising past here, and wrote in his diary: "White salt is produced on the east bank of the river, while red salt is produced on the west bank. Merchants and dealers from all directions came here for business. The salt-making method does not rely on fire. As the rocks on both sides of the river are as steep as walls, the natives have to build salt shelves along the cliffs, cover their tops with mud (each side higher than the enclosed flat area), and pour in brine water. Under the action of sunlight and

1 Li Shuicheng. "Vigorously Strengthening the Exploration and Protection of Ethnic Salt Cultural Relics: A Guide to the Column 'Salt Archaeology' Column" [J]. *Cultural Relics in Southern China*, 2010 (01).

2 Zhang Yulin: "A Random Account of the Historical Evolution of Batang", published in the Journal of Kangding Nationality Teachers College, 1990 (1).

wind, salt is gradually precipitated, and then swept and stored downstairs for sale. The tops of these shelves are known as salt pans, and between these salt shelves there are salt pits, which, like paddy fields but slightly deeper, are used to store brine water. When the mountain snow melts in summer after a warm spring, the river will flood and submerge the brine wells. In order to make salt as usual, salt makers would put the saltpan mud into the pits, and the mixture is very much similar to the water from brine wells. Orderly arranged like thousands of fish scales, the annual output value of these salt shelves may amount to millions." This precious material became the earliest record of Yanjing in Chinese classics.

The earliest systematic introduction to Yanjing, *Local Geographical Records of Yanjing, Batang*, which was written by Duan Pengrui, the salt officer there, in the first year of Xuantong (1909), comprehensively sorted out the social, economic (salt production), and cultural development of Yanjing from the perspective of "location, history, boundaries, situation, ethnicity, customs, salt pans, and Lama temples." And such records as "wooden shelves are put up by the natives one on top of the other on either side of the river; the salt pans, separated with boundaries, are very much like the inland paddy fields; brine tanks are dug nearby, which are usually filled with brine water, in order to continue salt making through solarization when floods have submerged the well mouths in summer and autumn" could clearly show the continuation of the traditional production methods of the salt industry for more than 200 years. At that time, there were a total of 2,763 salt pans, 689 brine storage tanks, and 52 brine wells on both sides of the Lantsang River in Yanjing, and the annual output of salt was about 18,000 *duo*[1], each weighing 70 kilograms.[2] In that way, the annual output of salt in Yanjing was 1.25 million kilograms, or 1,250 tons.

In Liu Zanyan's *Records of Salt Wells During the Republican Period*, it was recorded that: "In the salterns in Yanjing, there are springheads under both sides of the Lantsang River, enclosed by piled stones; wooden shelves are put up along the mountain slopes, the tops of which are shaped like paddy fields and laid with loess of 1 inch thick; brine water carried through manpower is poured into such

1 A load carried by a pack-animal.

2 See Duan Pengrui's *Local Geographical Records of Yanjing, Batang*, published by National Library of China, stereotype edition of the second year of Xuantong (1910), p. 13.

'fields', and salt would be precipitated by wind. Each salt maker can make over 15 kilograms of salt through solarization per day. Reddish and yellowish in color, it has a strong taste, and sold in dozens of counties in the southern Kangba district and the region extending west..." In the twentieth year of the Republican period (1931), the prosperity of salt production remained unabated, and according to Jinfei, "on the east bank there were 1,242 salt pans and 802 brine tanks; on the west bank there were 2,724 salt pans and 474 brine tanks."[1] Whereas the statistics in Cui Kexin's *Investigation of the Geology and Salt Production in Yanjing County* showed that such basic situation as salt pans, etc. in Yanjing in the thirtieth year of the Republican period (1941) was as follows: There were 163 salt-maker families, 23 brine wells, 2,329 salt pans in good condition (including 1,300 in Jiada Village), and more than 230 unrenovated salt pans (including more than 200 in Jiada Village). As to salt production, the number of salt pans in Jiada Village was counted as 1,500, and the number of salt pans elsewhere was 1,080. And the daily output of each salt pan was: 1/3 Tibetan *dou*[2] in Jiada Village; 3/2 Tibetan *dou* elsewhere (9 kilograms). Excluding the 156 rainy days in 6 months, during which salt cannot be made, the total annual output is 479,332 Tibetan *dou* (approximately 2,876 tons).

From the liberation of Yanjing on October 12, 1953, to 1959, the salt industry production here was mainly organized by the county government, but still based on handicrafts under individual ownership. Detailed record of salt production in the early days of Yanjing liberation has not been found yet. However, *Investigation on Commerce and Handicraft Industry in Tibet*, based on the tax situation during this period, estimated that within two years and five months, a total of 64,748 *duo* of salt were produced, with each *duo* only calculated as 75 kilograms, totaling 48,606,100 kilograms, and the annual output was about 2 million kilograms[3], i.e., about 2,000

1 "Investigation in Yanjing County" by Jin Fei, published in *Political Affairs of the Border Region*, 1931(8).

2 "dou" is a measure for grain in China; here, according to 6 kilograms per Tibetan *dou*, it approximately equals 2 kilograms.

3 See *Investigation on Commerce and Handicraft Industry in Tibet*, co-edited by Institute of Ethnic Studies, Chinese Academy of Social Sciences, and Institute of Social Economics, China Tibetology Research Center, and published by China Tibetology Publishing House, 2000 edition, p. 8.

tons.

The industry in Markam County started in the 1960s. According to *Markam County Annals*, in 1960, Yanjing salt makers, by digging more brine wells and building more salt pans along the east and west banks of the Lantsang River, enlarged the production scale and established a state-owned salt factory. With an annual production of more than 15 million kilograms, the salt, in addition to serving as a county supply, was also sold to Batang in Sichuan Province and other places. In 1963, Yanjing Salt Factory was abolished and transformed into a salt industry commune. Since then, all salt makers made, collected, and sold salt collectively, and at the end of the year, food was distributed according to their work points, which was not much different from the inland. Moreover, during the time of the people's commune, Jiada Villagers were divided into three groups, the first of which specialized in salt making through solarization, and has been retained until now. In 1965, local salt-maker families were organized to establish a producers' cooperative. In 1974, after the realization of people's commune in the whole county, the salt industry commune was changed into a salt industry team. In 1978, the county government, in order to adapt to market demand, established a salt iodization site in Yanjing, which led to the development of salt production to some extent during this period. However, due to the inability of technology to keep up and low yields, the salt iodization site closed down afterwards. In 1985, the system of contracted responsibilities on the household basis with remuneration linked to output was implemented in salt production. In 1986, there were already 66 households and 274 people who contracted for salt production in the county. By 2000, two townships (i.e., Nakhi Ethnic Township and Quzika Township)[1] were engaged in salt production in Yanjing County, with 320 households and over 1,500 people; there were 2,655 salt pans, over 300 brine tanks, and over 60 brine wells, the annual salt output totaling 750 tons.[2] In 2001, the salt output of the whole township reached 816 tons, including 344 tons in Jiada Village, 250 tons in Nakhi Village, and 222 tons in Upper Yanjing

1 During this period, Jiada Village on the west bank of the river was under the jurisdiction of Quzika Township, so there were salt pans in both Nakhi Ethnic Township and Quzika Township. Later, Jiada Village was put under Naxi Ethnic Township.

2 See *Markam County Annals* compiled by the Compilation Committee of Markam County Local Chronicles, published by Bashu Publishing House, 2008 edition, pp. 365-366.

Village. Since 2007, salt makers have no longer struggled to make a living as before; they have gradually switched to pumps for drawing brine, and the era of carrying brine on their backs has passed by degrees. And people would drive motor vehicles, instead of horse caravans, to collect salt in these villages.

Judging from the production and sales of Yanjing salt, the industry has basically maintained a stable state from the 1990s to the beginning of the 21st century. Salt, having gradually breaking away from the control of the system, was at a stage of development in which salt makers produced and marketed all by themselves. The protection of cultural heritage and the rise of tourism since the beginning of the 21st century, has also become important factors for the inheritance of the traditional salt-making technology through solarization in Yanjing. Currently, there are more than 300 salt makers in Yanjing Township, and salt production is a family-based workshop type. Except for some of the salt makers in Jiada Village, who live near the salt pans, most of the rest live in the mountains or in Yanjing Town. Due to the low technical and undemanding nature of salt production, there is no fixed type of work here, and every producer participates in such processes as carrying brine, building salt pans, and making salt through solarization, etc. A salt-maker household may produce 2,000 to 2,500 kilograms of salt at most per year, or 1,000 to 1,500 kilograms at least.[1]

The history of Yanjing is a cyclic process, in which people inherited the continuous production, distribution and consumption of salt. Today, the salt pans in Yanjing County are the only manual primitive salt-making site through solarization that survived on the Ancient Tea-Horse Road. This primitive folk activity of well salt production, having been maintained to the present day, continues to have an impact on the production and living of local ethnic minorities. With high cultural relic values, it has not only made a heavy mark in the history of China's well salt science and technology, but also provided valuable primary data for scholars at home and abroad to further study the salt industry in Tibet.

2) The Ethnic Corridor in Eastern Tibet

The concept of "ethnic corridor" is an important part of Mr. Fei Xiaotong's

---

1 *Yanjing Township: A Major Salt Town on the "The Ancient Tea-Horse Road"* by Tao Hong, published in China Cultural Heritage, 2005(5).

theory of pluralistic integration pattern of Chinese nation. The so-called "ethnic corridor", "refers to the route along which a certain nation or ethnic group migrates or flows outward along a certain natural environment such as rivers or mountains for a long time, and in this corridor, there must have been preserved lots of historical and cultural deposits of the nation or ethnic group"[1]. At a time when transportation was extremely inconvenient, Yanjing, the only salt production site within a radius of hundreds of miles, through its salt trade and exchange, as well as the formation of salt paths, not only spurred the local salt industry, increased the population along the trade routes, but also greatly promoted the interaction and integration between Yanjing residents and the surrounding ethnic groups.

The trade between Yunnan and Tibet has a long history. And the valley area of the Three Rivers Basin on the eastern edge of the Qinghai-Tibet Plateau became an important passage for ethnic exchanges during the Tubo period. In June 2009, an archaeological team composed of several institutions from Shaanxi and Tibet discovered the cliffside images in a little Lakang (Buddhist Hall) in the Brag sgo shod valley of Lower Yanjing Village, and concluded that it was a cultural relic of the Tubo period. This also indicated that there were human activities in Yanjing during the Tubo period at least.[2] This was once the thoroughfare from the Tubo Kingdom to the Nanzhao Kingdom, as well as the only route for Yunnan tea to be transported to the holy city of Lhasa.

In the Tang Dynasty, the Tubo regime expanded eastward, and in order to contend for resources, entered western Sichuan. Yanjing thus became the passage through which Tubo forces moved southward and was culturally influenced by it. However, since Yanjing was located in the border area of the three major forces, i.e., the Tang Empire, the Nanzhao Kingdom, and the Tubo Kingdom, the impact of the former two could not be ignored either. In particular, Yanjing had a relatively close interaction with the northern part of Nanzhao Kingdom, or the current northern

1 See the article "The Southwest Silk Road and the Ethnic Corridor" by Li Shaoming.

2 See "The First Comprehensive Archaeological Investigation and Record of the Tubo Stone Carving Remains in Eastern Tibet: Two Newly Discovered Tubo Stone Carving Remains in Markam County and Chaya County, Qamdo Prefecture, Tibet" by Xi Lin, Zhang Jianlin, Shar rgan dbang vdus, and Tian Youqian, published in page four of *China Cultural Relics News* on November 13, 2009.

Yunnan ethnic groups (Nakhi, Bai, etc.). After the Tang Dynasty, the Ancient Tea-Horse Road began to take shape gradually, and made progress due to the establishment of the Tea Horse Division in the Song Dynasty. Later, the ancient salt-horse road was absorbed by the Ancient Tea-Horse Road for trade between Sichuan, Yunnan, and Tibet. According to historical records, Yanjing already became the first courier station after entering Tibet from Yunnan through the Ancient Tea-Horse Road, as well as an important material distribution center in the Hengduan Mountains during the Tang Dynasty. Via this ancient road, travelling traders sold Yunnan tea to Tibet in exchange for animal hides, cordyceps, and medicinal herbs, etc. Salt makers, on the other hand, mainly bartered for food, oil, cloth, iron, and other necessities of life. The prosperity of commercial trade and goods transportation on the Ancient Tea-Horse Road brought about the mobilization of people and the growth of settlement scale, and there was a diversified division of labor and cooperation on the basis of the original agricultural economy, resulting in the influx of people from various industries, such as horse herders, porters, cobblers, traders, shopkeepers, *maguotou* (leaders of the gorse caravans), and Lamas, etc. As Yanjing is located in the border area of various major regimes, it has been constantly involved in wars and power games since the Tang Dynasty. The change of regime led to chaos for a time, which was not unrelated to the fact that Yanjing, situated in the ethnic corridor, was the necessary path for the integration between Han and Tibetan cultures.

In the 1620-60s of the Ming Dynasty, Chieftain Mu began to expand northward, and his influence soon extended to Yanjing and other places. It was an important period of great changes in the history of Yanjing, not only reflected in the possible updating in salt-making techniques, but also in the adjustment of the distribution pattern of local ethnic groups caused by Nakhi people's arrival and occupation. At this time, the Nakhi ancestors, compared with the original ethnic groups in Tibet which had long been under the feudal serfdom and the influence of theocracy, were more developed in politics, economy and culture, and relatively advantageous in agricultural and industrial technology as an enterprising ethnic group, thus bringing some impact on the local society since they arrived in Yanjing. In particular, both the improvement of salt-making skills and the traditional management style of Chieftain Mu had a profound impact on the local society of Yanjing. Henceforth, salt, continuously circulating within the border area of the three parallel rivers between

Sichuan, Yunnan, and Tibet, not only promoted the interaction and integration of multiple ethnic cultures such as Khamba, Nakhi, Yi, Pumi, and Lisu, but also created the unique social organization form and cultural type of the concentrated population.

Since the Qing Dynasty, the sales of Yanjing grainy salt were relatively active, maintaining frequent interaction with surrounding ethnic groups. Merchants from various regions entered Tibet to set up businesses and horse caravans. At this time, Yanjing was still in the traditional stage of barter, bartering with salt. As the exchange of its salt with the grain of the surrounding areas was gradually intensified, a salt-grain trade routes began to take shape by degrees, i.e., a major longitudinal one (north-south river valley) supplemented by lateral one (ravines on the east and west sides). To the south, it led to Deqin mainly through the Lantsang River canyon, then reached Weixi or Benzilan, and then went deep into Lijiang and Dali. This was one of the routes chosen by Chieftain Mu of Lijiang in the Ming Dynasty, with the rise of its force, to expand northward into the eastern part of Tibet by military action, basically going up along the Lantsang River and the Chin-sha River canyons.[1] The northern route, going east through the Zhaguxi Gorge, led to Nandun, and from there proceeded eastward across the river, would enter Batang and Litang; turning north, it would reach Ningjing (Jiangka), and proceeded northward to reach Zogang and Qamdo. To the west, it crossed the river, and after passing Laweng Temple and going over the mountain, reached Bitu; Then, further west, it led to Chayu and other places. Since the Qing Dynasty, due to the relationship of administrative subordination between Yanjing and Batang, as well as the fact that horse caravans were often organized from Batang for Yanjing salt transportation, the eastern route was basically unimpeded all along, and this situation lasted until the end of the 20th century.

During the Guangxu period of the Qing Dynasty, Adunzi was set up as a commercial port in Yunnan, and all local products in the county were transported to Adunzi for sale or exchanged for necessary groceries. At that time, the goods from Markam to Adunzi were mainly medicinal herbs, leather, and salt. Besides, there were more than ten Tibetan and Han shops in Yanjing. By transporting the salt, tea, copper, iron, and groceries purchased from Yanjing to Chayu County in exchange

---

1 See *The Social History of the Tibetan People in Diqing* by Wang Hengjie, China Tibetology Publishing House, 1995 edition, p. 53.

for medicinal herbs, and selling them on site or in Adunzi, they made Yanjing an increasingly prosperous market.

Meanwhile, the salt resources in Yanjing not only brought economic benefits, but also promoted cultural interaction between different ethnic groups, which was first manifested in the simultaneous use of Chinese and Tibetan in the local area. For example, in *Regulations on Salt Taxation in Yanjing*, it was recorded that: "An office clerk proficient in the Manchu language, and a Tibetan interpreter responsible for going to imperial court or visiting a friendly nation as an envoy would be appointed, each receiving a monthly salary of ten taels of silver." Secondly, the influence of the salt industry in Yanjing had deeply reached into the Drung, Nu and other ethnic settlements by the Nujiang River. From the perspective of the Lhoba and Mishmi people, they urgently demanded salt from Yanjing. And in the first year of Xuantong (1909), Battalion Commander Cheng Fengxiang, after arriving in the Zayu area, found that in the locality "the needed goods were mainly salt, and two or three *pi* (equivalent to 1 liter) of grain is often exchanged for one *pi* of salt"[1]. According to a survey conducted by the Institute of Ethnic Studies of the Chinese Academy of Social Sciences in the Mishmi people inhabited area of Zayu County, Tibet in 1976, it was found that close to Yanjing "salt was very lacking in the Mishmi people inhabited area, and they'd exchange the coptis they had dug up or other valuable medicinal herbs obtained occasionally for salt in the Tibetan inhabited areas"[2]. By means of bartering, these Mishmi people often exchanged coptis, gastrodia elata, fritillaria, musk, and bear gall, etc. for knives, iron pots, salt, yak butter, and other daily necessities with the Tibetans. Apparently, salt had become an important commodity between the Mishmi people and Tibetans. During the Qing Dynasty, when Xia Hu entered the Gongshan area by the Nujiang River, the exchange of salt was also recorded: "People in Chang (Gongshan County) all consumed sand-like salt, which was produced in Yanjing County, Xikang Province, and transported for sale by the uncivilized minorities in Tsarong. People generally traded grain for it,

---

1 *A Memorial to the Throne by Zhao Erfeng from the Sichuan Border*, compiled by Wu Fengpei, and published by Sichuan Ethnic Publishing House, 1984 edition, p. 227.

2 *Survey of the Social History of Mishmi people*, compiled by the Institute of Ethnic Studies of Chinese Academy of Social Sciences, and published by Yunnan People's Publishing House, 1990 edition, p. 59.

rarely buying it with silver coins. Due to the limited production of grain in Tsarong, salt was transported to exchange for it, which was then brought back home for their own consumption."[1] And in somewhere else it was also recorded that: "In the Zayu area, native merchants from Menkong would still transport well salt to exchange for coptis, which proved that well salt was produced there. Some local minorities had been dispatched to Yanjing to purchase it, in case of need for reward, and would return to Zayu soon."[2] Therefore, Yanjing salt did play an important role in promoting the interaction and integration between various ethnic groups in the Hengduan Mountains region of eastern Tibet.

Its most prosperous years were probably from the middle and late Qing Dynasty to the Republican period, when merchants and horse caravans came here in droves. These caravans, by transporting the salt out of Yanjing and bringing back the production and living materials from outside, greatly promoted the prosperity of Yanjing economy. The trade, with salt as an essential commodity, was achieved between Yanjing and its surrounding areas, of which Adunzi in Yunnan Province and Batang in Sichuan Province were two important market towns that had been maintaining long-term exchange relations with it. And with its trade strengthening relationship to such surrounding areas as Adunzi and Batang, Yanjing also exerted a significant influence on them. Yanjing sand-like salt, having been sold to Yunnan Province for long, was also described in *Newly Compiled General Annals of Yunnan* of the Republican period. The sales scope of Yanjing salt had expanded to the Weixi area in Yunnan, with Deqin as an important transfer station...Yanjing was not only geographically adjacent to Batang, but also used to be under the jurisdiction of it. Driven by these two relationships, Yanjing and Batang always maintained a long-term interactive relationship. And Yanjing sand-like salt were mostly sold to Batang. Economic development has led to the prosperity of culture. Along with the entry of outside materials, outside culture began to take root here, thus local culture also gaining new annotations in the economic opening up. Such primitive folk-custom of

---

1 *Changputong Annals*, compiled by Changputong Administrative Councilor's Office. See *Selected Literary and Historical Materials of the Nujiang River (Volume 18)* edited by Li Daosheng, 1991, p. 14.

2 "Brief Annals of Zayu", see *Collection of Tibetan Local Chronicles* (Volume 2), China Tibetology Publishing House, 1997 edition, p. 142.

well salt production, preserved until today in the 21st century, will continue to have an impact on the lives of the surrounding ethnic minorities.

Generally speaking, Yanjing salt, relying on its unique geographical location in the border area of Sichuan, Yunnan, and Tibet, has promoted the interaction between Yanjing and the surrounding areas, facilitated the exchanges between such ethnic groups as the Tibetan, Nakhi, Bai, Lisu, etc. on the Yunnan-Tibet Ancient Tea-Horse Road, and enhanced the understanding and unity between Lhoba people, Mishmi people, Tibetans, and Han people in the Zayu area, forming a unique regional culture of the "Ancient Salt-Horse Road". Some farmers and herdsmen, who have spent their entire lifetimes within the narrow communities where they engage in productive and living activities, know very little about the outside world. The economic activities of salt-grain exchange, while enabling mutual communication and broadening their horizons, have played a significant role in cultural dissemination and ethnic identity among inhabitants of different parts of the Tibetan Plateau. Moreover, they have exerted a significant impact on road construction, village development, and cultural education in the locality, and to some extent improved the living standards of the salt makers.

3) The Village Layouts

After the discovery of saline springs in Yanjing, people continued to gather here around the production, circulation, and exchange of salt. Most salterns, especially larger ones, were closely connected to villages, which further highlighted the phenomenon of population aggregation. Therefore, on the whole, these salt-making villages, mainly engaged in salt production, by inheriting much of the salt production modes in the past, have already formed a certain cultural landscape. In the academic circle, it was pointed out that "'landscape' is a geographical term with three layers of meaning: firstly, land and visible objects on it; secondly, visibility or ornamental value; and thirdly, the unity of natural and humanistic attributes."[1] And "cultural landscape is the fruit of human activities, the earth's surface traces of the interaction between humans and nature, as well as an identity that culture gives to a region,

---

1 Zeng Daxing. *Introduction to Literary Geography* [M]. Beijing: The Commercial Press, 2017: 230.

which may visually reflect the cultural characteristics of this region."[1] As a type of cultural landscape, the salt-making landscape here is the fruit of evaporating brine in the sun over the years. It's a characteristic bestowed on these salt-making villages by salt culture, reflecting their unique cultural features.

Yanjing is an ethnic township dominated by Tibetan and Nakhi nationalities, including Yi, Bai, and Lisu, etc. Located on a stretch of flat hilltops and hillsides, it's surrounded by lots of mountains, with the Lantsang River surging through from north to south. On either side of the narrow river valley are high and precipitous mountains, the east bank very steep, whereas the west side sloping comparatively gently. As water sources are abundant here, the mountain streams have been channeled into numerous canals, which flow around the houses, through the farmland, and finally into the Lantsang River. The entire township, covering an area of 11,654 square kilometers, has jurisdiction over four administrative villages: Jiada, Nakhi (also known as Lower Yanjing), Upper Yanjing, and Juelong, which, located on the east and west banks, are facing each other across the Lantsang River. Among them, Jiada Village is on the west bank, while Nakhi Village, Upper Yanjing Village, and Juelung Village are on the east bank (see Figure 1-4), with a suspension bridge connecting the salt pans on both sides. The location of the first three villages forms a triangle shape, but Juelung Village is in a valley about 10 kilometers away from the township government (Lower Yanjing village).

Among the four villages mentioned above, except for Juelong Village hidden in the valley, Lower Yanjing Village, Upper Yanjing Village, and Jiada Village are all near the mountains and by the river, and the villagers have been engaged in salt production for generations. The salt resources are very abundant here, with a total of 83 saline springs and a daily flow rate of 140~288 cubic meters. In each of these three salt-making villages, there is a villagers' group called the salt-making group, which relies entirely on salt making for a living. The salt pans of Upper Yanjing Village, located in the north near Quzika Township, are the farthest from its village, and the terrain is quite steep; the salt pans of Lower Yanjing Village are in the south; the salt pans of Jiada Village are closest to its village, and the terrain is fairly flat.

1 Zhou Shangyi and Kong Xiang. *Cultural Geography* [M]. Beijing: Higher Education Press, 2014: 301.

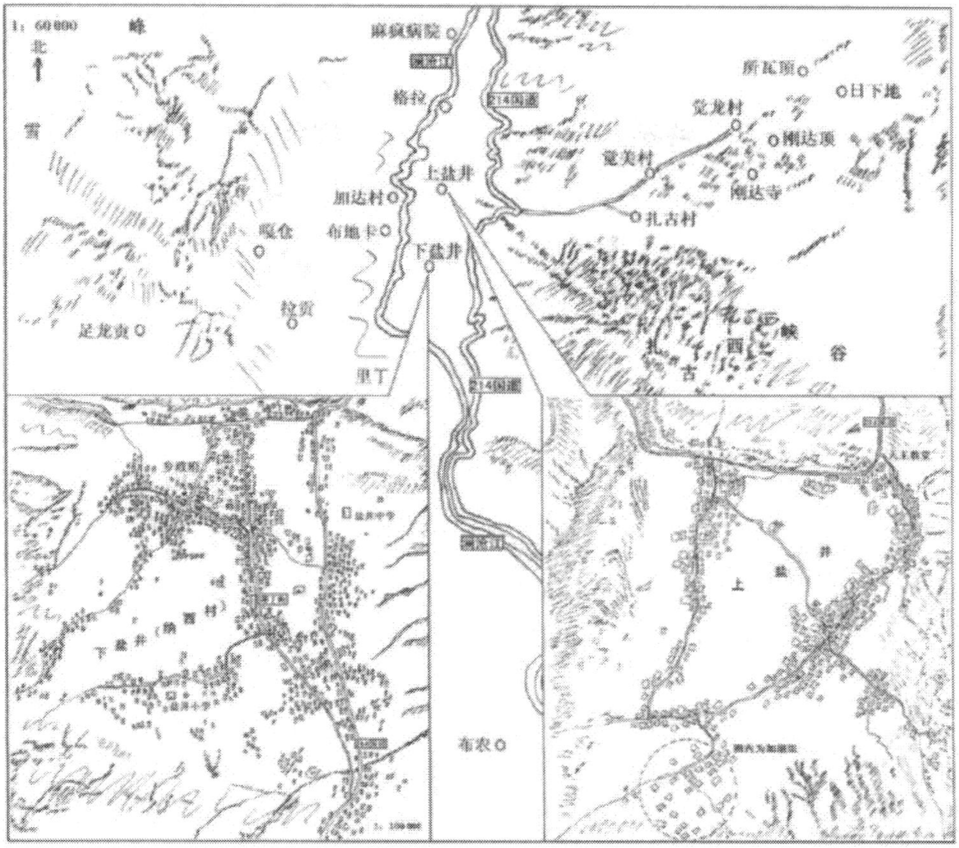

Figure 1-4 the village layouts of Yanjing

Looking north from the east bank of Lantsang River, the salt pans of Lower Yanjing Village are unfolded at your feet. And looking east from the west bank, the salt pans of Upper Yanjing Village, hanging on steep slopes and cliffs one on top of the other, or arranged like the teeth of a comb or the scales of a fish, appear to be very spectacular, indeed.

According to the survey result at the end of 2017, there were a total of 3,463 salt pans in the three salt-making villages. In terms of the number, Jiada village had the most, accounting for 80 percent. The basic situation of the three villages are as follows:

a. Lower Yanjing Village

Lower Yanjing Village, also known as Naxijing Village, is the seat of the Naxi Ethnic Township Government. Located next to the National Highway 214, the transportation here is very convenient. With an average altitude of 2,600 meters, the annual average temperature is about 10 °C, and the annual precipitation is 350~450 mm. Consisting of 7 natural groups, the village has a total of 291 households, 1,365 residents. The total arable area is 856.70 *mu* (a unit of area, equivalent to 0.0667 hectares), the grain crops mainly including highland barley, wheat, and corn, etc., and the cash crops including pepper, peaches, apples, pears, pomegranates, plums, and oranges, etc. In addition to considerable value placed on education, there are also restaurants, hotels, and retail stores, and tourism is well developed in the village.

The salt pans here, located on the left bank of the Lantsang River, belong to the Zongge Group of Naxi Village, Naxi Township, with an altitude of 2,315~2,329 meters. To its west is the Lantsang River, to its east are either hillsides or steep cliffs, to its south is a simply constructed road leading to the Jiada Suspension Bridge and Naxi Township, and to its north is a gully facing the salt pans of Upper Yanjing Village. There are 154 salt pans here, with a total area of about 30,000 square meters, which is 474 meters long from north to south, 115 meters at the widest point from east to west. And the height difference between the top and bottom is 50.47 meters. They have been built along the banks of the Lantsang River in accordance with the mountain trend, with staggered heights. Among them, the southern salt pans are mainly distributed on gentle slopes, with the widest point exceeding 110 meters; the northern ones are all built on the edges of steep cliffs, with the widest point only about 50 meters and the narrowest point only about 20 meters. The uppermost salt pans are about 120 meters away from the riverside, and about 70~100 meters above the river surface.

b. Upper Yanjing Village

Upper Yanjing Village, located also close to National Highway 214, is about 2.5 kilometers away from the seat of the township government, with an average altitude of 2,600 meters. In the village, there are a total of 157 households and 772 residents, mainly Tibetans, plus some Naxi people. The village has 5 natural groups under its jurisdiction, and the arable area is 530.80 *mu*, with a per capita area of 0.72 *mu*. The forest coverage rate of Upper Yanjing Village is 60.7%, mainly planted with pomegranates, peppers, walnuts, poplars, willows, and elms, etc., while the crop farming is mainly composed of highland barley, wheat, and corn, etc. The single-type industry of farming plus timber trees and fruit trees has developed into a multi-level and multi-chain industry, forming a new rural industrial development mode of "enterprises + bases + farmers". Moreover, the wine produced in Upper Yanjing Village, characterized by its pure flavor, comfortable taste, and obvious health care effects, is known as the "Bordeaux" of Tibet.

In the village, there are a total of 13 salt-maker households, with 587 salt pans and an annual output of about 5,000 kilograms, which has been a chief source of economic income. These salt pans, located on the left bank of the Lantsang River, with an altitude of 2,305~2,346 meters, belong to the Salt-making Group of Upper Yanjing Village, Naxi Township. To its west is the Lantsang River, to its east are mountain slopes or steep cliffs, and to its south is a gully facing the salt pans of Lower Yanjing Village. The total area is about 28,000 square meters, 364 meters long from north to south, about 77 meters at the widest point from east to west, with a height difference of about 41 meters between the top and bottom. The uppermost salt pans are about 120 meters away from the riverbank. As the terrain here is quite similar to that of Lower Yanjing Village, salt pans have also been built along the banks of the Lantsang River in accordance with the mountain trend, arranged in a neat and orderly manner at different heights.

The only Catholic church in Tibet, located at the southeast corner of the village near the national highway, has more than 500 believers, accounting for 75% of the total village population. Hence it is a village featuring the integration of Chinese and Western cultures. In December 2012, Upper Yanjing Village was included in the first batch of traditional Chinese village list. And in September 2014, it was awarded the title of "National Ecological and Cultural Village" for the year 2014 by the China Eco-Culture Association.

c. Jiada Village

Jiada Village, located on a terrace (alluvial fan) on the west bank of the Lantsang River, has a total of 205 households, and 1,236 residents. There used to be a temple (Lagong Temple, now relocated to Quzika Township), and there is an arable area of 589.60 *mu*, with an average altitude of 2,250 meters, belonging to a type of valley climate. The northern part of the village are high and precipitous mountains, thus forming a distribution pattern of farmland in the south and salt pans in the north. Due to the relatively open and flat terrain of the river valley terraces in this section, the land is relatively fertile and suitable for growing highland barley, sweet buckwheat, wheat, corn, grapes, pears, walnuts, green peppers, lettuce, potatoes, radishes, cabbage, pomegranates, and apples, etc.

farmland in the south and salt pans on either banks of the river in the north

The salt pans here, belonging to the Jiada Group of Jiada Village, Naxi Township, are located on the right (west) bank of the Lantsang River, with an altitude of 2,315~2,328 meters. To its east is the Lantsang River, to its west is a simply constructed earth road leading to the Jiada Suspension Bridge, to its north are steep sloping fields, and to its south is the White Pagoda, facing Jiada Village across a large gully. There are 2,775 salt pans, with a total area of about 35,000 square meters, which is 445 meters long from north to south and approximately 95 meters at the widest point from east to west. The uppermost pans are about 95 meters away from the riverside, and about 30 meters above the river surface. Compared with the two villages mentioned earlier, the salt pans here are not so much affected by the mountain trend and thus appear to be more open.

salt pans at the Jiada Village section of the Lantsang River

At the physical space level of the villages, the architectural culture landscape and the entire settlement, born of salt, have integrated rich salt culture into their development process. The unique structure stemmed from the joint effect of the Ancient Tea-Horse Road and its location environment; the unique layout originated from the corresponding relationship between saline springs and farmland resources and the production and living of the concentrated population; at the entity level of residential buildings, the unique types came from the combination of special users and local construction practices. And all these particularities came from the construction suitability between material space and people's production and living. Therefore, the architecture here has a typical Tibetan ethnic style and strong regional characteristics, its architectural structure and layout showing diverse changes, which, with a layered look, conforms well to the natural spatial pattern.

The houses of Tibetan residents in Yanjing are generally 2 to 3 storeys; each with a square plane, gradually shrinks from the bottom to the top, forming a terrace. Many residents built their houses against the hills, which, just like ancient castles, are very spectacular. All these houses have a large courtyard, and the gates were also built tall and sturdy. The first floor, with windows installed opening outward from the courtyard, may serve as a livestock shed, fodder room or garage. The second floor is for the whole household to live on. And the third floor is usually used for stacking grain and other things, etc. What is more, the top, solid and flat, which can be used for drying grain, is also a broad balcony. Standing there, one could gaze around the beautiful green fields of the countryside, as well as the snowy mountains and rivers in the distance. An incense burner in the shape of an ox's head would be placed on the broad balcony, with herbs and cypress branches, etc. burning inside, to piously worship the gods and Buddhas. And an oratory dedicated to worshipping gods and Buddhas is generally set up in each Tibetan house, too. Spacious and magnificent, one of its walls would be occupied by a huge Buddhist niche with exquisite colorful paintings and carvings. In such an oratory, Thangkas and dharma-vessels passed down by their ancestors, as well as auspicious tokens left by eminent monks after their stay, are also enshrined and worshiped. The meditation bed, specially made with extra care for lamas, are covered with a thick and gorgeous Tibetan carpet. Local residents are willing to spend money on their houses, and the doors and windows made of giant timber and exquisitely carved are always amazing. Of course, their oratories are particularly beautifully decorated. For example, some people may decorate their homes in varied colorful splendor, and even the beams and ceilings are covered with paintings and carvings. The copper pots inlaid with pearls and jade, dharma-vessels, jade wine goblets, and golden Buddhas...are simply more dazzling. Yet, the architectural structures of these folk houses also have their respective regional characteristics. They could be roughly classified into such types as stone and wood, mud and wood, wood, and commixture, etc.[1] But in general, these structural differences are changes in adaptation to the local environment.

For the Naxi ethnic group in Jiada Village, there are two types of residential buildings. The first type, generally a small independent building with two or three floors, is built with a family as a unit, for the family members living in it together; the second type, a two-storey courtyard-style building, is built with a large floor area and a relatively large enclosed courtyard, mainly for horse caravans or migrant workers engaged in salt production. The former type is generally a regular rectangular flat-roofed building in form. With reduced area of structure, a roofdeck would thus be left on the top, generally for drying grain, whereas the covered area behind it is a space for storing grain. The latter type, a flat-roofed, trisquare-shaped two-story building in form, is relatively unique. The retaining wall on the ground floor is made of stone, while that on the second floor is made of wood. Both types of residential buildings, including□the monastic architecture in the west of the villages, have prominent regional characteristics in construction; that is, the use of relatively short timber as the supporting structural material, wooden columns and flat purlins as the post and beam structure, and the local stone or rammed earth as the wall enclosure material. Thus these residential buildings, characterized by thick walls, low floor heights, and small windows, fully reflect the regional construction consensus on the use of local climatic conditions and local materials.

From the following pictures, the previous living conditions in the Naxi ethnic folk village--Jiada Village might be discernible. The dwelling environment has been preserved intact, allowing tourists to visit, even with someone offering free explanations.

---

1 "Traditional Tibetan Residential Architecture: the Kham Residential Buildings" by Zanhua. See Ancient Homestead, www.gujianchina. cn, March 13, 2020.

As for the Tibetans here, although they still retain the tradition of "covering houses with soil and preferring storied buildings", and most of their houses are Tibetan-style blockhouses, due to the influence of Han culture, they have gradually changed their architectural structure and style. Some villagers, by adding a protruding and roofed balcony to the second floor of the earth building, not only have facilitated their daily lives but also changed the Tibetan habit of spending their time inside the blockhouses. Many Tibetans have even built their courtyards with cement pouring method, and planted flowers and fruit trees, thus greatly expanding their living space. In recent years, with the help of the national targeted poverty alleviation policies, many Tibetans have built separate one-storey cement houses, and installed solar energy equipment donated by the state for daily washing.[1]

Nowadays, in the vicinity of the villages, we can hardly see vast grasslands, and there are no herds of yaks and sheep, either. Instead, there's a large stretch of green terraces from high to low, and the crops in the field are growing quite well, which indicates that the land here is fertile and the rain is abundant. And local villagers have

1 Han Yan. "Presented in *Weixi Jianwen Ji*: Muliti-ethnic Integration and Development in Northwest Yunnan". *Journal of Ethnology*, 2022,13(09).

already transitioned from herdsmen to half-herding and half-farming men (there are still yaks and sheep in the mountains behind), or even more farming than herding. Looking around, almost every household is dwelling in a small three-storied western-style building, with a flat roof and multiple windows. The Tibetan style is no longer that obvious, and there are not too many special designs. Priority seemingly given to practicality, they are scattered all over the green fields like stars in the sky.

Upper Yanjing Village, seen from a distance, has become an atypical Tibetan village, which looks very un-Tibetan and not much different from villages in the inland region. Most families have gradually moved into modern buildings equipped with modern electrical appliances and running water; many have bought cars, or at least motorcycles or pedicabs. Even if placed in the inland, it can be regarded as a well-off little village. At the village entrance, there's a white stone tablet that reads: "Chongqing Aid Construction, the new rural area of Upper Yanjing Village". Presumably, Chongqing cadres sent to aid Tibet, by bringing with them Chongqing's funds, Chongqing's new thinking on development, and Chongqing's way of life, have enabled the local people to live in better houses, grow better crops, and lead a better life.

Thus it can be seen that the cultural landscape of Yanjing villages is not static, but changes with the changes in people's living and production. Anyway, despite the impact of outside civilizations, these villages have still maintained their distinct characteristics of simplicity and naturalness on the whole.

3. Production and Lifestyle

Marx said: people make their own history, but they neither do it at will, nor under conditions of their own choosing; instead, it's under the conditions that they have directly encountered, or that are already established, or inherited from the past.[1] In Yanjing, the unique resource endowments and environmental conditions not only have constituted the extraordinary appearance of the "millennium ancient salt pans", but also created the distinct social production and living mode for local people.

1) Livelihoods of the Residents

The trend of diversified livelihoods is quite evident in Yanjing. Among the four administrative villages of the township, Upper Yanjing, Lower Yanjing, and Jiada mainly engage in salt industry in the early days. Besides this, the livelihoods of these villagers also include farming, animal husbandry, gathering, and planting, etc., all showing diverse characteristics. However, Juelong Village, located northeast of Yanjing, mainly engages in farming, animal husbandry, and gathering, for it's far away from the Lantsang River valley and thus lacks brine resource. In spite of the fact that in Lower Yanjing, the seat of the township government, villagers have gradually given up salt making and turned to other service industries (which is closely related to the development of Yanjing as a 4A-level scenic area), from the perspective of the entire Yanjing area, salt industry has always been a major means of livelihood for local residents.

Salt is inseparable from Tibetan life, and as a Tibetan proverb goes, "Tea without salt is just like water; a person without money is just like a ghost." The formation and development of Yanjing, based on the well salt industry, have been going hand in hand with the development and utilization of well salt resources. Salt makers learned to maintain their livelihoods by inheriting traditional production techniques, and meanwhile established the status of the region through salt trading. They have played an important role in history and thus influenced the historical process of the region. This might be confirmed in the literature of the Qing Dynasty, "Generally speaking, for Yanjing local minorities, the most important livelihood is salt making, followed

1 *The Complete Works of Marx and Engels* (Vol. 8), People's Publishing House, 1995, p. 121.

by farming and then commerce. And the livestock here are far inferior to those to the north of Batang."[1] It shows that the three industries, i.e., farming, animal husbandry, and commerce, are not as developed as the salt industry in Yanjing. Therefore, the primary job of local Tibetans and Naxi people is salt making.

In the traditional sense, there are two types of salt-maker households in Yanjing: specialized and semi-specialized. The former, having no farmland but only salt pans, are mainly engaged in salt production all year round, and may stop it only when the conditions for salt making are extremely harsh, such as continuous heavy rain, snow, and river floods. According to reports, now there are more than 60 specialized salt-maker households in Yanjing who live entirely on salt making. In Jiada Village, there's a group (Salt-making Group) consisting of 24 such households and 149 people; in Upper Yanjing Village, the Salt-making Group is made up of 12 such households and 52 people. For semi-specialized households, they have both farmland and salt pans, and thus have to shuttle back and forth between salt making and farming. When it comes to busy farming seasons, such as when corn, wheat, highland barley, and other crops have to be planted or harvested, villagers would stop salt making and engage in farm work.[2] There are more than 200 such households with both salt pans and farmland.

The income of salt makers may vary from one another depending on the size and operation situation of the salt pans. In Upper Yanjing Village, the average number of salt pans for each salt-maker household is about 7, while for specialized salt-maker household is nearly 12. In Jiada Village, the average number of salt pans for each salt-maker household is around 18. And in the entire Yanjing area, this figure for each specialized salt-maker household is around 22. The annual output of a salt-maker household here can be as much as four or five thousand *jin*, or only two or three thousand *jin*. Comparing the situation of Jiada Village with that of Upper Yanjing Village, it can be seen that most salt pans are concentrated in the hands of specialized salt-maker households. The output of salt varies slightly between the east and west banks of the Lantsang River. In Upper Yanjing Village and Lower Yanjing

1 Duan Pengrui's *Local Geographical Records of Yanjing, Batang*, published by National Library of China, stereotype edition of the second year of Xuantong (1910), p. 11.

2 Li Hechun. *Impetus and Shackles: Salt and Salt Taxes in the Lantsang River Canyon*. Guangzhou: Sun Yat-sen University Press, 2016: 210.

Village on the east bank, as brine can be evaporated in the sun in a short period of time with high yield, each salt pan may produce 30 to 40 *jin* of first-layer salt. In Jiada Village, 20 to 30 *jin* of first-layer salt can be scraped from each pan, while the second-layer of salt is generally around 30 to 40 *jin*. By this calculation, excluding the three-layer salt, the yield of each salt pan would be around 50 to 60 *jin*. And differing in color, the salt may be of two layers: the first is mostly white, and the second is red. The wholesale price of white salt is 3 *yuan* per *jin*, whereas red salt is much cheaper, only 1 *yuan* per *jin*. As each salt pan can bring in 900 to 1,000 *yuan* a year, most salt makers' annual income is about 10,000 *yuan*, with the lowest being only 4,000 to 5,000 *yuan* and the highest being around 18,000 *yuan*.

Salt makers often exchange salt with residents in surrounding areas. It would take place not only among the neighboring ethnic groups, but also between people engaged in different means of livelihood in the locality, so as to obtain daily necessities. As a result, the production and sale of salt have become the mainstay of local Tibetans and Naxi people to maintain their living, survival and development. And the development of the salt industry economy has also made Yanjing Township a fairly prosperous area in Tibet.

In addition to salt making, there are certain natural conditions for agriculture in Yanjing, too. As we know, the agriculture of Markam County is mainly concentrated in mountain-valley areas, plateau farming areas, and plateau farming and pastoral areas. The vast majority of Yanjing area belongs to the mountain-valley type, with its land distributed on both sides of the Lantsang River. Due to an altitude of 2,100 to 2,600 meters, the climate here is mild, with long hours of sunshine, and thus can basically ensure double cropping a year. There are 780.25 *mu* of single-cropping land and 2,172.5 *mu* of double-cropping land, but the per capita arable land area is less than 1 *mu*. In the locality, the staple crops include highland barley, wheat, corn, beans, and vegetables, etc. During the Greater Spring (from May to September), the major yields are wheat and highland barley, and with the development of modern farming technology, such as the provision of better seeds and fertilizers, farmers have started to grow corn on a large scale. But during the Lesser Spring (from October to April of the following year), the major yield is only wheat. The altitude of plateau farming areas is relatively higher, generally between 2,500 to 3,200 meters. Due to the influence of climate conditions, the major crops are highland barley and wheat, roughly one cropping per year. Animal husbandry accounts for a certain proportion here, and the altitude of pastures is 3,000~3,500 meters. Parts of Juelong Village and Jiada Village belong to this type.

According to Rinchen Dondub, a villager from Jiujia Village (seemingly another name for Juelong Village), the grasslands of Juelong Village are most widely distributed in the entire Yanjing area, mainly on the east bank of the Lantsang River and the mountain tops with a relatively higher altitude. It may take seven to eight hours to walk from the village to these grasslands. In the past, every household used to graze their animals on the pastures, but now only one in Jiujia Village still does so. Since the village is surrounded by farmland, seasonal transitions between productive activities are widespread. From May to October, cattle and sheep are driven up the mountains, and by mid-October, when the farmland harvest is already completed, they'd be driven back again to the village. The livestock here mainly include horses, mules, donkeys, cattle, and goats, among which, the cattle are mostly cows and yellow cattle, with few yaks.[1]

Local residents have also realized that owing to limited natural conditions and the lack of large areas of arable land and open pastures, they'd have to rely on other means of livelihood to obtain material resources. Therefore, while not delaying salt making and agricultural production, people would often go and collect matsutake, or dig Cordyceps sinensis, fritillaria, coptis, etc. in the mountains. As the main production area in Tibet of wild edible fungi, including matsutake, auricularia auricula, tripe mushroom, morchella, and Saarcodon imbricatum, etc, Markam County, rich in resources, is characterized by their high yield, large variety, and good quality, and has been exporting them to Japan and Southeast Asia. Among these fungi, matsutake has been regarded as the most precious, which is not only very delicious in taste, but also good for human health. Since the mid-80s, mountain mushrooms, represented by matsutake, have been worth hundreds of times in overseas markets, and matsutake has thus become a major source of wealth for Yanjing residents, too. Matsutake, growing in mountain forests with an altitude of over 4,000 meters, are collected every July or August. At that time, the villagers of

1 Li Hechun. Impetus and Shackles: Salt and Salt Taxes in the Lantsang River Canyon. Guangzhou: Sun Yat-sen University Press, 2016: 38.

Yanjing will take some solid food and stay in the mountains for four or five days. The price of matsutake is set by grade. Generally speaking, those with a big pileus not yet unfolded is rated as first grade, and priced between 40 to 50 *yuan* per *jin*; those with a slightly unfolded pileus is of the second grade, priced at around 30 *yuan* per *jin*; those with a fully unfolded pileus is of the third grade, priced at 10 to 15 *yuan* per *jin*. Matsutake of the third grade, with affordable price, is mainly purchased by locals for their own consumption. And Cordyceps sinensis, even more expensive than gold, has been a gift from the plateau for the local villagers. Cordyceps sinensis, together with ginseng and pilose antler, is known as the "Three Treasures of Traditional Chinese Medicine". Generally growing in grassy marshland with an altitude of 3000~5000 meters, it is collected in April and May. On the internet, well-packaged dried cordyceps is sold by gram, while in the mountains, the cordyceps freshly dug up would be sold by stick, and a high-quality stick might be sold for hundreds of *yuan*. Hence, more than a month's arduous gathering can often earn enough for a year's expenses. Simply due to its high price, every gathering season, all the villagers would go out and flock into the mountains, crawling around on all fours in search of the thin, black-brown "golden grass". In addition, the fritillaria in local mountains, because of its pure and pollution-free growth environment, has even higher health care effects and medicinal value, and is thus very popular among merchants from outside, often selling at a better price.

Yanjing is also one of the grape-growing regions in Tibet. As early as 1855, when Western Catholicism was introduced to Yanjing, two French priests brought grape seedlings from Europe, as well as viticulture and wine-making techniques. In recent years, local wineries have been built, driving the development of grape planting industry in Yanjing. Now the local government are actively guiding villagers to plant grapes, by providing an annual subsidy of 1,500 yuan per *mu* of vineland in the first three years of cultivation. A wine viticultural association has also been established in the village, and the winery has signed contracts with the association and farmers. At present, agreements have already been signed between the government, the winery, Muxu Township, and Quzika Township. By the end of 2013, there were about 270 grape grower households in the township, with a vineland area of over 600 *mu*. Since 2014, Zangdong Treasure Wine Company, a leading red wine enterprise, has been supported by the local government. On December 11 of the same year, the implementation of geographical indication product protection for "Yanjing Wine" was approved by the General Administration of Quality Supervision, Inspection and Quarantine of the People's Republic of China, and "Yanjing wine" has thus become a specialty of Markam County in Tibet as well as a national geographical indication product of China. In 2018, there was a total of 3,018.3 *mu* of vineland area in Yanjing Naxi Ethnic Township. In 2020, 33.4 hectares of land had been transferred to establish a grape planting base, and through the "enterprise + association + base + farmers" model, the local people had been driven to plant 166.8 hectares of wine grapes. As a result, Zangdong Treasure Wine Company's annual output value reached over 30 million *yuan*, with more than 700 farmers and herdsmen employed temporarily every year.[1]

Since the beginning of the 21st century, with the development of electric industry, women have been freed from the intense labor of carrying brine on their backs, and the traditional means of livelihood based on salt industry is also changing accordingly in Yanjing. As much salt-making labor of is saved, salt makers have more time to engage in other livelihoods, such as catering and construction, etc. For example, relying on tourism development, Tibetan farmhouse entertainments have been built in Jiada Village, Naxi Village, and Upper Yanjing Village to attract tourists, so that they can experience Tibetan customs and promote consumption. In Jiada Village and Naxi Village, there are more such Tibetan farmhouse entertainments, i.e., 28 and 25 respectively, compared with 19 in Upper Yanjing Village. The formation and development of emerging industries have provided new ways for the Naxi people in Yanjing to become rich by entering the market from farmland, and shifting from a single grain and salt production mode to commercial and transportation activities.

2) Family-workshop Type Production

The salt wells and salt pans in Yanjing Township were privately owned before 1970. Except for the period from 1970 to 1982 when they were owned by the people's commune, since 1982, the mode of contracted production to households has

1 Hu Jun et al. "Definition of Grape Growing Period and Analysis of Meteorological Conditions in Yanjing of Mangkang County of Qamdo City in Tibet". *Chinese Agricultural Science Bulletin*. 2023, 39(01).

been adopted, and thus it is still privately owned. Therefore, the salt production here is mostly organized in the form of family workshops, which have been producing and marketing all by themselves for a long time. Despite the relatively low-tech content in salt production, the customs here have existed for thousands of years: there's a clear division of labor between men and women; women are responsible for salt making, while men responsible for salt peddling and salt selling. As for when this division of labor by gender began, it is almost impossible to be ascertained now.

Although the work of salt making is all done by women, there would always be someone responsible for selling salt after it is produced, and this is no easy task. With horses and mules, men need to transport the sun-dried, hill-like salt heaps to more than ten neighboring prefectures and counties for sale. They would go through long and arduous journeys, braving the wind and dew.

a. Woman's Work

Since more than a thousand years ago, the ancient and primitive way of life here has been carried on by local Tibetan and Naxi women through their lives and beliefs. What they have been dedicated to guarding is a magical substance that can change the taste buds of us all--salt. Even from their birth date, it was destined that they would never part with the salt pans. With the craft of salt making learned from their mothers, they, following in the footsteps of their ancestors, have been engaged in this traditional livelihood at the bottom of the canyon for generations. In the salterns, women are the absolute protagonists, taking on almost all the labor, so it's not an exaggeration to say that this canyon is a "canyon of women". Whether in the salterns of Jiada Village, or Upper Yanjing Village and Lower Yanjing Village, the vast majority of salt makers are women. These "salt women" have also become part of this unique landscape.

In the entire salt production activity, except for the renovation of salt pans and the construction of brine wells and brine storage tanks, which would be men' responsibilities, the rest are basically undertaken by women. These women, constantly busy rushing about between the salt pans, have too much work to do throughout the day, sometimes carrying soil or wood, sometimes patting the salt pans with tools...According to their explanation, it's because men are careless and may smudge the glistening salt, which thus cannot be sold at a good price.

Salt making is the most common job for Yanjing women, and there are often one or two of them in the family who cooperate in salt production. Compared with women of Jiada Village, which is a mere stone's throw from their salt pans, where they may go to work unhurriedly at any time while humming a tune, or return for lunch during the rest interval--even find time to make some yak butter tea for themselves, women of Upper Yanjing Village and Lower Yanjing Village have to overcome more difficulties in their life and work. During the annual salt making season, they would get up in the early morning when the sky turns slightly red. After having breakfast and prepared solid food for lunch, they set off from the high part of the river valley, walking for nearly an hour, then along the steep narrow meandering footpaths and bypassing the maze-like salt shelves, before descending to the bank of the Lantsang River. The whole morning, they are either collecting the dried salt, or repairing salt pans, or carrying brine to the storage tanks. Around 12 noon, someone would whistle to everyone that it's time to have a break, and of course it won't be too long. These women then gather in twos and threes in the salt shelves below the salt pans, where pots, kettles, bowls and chopsticks, as well as some firewood are kept most of the time. First of all, they would light a fire to boil water; when the water is boiling, the first thing to do is to make yak butter tea; and when the tea is ready, everyone sits in a circle and places their lunch in the middle, mostly local cakes, salted vegetables, or stir-fried potato slices or hot peppers. Drinking yak butter tea while eating these solid foods make for an easy lunch. After lunch, they collect their own things, put them in bags, and return to their respective salt pans to continue working. In the afternoon, they usually collect salt. when this is done, they'd begin to pat (repair) the salt pans. And after a while, when the salt pans are dry, they pour brine into them. From a distance, these women are like worker bees in a hive, shuttling and working between salt wells, brine tanks, and salt pans, very hardworking. Sometimes, they'd work in the salt pans until 6 or 7 pm; when they return home, it is already 8 pm. Dragging themselves home, they still have a host of chores waiting to be done; their parents, children and husbands may be awaiting them to cook supper for the whole family. After supper, they still have no leisure time, for they not only have to clean up the dishes and tidy the kitchen, but also need to mix fodder and feed the livestock.

Due to the fact that women undertake almost all the work in the salterns, their bodies and minds are under extraordinary stress, and may look particularly haggard in their mid-forties. Just take carrying brine as an example, they start from the cardamom years until middle age, day after day, year after year. In salt pans with rugged paths and steep terrain, women in threes and fives can often be seen carrying brine on their backs, climbing upward with difficulty. And as the sun shines into the Lantsang River Canyon later than outside, the river water is freezing cold in the morning. But they never complain about these things. They skillfully scoop up brine from the steaming salt wells, and by carrying it on their backs, climb up the steep slopes and wooden steps to pour it into their own brine tanks, hoping to store as much as possible before the rainy season arrives. They moved mechanically through the salt shelves that are staggered in height, disappearing from one opening and emerging from another. In this way, they have to carry at least fifty to sixty, or even more than one hundred times a day. Early in the morning, as the brine is especially abundant in the salt wells after flowing overnight, they'd carry one bucket after another without stopping. Near noon, only when the brine has been dried up, they are able to take a short break while eating lunch and waiting for the brine water to seep out again. In the afternoon, they have to finish carrying all the brine water from the wells before returning home. Within a single day, each of them may carry as much as 1 ton of brine! The surface of a brine tank is as flat as a mirror, reflecting the golden hilltops, as well as these women's bent silhouettes.

As a result of working in the salt pans for a long time, these women's hands, feet and lower legs would often peel, or even crack and bleed. And the situation is even worse in winter, for whenever they touch the brine again, the pain is simply unbearable. In the locality, as those anti-cracking drugs cannot be found anywhere, the best way is simply to apply lard, beef tallow, mutton tallow, or yak butter to the affected area, which may help to relieve skin cracking. Whether during pregnancy or menstruation, they have to clench their teeth and keep working. Sometimes, if they do not pay attention themselves, or their husbands are not considerate enough, their bodies would be even more overwhelmed. What's more, because their feet are always soaked in brine, these women are especially prone to rheumatism. They may not feel it when young, but as they age, their legs and feet will become less and less agile.[1]

1 Li Hechun. Impetus and Shackles: Salt and Salt Taxes in the Lantsang River Canyon. Guangzhou: Sun Yat-sen University Press, 2016: 67.

Yanjing women have clearly been accustomed to this laborious lifestyle. The mountain wind and sunshine have prematurely carved the traces of time on their faces; due to the excessive labor, sweat may bead their foreheads every now and then. The so-called "beauty of coarse clothes and unkempt hair" is only relative to the white-collar women in office buildings; as a matter of fact, women here might also be dressed up clean and refreshing: the fluttering skirts and the flashing silver bracelets on their wrists make their weight-bearing images appear natural and unrestrained in a impressionistic style. The flat and bright salt pans are like mirrors, reflecting their dignified and robust beauty. Although their smiles were as bright as peach blossoms, yet their faces were covered with fatigue. Sometimes they'd stop their work and stare blankly at the unceasingly flowing river; occasionally, they may also sing a song loudly, and there is one song that goes like this: "How I wish life was easy and happy, but life always kept me busy..." They calmly shuttle between salt and wells, and harvest the gift of nature with their sweat, solidifying their hard and busy figures in this ancient and desolate scenery on the banks of the Lantsang River. These women, born for salt, pour their lives into the square salt pans; while enjoying the gifts from nature, they are also savoring the mixed tastes of life. Hence salt has become a lifelong bond for the women on this land.

Today, with the development of electric industry, electrical poles have already been erected everywhere by the salt pans. With pumps for drawing water, these women don't need to undertake the laborious work of carrying brine on their shoulders any more. The brine can be drawn directly from saline springs into the brine tanks, and after one or two days, when the concentration is saturated, it is then filled into the salt pans of different heights. As women's work mainly focuses on evaporating brine in the sun and repairing salt pans now, their clothes are no longer dirty and torn; instead, they can finally dress clean and decently. With much time saved in conveying brine, these women even have some free time; they may participate in other livelihood, or have some leisure activities of their own. However, every time recalling the hard days of the past, they will inevitably feel sick at heart.

b. Cooperative Work between Men and Women

In the Yanjing area, men are mostly engaged in farming, trade, and other occupations, because the locals believe that making salt is the work for women, and that men working in salt fields is a shameful thing and will be ridiculed. Of course, men will also work in salt pans when the family labor force is not enough or the conditions do not permit.

There are only three occasions where men might be seen here: first, in construction work, such as the maintenance of brine wells, brine storage tanks, and salt shelves; second, when women have collected salt under the salt shelves, men will appear leisurely on horsebacks, driving mules, on the paths leading to the salt pans, pack the salt in bags and put on the animals' backs before driving them away from the mountains; third, in the past decade or so, thanks to new technologies such as the internal combustion engines, electric wires have been linked to these salt pans, and poles have been erected everywhere, mechanics and electricians being all male. On these occasions, women would take a back seat, and men step forward to play the leading role.

However, during the investigation in Yanjing, we also found that the division of labor between the two is no longer so obvious, and men often help women repair salt brine tanks and carry salt. The traditional phenomenon, "for the salt-maker households, men are responsible for salt transportation while engaging in farming; but such work as carrying brine and evaporating it in the sun, etc., is generally undertaken by women"[1] has also undergone some changes. Therefore, locals believe that it's only a matter of different labor division for women to be devoted to making salt, and it seems that men and women do not care what kind of work they'd undertake.

After the transformation of the collective economy, the production and operation of the salt industry have become entirely family-based. The family economy, integrated with the social economy, is showing the characteristics of both independence and interchange simultaneously. Families in Yanjing basically have three tasks: farming in the fields, grazing their animals in the mountains, and making and selling salt. There's labor division as well as cooperation among family members; women are mainly responsible for the work in the salt pans, while men mainly for salt transportation and exchange, with other work shared by both. In addition, many families also take transportation service and short-term part-time job

1 "Investigation in Yanjing County" by Jin Fei, published in Political Affairs of the Border Region, 1931(8).

in the vicinity as a partial source of income. It's difficult to calculate the labor force in various industries accurately, and seasons and opportunities are levers that regulate labor distribution.

In Yanjing, men are indispensable for the construction work of salt pans, brine storage tanks, and brine wells, etc. as well as electrical installations. The reason is that these tasks may require able-bodied laborers in many cases, and some also call for certain skills. When constructing salt pans, there's naturally a relative division of labor between men and women, too. Men's tasks may include conveying timber, demolishing salt pans and putting up wooden shelves; especially for the latter, experienced male carpenters would be needed. In contrast, women are engaged in some secondary work, such as sorting out the wood that has been removed by men, and carrying sand and mixing mud, etc. during repairing the salt pans. When constructing a brine storage tank, men are mainly responsible for conveying stones and cement, and building its four walls with stones, whereas women would assist in mixing cement and mortar, and may also need to deliver stones and other materials in the process of wall building. And when constructing a brine well, the general process is relatively more complex, and the work undertaken by women is basically the same as the above.

To get a full picture of the male and female activities here, we must also look beyond the salterns. For example, when the weather is not conducive to salt making, do women still stay in the salterns? If not, what kinds of production are they involved in? And in the meantime, what are the men doing?

Every year from the end of June to the beginning of October, due to the rainy weather in the locality, the work in a saltern would be stopped for a while. With women joining in other industries, the entire saltern will become quite deserted. They'd join in the gathering crowds one after another. From May to June, it's the season for digging cordyceps sinensis, which may last for forty to fifty days. From August to September, it's a month for picking matsutake. And subsequently, October is the season for digging Fritillaria, which also lasts for almost a month. So, starting from May to the end of October, it's an important period for local villagers to engage in gathering activities in the mountain forests. At this time, generally regardless of gender, residents from each village are rushing up the mountains. What's more, in the mountains where gathering activities are conducted, some people, in order to meet the gatherers' daily needs, even provide various services, such as keeping motorcycles, opening retail stores, and repair shops, etc.

Tashi Yangzong, a Lower Yanjing villager of the Naxi ethnic group, spends most of her time making salt at home, for her family owns 11 salt pans. But in the season of matsutake, as it's not suitable for salt production because of too much rain, she, together with her sister-in-law and younger cousin sister, would let go of their household chores and go to pick matsutake in the mountains. She told the investigators that as the matsutake is very far away, they usually have to start off at 6 am in the morning, and arrive at the destination at 3 or 4 pm in the afternoon. Three or four of them would go together, carrying solid food and luggage, and it may take five or six days. After searching laboriously for 4 days, she finally picked 15 *jin* of matsutake, earning a total of over 460 *yuan*. However, with matsutake becoming less and less, and people looking for it more and more, her husband has also gone to work in Markam County town. Every time, as soon as she returned to Yanjing, the first thing Tashi Yangzong would do was to sell matsutake. As there are four or five shops for collecting matsutake on the street, this process could be easily finished. After the it was sold out, Tashi Yangzong had only one day to find enough greenfeed for pigs and cattle feed for the next five days or so, for early the next morning she would have to go up the mountains with a few sisters to pick matsutake again. Her destination was a high mountain on the east bank of the Lantsang River, where only residents from Lower Yanjing Village, Upper Yanjing Village, and Juelong Village are allowed to search for it, while residents from beyond the river are absolutely prohibited. Every day, about seven or eight pm at this time of year in Yanjing, villagers who have picked fresh matsutake would be busy carrying it down the mountains for sale. There are two major collection spots on the street: the vacant lot in front of the Yantian Hotel, and the vacant lot behind the local Agricultural Bank. The merchants who come here to purchase matsutake are mainly from Shangri-La and Lijiang. Usually, three or four of them would work together; some are responsible for collecting matsutake, while some drive the motor vehicles. At 8 pm, when everything is done, they'd immediately transport it overnight to the areas of Shangri-La and Dali, etc.[1]

---

1 Li Hechun. Impetus and Shackles: Salt and Salt Taxes in the Lantsang River Canyon. Guangzhou: Sun Yat-sen University Press, 2016: 69.

In Tashi Yangzong's view, although it's very hard to search for matsutake, every time she would bring some back home, anyway. And compared with making salt, it is far more profitable. Working laboriously in the salterns, the salt produced sells for no more than 1 yuan per *jin*, and the average income per day is only dozens of *yuan*. But by searching for matsutake, one can generally earn about 100 *yuan* a day. Therefore, in the season of searching for matsutake, everyone would drop what he/she is doing (or try to finish it early), and be fully devoted to this activity. As soon as the gathering season is over, these women will return to the salterns. Homes, mountains and salterns are the three spaces in which they engage in activities. Before the roads were built, men used to drive mules and horses for salt transportation. But now, some people often come here to purchase salt in motor vehicles. In Upper Yanjing Village and Lower Yanjing Village, since the salterns are not yet accessible by road, they still need mules and horses to transport salt, and gender division of labor remains the same as before.

In Yanjing, in addition to gender division of labor, there is also an age division of labor. Healthy individuals aged 18~55 are young and middle-aged laborers, who almost undertake all kinds of work in local production activities. Primary and secondary school students are not counted as normal labor, though they may sometimes offer a helping hand; and university students, still in the learning stage, whose registered permanent residence has been from their hometown, are not counted in the labor force, either. Furthermore, after the age of 55, people gradually withdraw from the category of able-bodied laborers. It is not, however, as some scholars have suggested, that villagers would not engage in any work other than household chores at the age of 60; on the other hand, depending on the actual situation of the family, some people in Yanjing may remain active in farming even after 60.

3) Cooperation between Families

As mentioned earlier, in Yanjing Township, a family is the basic economic unit. Normally, as salt making activities are regulated by the production cycle, drawing brine, carrying brine, filling salt pans with brine, and collecting salt, etc. are mostly done by each household on its own, and thus there's no need for cooperation. But salt makers still attach great importance to the team spirit of unity and cooperation, and collaborative support. In several cases, there is also cooperation between families.

Firstly, in the event of floods or unexpected events, it's necessary to urgently repair infrastructure such as salt pans; and to ensure the construction within a limited time, different families may cooperate quite closely. At this point, due to the urgent need to restore salt production without affecting its output, families whose salt pans have been severely damaged may need the help of their relatives or neighbors, in order to repair them in the shortest possible time. Another situation is that when building public brine wells, it's necessary to joint everyone's efforts. Through division of labor and collaboration, not only can each individual's strengths be given full play while making up for their own shortcomings, but also team spirit and atmosphere may generate strong motivation, greatly improving work efficiency.

Secondly, as different families have different labor times, different princile livelihoods, or different land types, during in a certain period of time, some farmer households may be busier, while others are comparatively idle. If the two groups adopt a strategy of complementary collaboration, i.e., the idle households help the busier ones first, and when they need labor force later, the latter will be able to offer help in return, it can improve the efficiency of labor utilization. Therefore, salt-maker households often collaborate in different forms, so that it is very effective to be organized: on the one hand, it's easier to get something done; on the other hand, by making use of the time difference between various means of livelihood, labor resources would be utilized more reasonably. Overall, salt making is conducted in the form of a single household, while the construction of brine wells or salt pans is in the form of collective activity. The process of building brine storage tanks varies among different families. Some would prepare materials first and then concentrate on laying stones, while others may invite their relatives and neighbors to construct it in a centralized manner, all helping each other by exchanging labor. If it is a brine storage tank shared by several households, then everyone should work together, and the division of labor is the same as above.

Thirdly, there's cooperation between families with farm mules[1] and those without farm mules. In Yanjing, there are two major ways of ploughing: one is to use

---

1 Mules and horses used for ploughing. In Yanjing, people mainly plough the farmland with mules, with only a small portion already mechanized, and yellow cattle are not seen here for this.

machinery, which is mainly seen in the relatively flat fields of Upper Yanjing Village and Lower Yanjing Village; the other is to use mules and horses, which may differ from the use of yellow cattle and buffalo in other parts of China. Hence, for families without farm mules, when engaged in farming production and land cultivation, they may need to communicate with those with farm mules and seek a solution, usually in the form of exchanging labor; i.e., each farm mule ploughing for a day is counted as one man-day, and so is the mule's owner responsible for the ploughing. In this way, if a farm mule, together with its owner, ploughs the land for a day, the other party need to work two man-days in return.[1]

In Jiada Village, when someone is building a house, harvesting corn, sowing seeds, or spreading manure, etc., many villagers will come to help with. During labor days, people eat three meals a day. The first meal is considered as breakfast, usually eaten at their respective homes around 6:30 am, and mainly consisting of wheat flour cakes and yak butter tea. Farm work starts from around 7 am, and generally the lunch time is around 12 am. Provided by the host, the food may include sliced pork that has already been cooked, and stir-fried potato with chili, with some yak butter tea afterwards. The lunch time is usually very short, less than an hour, and after that, people will resume work soon. The mode of harvesting corn is mainly by the family members, and sometimes their relatives, friends and neighbors would be invited to help, often by means of changing labor, and the working time is the same as above. This kind of mutual collaboration has been the survival rule of the villagers.

Moreover, when there is a shortage of manpower, they may even need to hire some handymen. Dengzeng, an old man from Jiujia Village, Yanjing, recalled that his father Meilan had always been traveling on the Ancient Tea-Horse Road for a long time, when it was still before liberation[2], and that his father transported the salt made by his own family to Zhongdian, Lijiang, and Dali in the direction of Yunnan by driving a horse caravan and hiring some handymen, so as to exchange for grain, cotton cloth, oil, and other daily necessities. At that time, they had more

---

1 Li Hechun. *Impetus and Shackles: Salt and Salt Taxes in the Lantsang River Canyon*. Guangzhou: Sun Yat-sen University Press, 2016: 70.

2 The People's Republic of China was founded on October 1, 1949, and Yanjing was peacefully liberated in 1950. The "liberation" time in this book generally refers to the peaceful liberation time of the local area, rather than the founding time of New China.

than 20 salt pans, and his mother was too busy to make salt alone; so they also hired a lot of handymen. But the number of handymen was not fixed; sometimes one was hired, sometimes two or four, and eight at most. The wages of these handymen were sometimes paid in salt, and sometimes in highland barley. Occasionally, other methods were also adopted. For example, if the handymen were responsible for the salt production in eleven or twelve salt pans, then the output from two or three of them might be paid to them as wages. Of course, the salt could also be converted into grain or Tibetan silver dollars.

Material production activities and production modes are the basis for the existence and development of human society. In the long practice of salt production, salt makers in the Yanjing area have already found out a set of production modes with unique and historically prescribed characteristics. As a long-standing livelihood, the existential reasonableness of salt making can be traced in the local physical geography and human history. The salt production process may seem ordinary, but upon careful analysis, it is found that the traditional salt making in Yanjing contains important ecological principles, which helps to solve the contradiction between human and land as well as material poverty.

Today, Yanjing is the only primitive artificial salt making site that has survived from earlier times in the world. The local residents worship everything related to salt, and everyone has their own simple salt industry ideology, which has thus formed the unique chaste and upright character of salt makers. Salt flavor, has not only long been integrated into the body of everyone in the local area, but also permeated every corner here.

CHAPTER TWO

Ancient and Traditional Salt-Making Technique

The history of salt making in China can be traced back to at least 5,000 years ago, almost synchronizing with the history of Chinese civilization in historical records. According to the sources, salt in ancient China is classified into sea salt, lake salt, well salt, and rock salt, etc; each type has its different techniques. Among these ancient salt-making techniques, well salt's is known to be the most complex and best reflects the wisdom and talents of ancient Chinese people.[1]

To this day, Yanjing, Markam County in Tibet still preserves the unique ancient salt making technique in the world. The salt production mode here, very rare in the world, is a most primitive type, i.e., by evaporating the brine under the sunlight. The tools are very primitive, the method, too. The perfect combination of wind and solar energy forms a unique condition for salt making. Its most typical feature is that the brine, having been carried into the salt pans, would crystallize into salt by wind and sunlight. As a way of survival bestowed by God, it is completely different not only from the traditional methods of scraping charcoal to extract salt or boiling the salt off the seawater in an open pot, but also from the vacuum salt production in China. For the purpose of clarity, it is ideal to start with the infrastructure, and then relate to the salt making tools and operation techniques.

---

1 Ling Xue et al. *A Preliminary Study on Technology of Well Salt Production Technique in Yanguan Town, Lixian County, Gansu Province*. Salt Industry History Research, 2012(02).

1. Infrastructure

The traditional salt making is always inseparable from a certain space and composed of a variety of elements. The saltern infrastructure is a necessity for salt production. Different salt production modes, under different natural and geographical conditions, require different infrastructure. These facilities, which demonstrate the traditional techniques of salt production, are an important component of salt culture, as well as a typical feature that distinguishes salt-making villages from other traditional villages. These engineering facilities are the material guarantee for normal salt-making activities, and their maintenance and improvement are of great significance to the production and living of local people.

1) Infrastructure Essentials

Due to the fact that salt making in Yanjing requires first carrying the brine drawn from the brine well to the brine storage tank for natural evaporation and concentration, and then pouring it into the salt pans as appropriate, resulting in the precipitation of salt particles after moisture permeation and exposure to wind and sunlight, such procedures as digging brine wells, putting up salt pans, and building brine storage tanks become indispensable. During the investigation in Yanjing, the author seemed to have walked into the historical scene, and the layout of the saltern, which was almost the same as the descriptions found in historical data, all accord with the characteristics of the terrain and climate. Salt infrastructure includes brine wells, public brine pools, and salt making operation zone, according to their functions. Brine wells, built by piling up irregular stone blocks, are situated by the river, and the public brine pools are close to them on the riverbanks. Salt-making operation zones, whose total number is over 100, are located on slopes or steep cliffs, and built layer by layer in accordance with the terrain, consists of two parts: private brine pools and salt pans. The following is an introduction in turn.

a. Salt Shelves

The salt shelves in Yanjing are known locally as wooden shelves. Due to the extremely narrow Lantsang riverside and the lack of flat land in the canyon, they, built one on top of the other and staggered orderly, were erected in a ribbon pattern on the steep slope from the ancient Tea Horse Road to the river. The purpose was to build the maximum number of salt pans in a limited space, while ensuring salt crystallization in each pan through exposure to sunlight and wind.

These wood-supported structures are collectively called the "salt shelves", each with a platform on the top for evaporating brine. Their construction is quite similar to the common stilted houses in the southwest mountainous region: set into a hillside, or isolated on all sides, their sizes and heights vary depending on the terrain. First, a foundation is built with stones, etc., and then whole round timbers are erected as a support, forming a frame. A salt shelf needs lots of such round wooden columns, whose lengths depend mainly on the gradient of the slope and whether the foundation of the salt shelf is solid. Due to the difficult terrain, long columns are set up in the front and short ones in the back. They are arranged in a network, with a distance of 1-2 meters between each other (see Figure 2-1). If the height difference between the front and the back is relatively large, and the front columns are not long enough, overlapping crossbars, which are mostly two or three connected sections, would be necessary. These round wooden columns are cut short according to the terrain, ranging from 1 meter to 5-6 meters. Near the shore, the columns are less than 1 meter, and the shorter ones stand directly on the ground, without wooden shelves. Such columns, in seven or eight rows at most, and three or four rows at least, are neatly arranged, thus ensuring well-distributed load-carrying capability. The columns having been erected, wooden beams are put up longitudinally or transversely on top of them, with a distance of 34 centimeters between each other.

Generally speaking, these wooden shelves range in size from 10 to 60 square meters, with each storey being 6-7 meters high (Nakhi Village, and Upper Yanjing Village) and the lower ones only 1-2 meters high (Jiada Village). A whole shelf, which may overlap up to seven or eight storeys, is tens of meters high (with the highest point 150 meters above the river surface), stretching 1.5 kilometers--a very magnificent and spectacular scene, indeed! These salt shelves are either adjacent to each other, or connected by walkways made of long planks, extending storey by storey from bottom to top. The upper and lower storeys, whose space design is very ingenious and flexible, are connected by a single-wood circular ladder. Between them there's a square hole with a width of about 80 centimeters, only big enough for one person to pass through.

Every block of wooden shelf has a small hole opened like this. Moreover, on the backside of each salt shelf, a hole with a diameter of about 20 centimeters would be opened by salt makers. And a salt basket is placed here, so as to strain off the water in the undried salt particles that have just been collected. The brine water drips into the brine storage tank below, and is thus recycled and utilized. In the long history of salt production, salt makers, deeply aware of the hard work of carrying brine, learned to cherish brine resources very much.

By putting up salt shelves near the spring on the riverside, they can draw brine from the salt spring with maximum efficiency. Salt makers, traversing between the columns, are busy with their tasks. Looked up from a low spot, the salt shelves very much resemble clusters of dark woodland, seemingly "thousands of fish scales arranged orderly" (see Figure 2-2). Walking below these shelves and shuttling between the columns, it feels like entering a castle built of a wood, or coming to an ancient labyrinth. If you approach these platforms for evaporating brine, and take a closer look, you'd find the clue: there are narrow crawlways leading down to the ground. The path under the salt shelves is extremely rugged and narrow. In some places, you can't even stand up straight, and your head will bump against the frame if you don't pay much attention.

Figure 2-1 salt shelves put up in steep terrain Village in the west

Figure 2-2 overlooking salt shelves of Yanjing

Since the platforms on the top of a wooden shelf are laid with wood blocks and soil, its leakage would be unavoidable. Some of the leaked brine directly drips into the brine storage tank below, while the rest drips into the soil or wood blocks below the salt pans. And over time, the brine water that has seeped under the shelves condenses into long stalactite-like columns of salt. At the slightest touch, these salt columns would break off with a crisp snap, running down your neck like ice or breaking into pieces in your palm. Lick it with your tongue, the salty flavor is simply unadulterated.

b. Salt Pans

Salt pans, also known as salt boxes or salt ponds, are actually divisions of the flat roof at the top of a wooden salt shelf, with a total area ranging from 9 to 50 square meters. In *Local Geographical Records of Yanjing, Batang*, Duan Pengrui mentioned that "the natives put up wooden shelves one on top of the other on either side of the river, and the salt pans, separated with boundaries, are very much like the inland paddy fields" (see Figures 2-3 and 2-4). And "Investigation in Yanjing County" also records: "The mountain being very steep, their width and lengths depend on the terrain where they are situated. One on top of the other, they are closely arranged like the teeth of a comb or the scales of a fish, their tops resembling fields with boundaries, known as salt boxes, or salt pans." According to local salt makers, the sizes of current salt pans are larger than before, but the shapes and construction method remain basically the same.

The salt pan structure consists of two layers: the first is made of fine wood, arranged closely together to form a plane, and the second is laid with wood blocks of different lengths, or mixed firewood and branches, and then coated with gravel and coarse sediment. The method of laying sediment is to separate it first into several heaps in the middle of a salt pan, then use a shovel to mix it thoroughly with some brine, and after spreading the mixture out, tamp and flatten it with a wooden hammer. Such a sediment layer is about 10 centimeters thick, with the edges around it 13-18 centimeters higher than the pan bottom, forming a rectangular "shallow sink" shaped with a flat bottom and raised edges, i.e., "the natives put up shelves in accordance with the terrain, and covered the salt pans on the top with mud, the bottoms being flat and edges higher" described in the late Qing literature. The salt pan bottoms must be made horizontal; otherwise, the inclined surface will cause uneven evaporation, resulting in brine loss and reduced production. Lastly, level and smooth the surface with clay powder already stirred into mud, and divide it into several squares with soil, patches of salt pans take shape.

The number of these pan patches on each wooden shelf range from three to four at a minimum, to about ten at a maximum, with the majority being four to five, to facilitate salt making. Most salt pans are rectangular, while fan-shaped, round, and trapezoidal ones, etc., are relatively fewer. The main reason is that such shapes are difficult to build and inconvenient for making and collecting salt. Usually, they are 3-6 meters long and 2-4 meters wide. Varying in size, the largest is no more than 10 square meters, or it might be difficult to support with round timbers, and small ones are 4-5 square meters, or even only about 2 square meters. Due to the terrain, the sizes of salt pans in Jiada Village on the west bank are larger, while those in Nakhi Village and Upper Yanjing Village on the east bank are comparatively small.

These pans, of different shapes and sizes, are built one on top of the other. Viewed from a high place, they resemble both miniature terraces and enlarged beehives. Some have just been filled with brine, sparkling like mirrors reflecting light; Some, the water in which has already evaporated, show a white layer of salt; some, which have not yet been filled with brine, expose their brownish-red (the salt pans of Jiada Village) or yellowish-gray (the salt pans of Upper Yanjing Village and Lower Yanjing Village) bottoms. Against the background of river water and farmland along the banks, these salt pans appear to be even more colorful, just like a huge "eyeshadow palette" with a slight gleam of red or yellow, or many mirrors collaged together, reflecting the sky light and cloud shadows, which is particularly eye-catching and striking in the desolate and empty canyon.

Most of the salt pans in Yanjing are in the form described above, but in Jiada Village, owing to its slightly flatter terrain, some are built by directly piling up stones.

Figure 2-3 overlooking salt pans of Lower Yanjing Village in the south

Figure 2-4 salt pans of Jiada Village on the west bank

The gaps between these stones, with a function for moisture permeation, can also result in salt precipitation. Such salt pans, 30 to 40 in total, are mainly distributed on the lower side of the road from Jiada Village to Quzika Township.

According to their locations, salt pans have long been divided into three areas, as is documented in old records: "On the east bank (in Yanjing), there are two areas for salt making, i.e., Puding and Yaka; on the west bank there is only one area, i.e., Jiada." [1] Now, Puding is known as Lower Yanjing Village, Yaka as Upper Yanjing Village, and Jiada as Jiada Village. The salt pans in Yanjing Nakhi Ethnic Township can be clearly divided into three parts according to the administrative villages to which they belong (see Figure 2-5). On the flat west bank, salt pans are dense and numerous, while on the relatively steep east bank, they are scattered here and there.

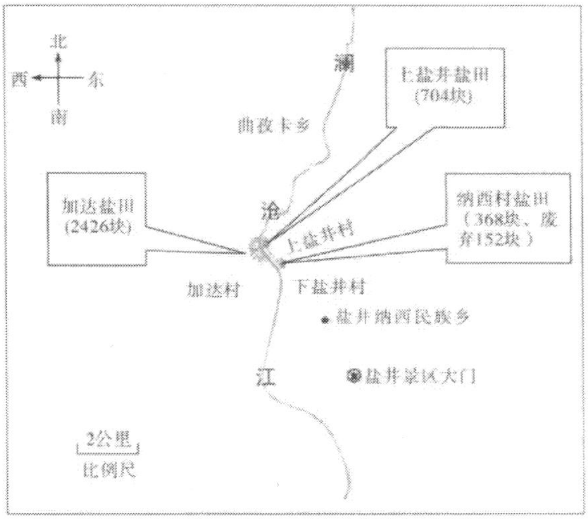

Figure 2-5 the distribution of salt pans

(1) Salt pans of Jiada Village, on the upstream west bank of the "S"-shaped bend, are the largest in number, density, and size.

(2) Salt pans of Upper Yanjing Village, on the east bank in the middle reaches of the "S"-shaped bend, facing the salt pans of Jiada Village on the opposite bank, appear to be extremely spectacular, for they have been built one on top of the other

---

1 Duan Pengrui. *Local Geographical Records of Yanjing, Batang*. National Library of China, stereotype edition of the second year of Xuantong (1910), p.13.

along the precipitous cliffs.

(3) Salt pans of Lower Yanjing Village, on the downstream west bank of the "S"-shaped bend, are not only the smallest in size and most dilapidated, but also constantly stricken by debris flows during the rainy season. Therefore, many salt makers either rent them out or give them to their relatives, or even simply abandon them. [1]

The number of salt pans also changes from year to year. According to the data in the first year of Xuantong (1909), there were "2,763 salt pans" at that time.[2]and the record during the Republican period shows, "on the east bank there were 1,242 salt pans ...; on the west bank there were 2,724 salt pans... a total of 3,966, ...this was the statistics in the second year of Xuantong (1910)." As to the exact number of salt pans in Yanjing, scholars hold different opinions since the 21st century. According to an investigation by Tao Hong and Huang Jian, "currently there are more than 300 wooden shelves and more than 1,500 salt pans in Jiada Village, more than 120 wooden shelves and more than 800 salt pans in Lower Yanjing Village, and more than 200 wooden shelves and more than 1,100 salt pans in Upper Yanjing Village"[3], much smaller than the numbers in historical records. Obviously, these were only estimated figures; statistics had not been actually conducted. Zhu Xia and Li Xiaocen's investigation suggested that there were about 3000 salt pans[4], an increase in number, which was somewhat different from the data of the above scholars. After that, Wu Chengli, based on his investigation, believed that there were about 3000 salt pans on east and west banks, including 1784 (more than 300 wooden shelves) in Jiada Village, 789 (more than 120 wooden shelves) in Lower Yanjing Village, and 1223

---

1 Li Hechun. *Impetus and Shackles: Salt and Salt Taxes in the Lantsang River Canyon*. Guangzhou: Sun Yat-sen University Press, 2016, p.40.

2 Duan Pengrui. *Local Geographical Records of Yanjing, Batang*. National Library of China, stereotype edition of the second year of Xuantong (1910), p.12.

3 "The Study of Mang Kang Tibet Salt Industry" by Tao Hong and Huang Jian, published in Salt Industry History Research, 2002(4).

4 See "Well Salt Production in Yanjing Township, Markam County, Tibet Autonomous Region" by Zhu Xia and Li Xiaocen, published in China Tibetology, 2007(3).

(more than 200 wooden shelves) in Upper Yanjing Village. [1]During an investigation in 2013, Dr. Li Hechun, linking it with the statistical data of each village, concluded that there were 704 salt pans in Upper Yanjing Village, 2426 in Jiada Village, and 368 in Lower Yanjing Village (excluding 152 abandoned), totaling 3498. If the abandoned salt pans were also counted in, the total number would have been 3650. It thus appears that salt pans on the west bank were more than those on the east bank.

During the People's Commune period in the 1960s, it was once proposed to lay salt pans with cement, but the evaporation effect was not ideal. Once filled with brine, such pans remained ponds after more than a month, and no salt could be precipitated. This incident showed that the indigenous method of salt pan construction, which had been passed down for thousands of years, seemed simple, but contained too much wisdom of their ancestors.

c. Brine Storage Tank

On the top of a salt shelf there are patches of salt pans, while on its foundation below are brine storage tank, so as to make efficient use of the space. Brine storage tanks, also known as brine tanks, or ponds, are called "salt pits" in historical records. As the name suggests, they are used to store brine. "Salt pits, dug between a couple of salt shelves, are similar to wet fields but slightly deeper, used for storing up brine water..."[2]Brine storage tanks are pit-type, mostly square, straight up and down, and a few smaller ones are inverted cone-shaped like "V" or pit shaped (see Figure 2-6). Their sizes vary a lot. According to the survey data, the depth is generally 1.0-2.5 meters, and the length and width are not fixed, either. Small brine tanks are generally 5 meters long and 3-4 meters wide, with a volume ranging from several cubic meters to tens of cubic meters. Some large tanks are about 8 meters long and 5 meters wide, and they were generally built by salt makers in the later period. Limited by the steep terrain, they could only be built in accordance with it, thus naturally the volumes are relatively smaller than elsewhere.

---

1 See "The Study on Salt Culture of Yanjing Nakhi Ethnic Township, Markam County, Tibet", Wu Chengli's Master's Thesis of the School of Sociology and Anthropology, Sun Yat-sen University, 2009, p. 66.

2 See "Journey to the Southwest of Kamu", in *A Compilation of Sichuan-Tibet Travel Records*, p. 444.

Figure 2-6 pit-shaped brine storage tanks

The brine storage tanks have three functions: First, when brine have been poured into the storage tank, highly-saline soil might be added to achieve naturally concentration within a short period of time, facilitating the full combination of brine and crystalline saline soil, and promoting the speed of salt formation; second, when brine has just been carried out of the brine well, the temperature is often up to 30°C~40°C, and in order to guarantee the quality of salt, it must be cooled to about 20°C in the storage tank before being transferred to the salt pans; third, the Lantsang River, whose water level rises sharply from May to August every summer, once overflowing, will submerge all the brine wells, and salt makers can only rely on the brine in the storage tanks in order to continue salt production and ensure the normal output. Thus, someone said, "before the river flooded, shrewd and diligent salt makers would try to dig tanks for storing brine water, in case the brine wells were submerged"[1]. In short, a brine storage tank serves as a transfer station and storage station for brine. Afterwards, if necessary, the brine in the storage tank could be scooped up and poured into the salt pans, or directly carried into them by manpower, or piped into them with a thin tube for evaporation.

Once the brine has been poured into the storage tank, it is necessary to add some substances with high salinity that may help to crystallize, such as wood removed

1 "Investigation in Yanjing County" by Jin Fei, published in *Political Affairs of the Border Region*, 1931(8).

from salt pans. These substances are highly-saline due to the long-term erosion by brine, and will be put into the storage tank when renovating the salt pans, as as to increase the concentration of brine and improve the quality of salt. Another example is the soil below the salt shelves, where brine continuously leaks and drips into the soil during evaporation, forming salt-affected soil. They are both salt makers' preferred additives for brine concentrattion. This method can not only promote salt precipitation by fully integrating the brine in storage tanks with the crystalline saline soil, but also improve salt quality by lowering the temperature of brine drawn from the brine wells to the ordinary water temperature.

Due to the fact that such stone-built storage tanks are prone to brine permeation, it is often necessary to inspect their surrounding areas carefully, lest all previous work come to naught because of the leakage of the hard-earned brine water. What is more, brine storage tanks are also vulnerable to the damage caused by river floods. Hence in addition to regular inspections, it must be renovated at the end of the year. Fortunately, since the application of cement pouring, people are no longer worried about the leakage in brine storage tanks. But after they have been used for a long time, a large amount of sediment of various wastes at the bottom need to be cleaned.

There are hundreds of brine storage tanks in Yanjing, one for each household. However, due to their different shapes and sizes, the exact number is almost unknown. During the second year of the Xuantong period in the late Qing Dynasty (1910), it was probably "six hundred and eighty-nine"[1]. During his investigation in 2013, Dr. Li Hechun, based on rough statistics, found that on the east bank there were about 120 brine storage tanks for salt pans in Upper Yanjing Village, and about 70 for salt pans in Lower Yanjing Village, and on the opposite bank nearly 100 for salt pans in Jiada Village, with a total of about 300 on both banks.

In recent years, some local salt makers have begun to build larger brine storage tanks with stones and cement, with a volume four or five times, or even eight times, that of the past. So currently, the largest storage tank has a volume of over 400 cubic meters.

---

1 Duan Pengrui. *Local Geographical Records of Yanjing, Batang*. National Library of China, stereotype edition of the second year of Xuantong (1910), p.13.

d. Brine Wells

The most basic thing for salt making in Yanjing is brine. On both banks of the river, there are saline springs gushing out. And in order to draw sufficient brine, villagers have built deep wells or pools on the shoresides adjacent to the river. The brine looks yellowish or greenish, in sharp contrast to the red river water. All brine wells are distributed along the river in accordance with the terrain, with the nearest being tens of centimeters to the river, and the farthest only more than 10 meters. The well bottoms are parallel to the riverbed, and the well mouths are almost parallel to the river surface--the higher ones are only about 3 meters above. In terms of dimension, they can be classified into small brine wells and large brine wells. The small ones, closest to the river, are square in shape, with a width of 0.6-2 meters and a depth of 1-2 meters; while large ones, generally more than 5 meters deep (even as deep as 15 meters), are mostly circular, and known as large mouth wells.

Cui Kexin, an investigator to Yanjing during the republican period, also described: "On the right bank, the diameter of the brine wells is generally about 1.5 meters, and the depth varies from 2 to 5 meters. But most of these wells are 2-3 meters deep, and the well mouths are all inlaid with wood in a rectangular or square shape, with the longest side being 1.5 meters, the shortest 0.5-1 meter. There is only one well whose circumference is also inlaid with wood. The farthest well is not more than 5 meters away from the river, the nearest only 1 meter. The brine wells located on the left bank, known as salt ponds, are very wide and shallow, unlike the shape of a well. The lengths of their sides are generally 3-4 meters, the depths not more than 1 meter. None of them are inlaid with wood. Except one or three wells are 5-6 meters away from the river, less than 1 meter above the river surface, the rest are 10-20 meters away from the river, more than 5 meters above the river surface.[1] Obviously, compared with the "Bamboo Salt Well" in the Song Dynasty and the small-caliber deep wells in the Qing Dynasty, Yanjing salt wells are not only much larger in caliber, but shallower in depth. What is more, in terms of size dimension, the latter is far lagged behind the sinking technology of the Pujiang salt wells in Sichuan in the Tang Dynasty. The sinking technique in Yanjing Township, lagged behind that of

---

1 "Investigation of the Geology and Salt Production in Yanjing County", by Cui Kexin, published in *Xikang Economic Quarterly*, 1944(8).

the Bamboo Salt Well in the Song Dynasty, still remains at the level before the Tang Dynasty at least.

As brine resources are rich in Yanjing, salt makers can dig and sink wells freely. But how to select proper locations is a matter of skill. Often, they are selected and construction begins in winter, and it is also necessary to invite experienced villagers to examine the riverbed and shore area carefully in advance. Preferably, select a spot where brine is abundant, easy to be drawn, and not prone to be submerged or washed away by floods, and dig a little by making use of natural rock fractures or vugs. When a mouth of saline spring has been discovered, if it is in the gravel stratum, salt makers would often directly dig it manually, using such tools as hoes and pickaxes. For in this situation, the work becomes relatively easy and does not take long, ranging from only one day to a few months. But if the mouth of spring is in the rock stratum, where drilling and blasting operation would be unavoidable, it will take a little more time. Generally, brine wells are 1-2 meters deeper than the brine aquifer. By renovating and enlarging a spring mouth on the riverside, one can clearly see the continuous gushing of saline spring from the crevices in the well below when standing on the side. The amount of gush can reach 20 liters per second, like the description in the record: "...the brine well is less than 1 *zhang*[1] deep, and brine gushes out like a spring..."[2]

Of course, the degrees of brine abundance vary a lot in different wells. According to an old record: "On the east bank there are two brine areas, where a well is about 2 *zhang* deep, with one or two streams of spring veins at the bottom. The brine, exuding a little slowly and varying in quantity in each well, is thick in quality and strong in taste...On the west bank there is only one brine area, where a well is only several *chi*[3] deep. The brine, exuding very quickly, is thin in quality and light in taste..."[4] At the mouths of some high-yield wells, "enclosing walls" are built with stones and mud, resembling cylindrical pillboxes. An access at the lower part of each

---

1 1 zhang is equal to 3.3333333333333 meters.

2 See "Journey to the Southwest of Kamu" in *A Compilation of Sichuan-Tibet Travel Records*, p. 444.

3 1 chi is equal to 0.33333333333333 meter.

4 "Investigation in Yanjing County" by Jin Fei, published in *Political Affairs of the Border Region*, 1931(8).

well, big enough for only one person to pass through and about 3 meters above the ground, is opened for drawing brine water during floods. According to statistics, there are a total of 7 such wells on both banks, including 1 in Jiada Village, 2 in Nakhi Village, and 4 in Upper Yanjing Village. All dug collectively, their interior has been treated simply with stone or cement, with a diameter of 1.5-2 meters and a depth of 5-10 meters. Except for the 7 circular old wells, the shapes of others are generally irregular. Their diameters and depths vary, too, the largest being 2-3 meters and the smallest only tens of centimeters in diameter, and the depths ranging from tens of centimeters to about 11 meters--mostly less than 5 meters. For example, there is a brine well in Jiada Village, whose mouth is 3.1 meters at the widest part and 2.7 meters at the narrowest part, but the depth is only 2.75 meters, with the minimum caliber being only 0.64 meters. Most of these wells are built with stone steps, against which a simple wooden ladder or single-log ladder (a thick log carved with foot grooves) is leaning beside. Such a ladder, almost upright, allows people to get down into the well for brine. They are generally raised by putting stones underneath, so as to avoid brine corrosion. In summer, with the outbreak of floods, one side of the leaning ladders is submerged, suggesting the process of brine formation, i.e., that the riverbed might have been cut by river water, causing water to soak into the stratum, and "dissolve salt in the salt bed into water"[1].

Figure 2-7. the brine well in Yanjing, Markam County

1 "Investigation of the Geology and Salt Production in Yanjing County", by Cui Kexin, published in *Xikang Economic Quarterly*, 1944(8).

Brine wells can also be classified into public brine wells and private brine wells. It goes without saying that the former are used by all salt makers here, while the latter are exclusively used by a certain household. The earliest brine wells have already been in use for hundreds of years, and more interestingly, they are usually named after local geographic names or family names. Currently, there are five brine wells (brine pits) named in this way in Jiada Village: Qumo, Bako, Drosago, Jidon, Moller. Among them, the one named "Qumo" is a public brine well.

Because well-digging is quite easy on the riversides, and such brine wells are frequently flooded in summer, their number varies from year to year, and from one historical period to another. In the first year of the Xuantong period in Qing Dynasty (1909), "according to investigation, there are more than 30 wells on the east bank of the Lantsang River, and 20 on the west bank"[1], a total of about 50; it is also said that there were 53 brine wells during the Xuantong period[2]. "There are 52 wells", as is recorded in *Local Geographical Records of Batang and Yanjing*[3]. This number was increased in the later *Investigation in Yanjing County*, which recorded: "There are 24 wells on the east bank, and 31 on the west bank; a total of 55."[4] In the 1930s, for a time there were 60 brine wells.[5] In 1976, when the Qamdo Detachment of the Qinghai-Tibet Plateau Comprehensive Scientific Expedition of the Chinese Academy of Sciences visited Yanjing Township, it was mentioned that "there are 24 relatively large brine wells, with depths ranging from one meter to several meters..." According to the statistics of Tao Hong and others, the number of brine wells on both banks

---

1 *Archival Historical Data of Sichuan-Yunnan Border Affairs in the Late Qing Dynasty (Middle Volume),* compiled by Sichuan Institute of Ethnic Studies and the editorial group of *Archival Historical Data of Sichuan-Yunnan Border Affairs in the Late Qing Dynasty*, published by Zhonghua Book Company, 1989 edition, p. 446.

2 See the section of salt pans in *Local Geographical Records of Batang and Yanjing.*

3 Duan Pengrui. *Local Geographical Records of Yanjing, Batang*. National Library of China, stereotype edition of the second year of Xuantong (1910), p. 13.

4 "Investigation in Yanjing County" by Jin Fei, published in *Political Affairs of the Border Region*, 1931(8).

5 Zeng Yangfeng. *History of salt administration in China.* Commercial Press, 1937 edition, pp. 83-84.

of Lantsang River are 79, including 58 on the east bank and 21 on the west bank.[1] According to Dr. Li Hechun's investigation in the past few years, there were 61 brine wells on the east bank and 17 on the west bank, totaling 78. Among the three salt-making villages, Upper Yanjing Village is the most densely distributed one of brine wells, with over 50 large and small wells in an area with a length of more than 100 meters and a width of more than 20 meters.

Although it was recorded in *The Well Salt in Yanjing County, Xikang* that: "In the rainy season, when the river rises and the original wells are submerged, the source of brine will also rise with the river water, and the brine might be drawn by digging a hole at a higher place...The number of wells can be increased at any time without restriction." Unfortunately, according to local salt makers, some brine wells have been abandoned because of river floods or being out of use for a long time. A few years ago, the Tibet Industry Department tried to develop local mechanized brine wells by investing about 200,000 *yuan*, but failed due to lack of electricity.

e. Road and Path Network

In salt production, although the accessibility condition near salterns are not an essential, its importance can never be ignored due to the impact on salt delivery. It mainly involves three types.

(1) Trails inside the Salterns

Simple roads generally lead to the saltern sides, while the internal areas are connected by little trails, which are wide enough only for people, mules, and horses to walk on. In salterns of Upper Yanjing Village, mules and horses can only reach the highest flat part of the salt pans. The trails between salt shelves are easy to walk on, but the section from the salt shelves to the brine wells would be quite difficult. Many spots are steep and narrow, only enough for one single person to pass at a time. In general, the saltern trails in Garda Village are relatively flat, and those in Upper Yanjing Village and Lower Yanjing Village are more rugged.

(2) Roads Leading to the Salterns

The salterns located on both sides of the Lantsang River are at different distances from each village. The salt-maker households in Upper Yanjing Village and

1 "The Study of Mang Kang Tibet Salt Industry", published in Salt Industry History Research, 2002(4).

Lower Yanjing Village on the east bank, in order to work in the salterns, have to walk a slightly longer distance, going down and climbing the mountains, which is not as convenient as their counterparts in Jiada Village on the west bank.

There are two roads from the Yanjing Township Government's location (Lower Yanjing Village) to the salt pans: One slopes down along the asphalt road paved for the development of salt pans in the scenic area, first leading to the salt pans in Lower Yanjing Village on the east bank of the Lantsang River, and then to those in Jiada Village on the opposite bank after crossing the reinforced concrete bridge over the river. This journey is about 4.5 kilometers, and the road runs down the mountain with numerous narrow bends, which can easily cause traffic accidents. If one chooses to go to the salt pans in Lower Yanjing Village in a motor vehicle, it will take about 15 minutes. And it will take about one hour on foot, or at least one and a half hours to return from the salt pans to where the township is situated. This road is mainly designed to facilitate passengers to reach the salt pans. The other road, or the main passage through which the salt makers and the horse caravans go to the salt pans, heads towards the right side of the township government and down the canyon along the farmland edge. This road, wider than ordinary mountain paths, is obviously suitable for horses and mules to walk on, and they would not bump against either side when carrying a load. All the way down is the river valley, with abandoned houses on both sides of the creek. According to the elderly of Lower Yanjing Village, there were originally eight or nine households here, but later, in response to the call of the township government, they all moved to the streets. Passing through the village and going straight down, one would arrive at the salt pans of Lower Yanjing Village; then from here, crossing the creek and climbing up the cliff, one would reach the salt pans of Upper Yanjing Village. As it is a downhill road, it takes half an hour to go from Upper Yanjing Village to salterns on the riverside; and then from there to the salterns of Upper Yanjing Village, due to the uphill road, which is rugged and narrow, it would take another half hour. Tourists often take this route to enjoy the sight of salt pans, but dare not go forward when reaching this section.

Upper Yanjing Village is linked to the salterns by little paths; it takes 30 minutes to descend and 40 minutes to ascend. If one walks along the road, instead of the little paths, it will take an extra hour by making a detour. The salterns of Jiada Village are right on the edge of the village, and the road is wide. Currently, the dirt road from Jiada Village to Quzika Township is still accessible to horse caravans and motorcycles, but ordinary vehicles are rarely seen.

(3) Roads Connecting Salt Pans to the Outside

Since the salt pans are distributed in the valley, with Jiada Village almost located along the Lantsang River, and the villagers of Upper Yanjing Village and Lower Yanjing Village living on the tableland by the riverside, access to and from the salt pans would be very necessary for these three salt-making villages.

In the old days, business travelers took the widely reputed "Ancient Tea-Horse Road". Until the 1950s and 1960s, when horse caravans were carrying salt to exchange for food, because there was no other road, they would still travel along the Ancient Tea-Horse Road, and cross the river by bamboo ropeways if there was no bridge. Interestingly, horses were also tied to bamboo ziplines, and they had long been accustomed to this exciting way of crossing the river.

In the 1970s, the National Highway 214 (also known as the Yunnan-Tibet Highway) ran through Yanjing from Deqin County, Yunnan Province, connecting the Sichuan-Tibet Highway in the north. Afterwards, it was built from Yanjing to Gatuo Township in the county town, becoming an important passage for Yanjing to exchange with the outside world, as well as the nearest route from Tibet to the coast. Previously, it took 9 days for horse caravans to come to Yanjing from Shangri-La, but after the road was opened to traffic, it would only take 6 days. Today, this 310-kilometre journey takes just seven or eight hours by motor vehicle. It used to take one day for a horse caravan to cross the Lantsang River by ropeways; since the bridge was opened to traffic in 1983, it only takes two minutes to cross the bridge now.

Considering different passages between the villages in Yanjing, it can be seen that the main route from Upper Yanjing Village to Lower Yanjing Village is National Highway 214. While from Upper Yanjing Village to Jiada Village, people used to take a mountain path, and afterwards, in order to develop tourism, a section of asphalt road from Yanjing Checkpoint to the salt pans of Lower Yanjing Village was built with the investment of 4 million yuan by Tibet Hongji Group Co., Ltd. Since the beginning of the 21st century, a reinforced concrete bridge spanning the Lantsang River from the east bank to Jiada Village has been built, and it is wide enough for two vehicles to pass in parallel. The suspension bridge, which was only used for horses and mules, has gradually been forgotten, and travelling by ziplines across the river has simply become a thing of the past. Under the aid-Tibet project by the Chongqing Municipal Government, the road surface from the riverbank to Jiada Village has been paved with cobblestones. In winter, people organize themselves and constantly renovate the road from Jiada Village to Quzika Township. As such a road connects the two places, making it more convenient to get around, villagers are willing to input labor to level the road surface. Starting from Quzika Township, one can also first cross the Lantsang River, climb up the "Z"-shaped road, and then reach Lower Yanjing Village through National Highway 214 in a clockwise direction. In this way, a circular transportation route took shape, linking Jiada Village, Lower Yanjing Village, Upper Yanjing Village, and Quzika Township. [1]

The continuous improvement of transportation conditions plays an important, sometimes even decisive role, in fully developing and utilizing local well salt resources and promoting the economic development of the salt industry. Moreover, it also has a significant impact on the production and lifestyle of salt makers and the transformation of their concept of life.

2. Maintenance of infrastructure

(1) Maintenance of Old Wells and Construction of New Wells

For salt makers, brine wells, as a source of brine, are also of utmost importance. On the one hand, they need to maintain the normal use of old wells, and on the other hand, they also have to find new brine to build new wells.

Above all, the biggest trouble in the salt production in Yanjing area is the damages of brine wells caused by floods. Once there is no brine supply, the production of the salt industry will be directly affected. Therefore, old wells need to undergo periodic manual maintenance. During the rainy season from May to September each year, the waves are rough and the river water rises by nearly ten meters; many brine wells on both sides of the river are in danger of being submerged by floods. When the river recedes, salt makers will have to spend two or three days

1 Li Hechun. *Impetus and Shackles: Salt and Salt Taxes in the Lantsang River Canyon*. Guangzhou: Sun Yat-sen University Press, 2016, p.47.

clearing the sediment from the wells. As a predecessor described: "The rise and fall of the river, from April and May to July and August, saw a surge of several *zhang* in water, with turbulent waves, and all the brine wells on either bank would be submerged...When the water recedes, the wells are completely blocked by sediment, whose cleanup work takes a long time. And this happens every year. "[1] Thus, salt makers took many precautionary measures, the most effective of which was to "encircle the wells with walls" beyond the embankment, i.e., to build blockhouse-like enclosures several meters high around some high-yield wells with stones, so as to ensure brine drawing during floods. (see Figure 2-8). Of course, the cost is very high. "Once the brine source is depleted, the situation is difficult to sustain, and the first step to make up for the loss is to protect the brine wells. Apart from building embankments, there is a method of encircling wells with walls. But as the raging waves surge, the sound vibrating the valley, with a differnce of several *zhang* between the ups and downs, it cannot be achieved without a huge sum of money. "[2]In addition, salt makers keep raising the height of large opening wells, so that when the river swells, the well mouths are not likely to be submerged. And when the door opened on one side of the wells is firmly blocked with wooden planks, the amount of sediment entering these wells would be reduced effectively.

As for the construction of new brine wells, it is also an indispensable project. But due to the active geological structure of Yanjing, earthquakes are prone to occur. This could be extremely damaging to such infrastructure as brine wells. According to *The Eastern Miscellany* and *The Chinese Newspaper Shenpao* (*Shanghai News*), an earthquake of around magnitude 6.75 occurred in Yanjing in 1921, which not only injured and killed many people and livestock, crushed houses, damaged salterns, but also caused the locations of saline springs to move forward. Therefore, after the earthquake, salt makers need to continuously increase the height of brine wells, or choose new sites to build brine wells. [3]

---

1 "Investigation in Yanjing County" by Jin Fei, published in *Political Affairs of the Border Region*, 1931(8).

2 "Investigation in Yanjing County" by Jin Fei, published in *Political Affairs of the Border Districts*, 1931(8).

3 *The Livelihood Model of People at the Two Foothills of Biluo Snow Mountain* by Li Hechun and Li Yafeng, published by Sun Yat-sen University Press, 2013 edition, pp. 178-179.

Figure 2-8 flooded brine wells during the sudden rise of river water

The construction of new brine wells requires a large amount of manpower, as well as material and financial resources. After the founding of the People's Republic of China, salt pans were redistributed many times, and brine wells gradually became public ownership here. As a result, the construction of brine wells has become a collective activity, which is no longer a matter for any village group or individual. The organization of labor force, centered around the village director, calls for voluntary work of all salt-maker households. Funds are raised to purchase the necessary cement, and the transportation of other required materials, such as coarse sand and building stones, etc., relies on organized labor. If government support can be obtained, villagers only need to put in a certain amount of labor. The brine well construction, as a continuous project that requires large-scale labor input, whether on the east bank of west bank, would involve nearly 100 villagers in the whole process (see Figure 2-9).

In April 2013, Jiada Village built a relatively large brine well after receiving a financial aid of of 200,000 yuan from the National Bureau of Religious Affairs. It is understood that this was due to the relatively flat terrain of Jiada Village and its location at the bend of the Lantsang River. When it rains, the lower brine wells would be easily submerged by the surging river water; therefore, a larger brine well was necessary to be built this time. In mid-March, a team of 70 to 80 people were busy building brine wells. The fund allocated by the government was mainly spent on cement; as to other materials required for the construction, such as sand and building stones, etc., the villagers would have to find a way to settle it themselves. In doing so,

Figure 2-9 collective construction of brine wells by Jiada villagers

the division of labor between men and women was relatively clear. Men were mainly responsible for concreting brine wells, mixing cement, and prying stones, while women were responsible for carrying stones and cement, and picking stones. When a man found a suitable stone by the river and pried it up with a crowbar, a woman would carry it on her back to the brine well with a rope. According to the director of Jiada Village, villagers started working from 8 a.m., broke for an hour after 12 at noon, and resumed work at 1 p.m. until 6:30 in the evening. It has been observed that this construction team of brine wells required at least 40 people per day, with a maximum of over 80. Except when it rained, everyone had been working for six or seven days straight. At such intensity, the whole construction work of the brine well lasted for an entire month. In general, people are willing to participate in the brine well construction, because it provides a guarantee for their salt-making in the future.

(2) Renovation of Brine Storage Tanks

The brine tanks need to be renovated every year. And when doing so, wooden columns that were previously used and old saline soil are put in, to better enhance the concentration of brine water.

Likewise, damages to brine storage tanks caused by earthquakes cannot be ignored, either. As mentioned earlier, the 1921 earthquake "injured and killed people and livestock, crushed houses, as well as brine tanks and saline lands, in large quantity." When the rainy season comes, these brine tanks are liable to be destroyed, too. For example, on August 21, 1980, torrential rains hit Yanjing and triggered flash floods and mudslides, washing away 170 brine tanks and 34 salt pans; as the newly-built brine wells were also submerged, salt production was forced to stop. [1] In order to be more durable and earthquake-resistant, the building materials of brine storage tanks are also constantly being updated. Before the liberation of Tibet, brine tanks were mostly built with stones; thus a large number of stones were needed, making them into "pit"-shaped. Since the production of cement with strong structural force, it is no longer necessary to build thicker sidewalls. As long as the concrete is poured into a wooden mold prepared in advance, and evenly shaken with a vibrator, a brine storage tank would take shape in two or three days when the mold is removed.

The construction of a brine storage tank also consits of several procedures. First of all, like building salt pans, it is to choose the proper site. This step calls for careful study, and the site should not be too far from the salt pans. If a salt-maker household owns more than a dozen salt pans, it is even more necessary to comprehensively consider the distance between the brine storage tank and each pan. If it is too far away, carrying brine from the storage tank to salt pans will waste too much labor and time. Besides, since brine storage tanks generally need to be built with a certain volume, it is often necessary to dig deep into slopes where the terrain is uneven, and if the soil there is mixed with sand and gravel, the work would be even tougher, only fit for hitting slowly with a pickaxe by those who are physically strong. If the slope is particularly steep, it's even more difficult to dig a rectangular brine tank (see Figure 2-10). In this situation, the method adopted by salt makers is to dig only a half, and the other half is concreted with stone and cement.

---

1 See *Markam County Annals* compiled by the Compilation Committee of Markam County Local Chronicles, published by Bashu Publishing House, 2008 edition, pp. 37.

Figure 2-10 a salt maker of Upper Yanjing Village is building a square brine storage tank

The second step in the construction of brine tanks is to carry stones, which usually come from the riverbank, and the salk makers, with baskets or ropes, could only carry one or two at a time. The selection of stones needs to be careful, too. One must try his best to choose stones with regular shapes and smooth surfaces; in this way, it will be relatively easier to form a masonry block and save materials. Some salt makers may adopt the concreting method mentioned above. Generally speaking, to build a brine storage tank with stones would consume a lot of labor; there are usually three or four people busy laying stones while pouring concrete into the gaps. Tashi Yangdzong's family once built a brine storage tank of this type, and according to her husband, it took as many as fourteen or fifteen days from digging the foundation to the completion of works. When researchers went to the salt pans with him, the brine storage tank had just been built for four or five days, filled with water, and he was concreting its exterior walls by contintuously pouring water onto them with a bucket. Later, we learned that it was because the brine storage tank had not yet been fully shaped, and if not watered, cracks may easily appear in the tank walls, resulting in water leakage. However, the salt makers told us, although such a brine storage tank consumes a lot of materials and labor, once built, it can be used for more than ten years.

(3) Maintenance of Salt Pans

As the main site for salt makers to carry out salt production directly, the importance of salt pans is self-evident. Salt shelves are not built once and for all. On the one hand, as the main structures are supported by wooden frames, after repeated use for a period of time, the salt pans on their tops, which have already been eroded by brine, need to be thoroughly renovated through disassembling in order to enhance the salt quality and production; on the other hand, due to such natural disasters as heavy rains, mudslides, and earthquakes, wooden shelves and salt pans built on slopes are liable to be damaged, too. For salt pans, "the rise and fall of the river water is most difficult to prevent...when a flood breaks out, even if the salt pans have been build at a high site, there is a risk of being washed away"[1]. According to literature records, the earthquake which hit Yanjing in 1921 also destroyed a large number of salt pans[2]. Hence, no matter what factors may cause damage to them, the salt pans should be renovated in time lest salt making be affected.

Normally, salt makers would walk around the shelves for inspecttion every time after collecting salt, and if any problem is found, they will deal with it promptly. At the end of the year, overall maintenance is still necessary to ensure the safety of salt workers in production. There are two types of renovation activities: temporary and periodic. Temporary renovation is carried out immediately if there's a leak of brine found in the salt pans or if the wood has decayed. While periodic renovation is generally carried out at an interval of about one year, or in the third month of the Tibetan calendar; the entire wooden shelf or part of the salt pans is uniformly repaired (see Figure 2-11), including the replacement of rotten wooden columns in the structure and rotten wood in the bottom layer of these pans. To reduce the amount of labor, before the repair work starts, the sandy soil of the salt pans is usually dug up and heaped in a corner of the wooden shelf first, for later reuse. Then, the floor slabs (wooden blocks) beneath is taken down and immersed in the brine storage tank

---

1 "Investigation in Yanjing County" by Jin Fei, published in *Political Affairs of the Border Region*, 1931(8).

2 Relevant reports from *The Eastern Miscellany* and *The Chinese Newspaper Shenpao* (*Shanghai News*) on May 25, 1921.

below for 2-3 days, so that the salt contained in the wooden blocks might be fully dissolved. This not only increases the concentration of brine in the tank, but is also conducive to corrosion prevention of the wood; the soaked wood blocks could then be reused to build salt pans. It is worth mentioning that the columns do not need to be soaked in brine water. Periodic renovation also includes overall reconstruction of the wooden shelf and salt pans every 10 years or so.

Figure 2-11 salt makers in Jiada Village are demolishing salt pans

The renovation of salt pans involves the following steps: Firstly, materials should be prepared and processed; trees are felled in the mountains and transported back to the saltern. When the timber is half-dry, it can be used to build a salt shelf. At this moment, with the help of a skilled craftsman, the timber is processed into columns or beams of the required sizes. And in accordance with the terrain, a main frame is put up, its top laid with a layer of logs or wooden blocks, and the gaps would be filled with branches or wood pieces if there are any. After that, a layer of coarse sand and gravel is laid and leveled, and finer clay is selected, watered, soaked for about ten minutes, mixed into mud, and then smoothed with a mud palm; the edges, laid with wooden blocks, are reinforced with mud. The renovated salt pans must be fully dried before new brine water is poured in for another round of salt making. But if they are not filled with brine water for two or three days, the salt pans will become too dry and crack. Up to now, the basic renovation of salt pans has been completed (see Figure 2-12). It does not take very long; generally speaking, four or five people can finish it in two or three days.

In a word, these salt pans are constantly constructed, demolished, and then constructed and demolished again and again, repeating the cycle.

Figure 2-12 women of Jiada Village are repairing salt pans

repairing the salt pans by patting

repairing salt pans

II. Salt Making Tools

As the saying goes, "A workman must (first) sharpen his tools if he is to do his work well." The importance of tools for productive activities is absolutely self-evident. The hand-made tools for salt making through solarization in Yanjing are not only closely related to the natural environment there, directly showing the local production level, but also reflect the infinite wisdom of generations of salt people. These salt making tools, simple and applicable, could be basically classified into three categories: tools for conveying brine, tools for renovating salt pans, and tools for collecting salt.

1. Tools for Conveying Brine

In the early days, people in Yanjing salterns conveyed brine by carrying it on their backs or shoulders, and the tools were brine scoops, pails and shoulder poles, etc.

1) Brine Scoops

A brine scoop (pronounced "Zhengzhen" in Tibetan) is a tool for scooping brine from brine wells into a brine pail. Generally sewn from birch bark, it's "a slightly round-cornered rectangle in shape, with a flat bottom and curved walls, and at the joints its parts are fixed with leather cords, birch bark shims, and inserted pins made of wood. The mouth is 20~22 centimeters long, 11~16 centimeters wide; the bottom is 19 centimeters long , 8 centimeters wide, with a height of 9 centimeters and a wall thickness of 3 centimeters"[1] (see Figure 2-13). They are used in collocation with brine pails, one for each pail. In a brine well, it's used to scoop brine water, and when the pail has been filled, it would be placed upside down floating on the water surface to keep it from spilling while being carried along the rugged paths. We also saw salt makers scooping up the river water with it for drinking; a small scoop has multiple uses here.

Figure 2-13 a real product photo and structure drawings of the brine scoop

People also used to scoop brine with gourd ladles and iron ladles. However, as they were easily oxidized and not durable, these iron ladles were abandoned soon.

1 The Institute of Cultural Relics Protection of Tibet Autonomous Region, Shaanxi Academy of Archaeology, and Sichuan Archaeological Research Institute. "Investigation report of Salt Pans in Yanjing, Mangkang County, Qamdo Prefecture, Tibet Autonomous Region". *Cultural Relics in Southern China*, 2010 (1).

Later, people switched to scoops made of plastic, which avoided either of the above shortcomings, and thus became comparatively popular in the locality. It can be seen that in different periods, the types of scoops used by people may differ from one another. By constantly exploring and trying in their production practice, they would make optimal choices based on practicality.

2) Pails

The brine pails is a type of unique tool used by Yanjing salt makers for brine conveyance. Mostly made of such locally produced raw materials as black goat wool yarn, and birch bark, etc., they are very good-looking in shape and design, durable, and full of a strong flavor of rural life. The style is rough and simple, but not lacking in delicacy and refinement. According to the material and shape, it can be classified into the following types.

The first type, pronounced "Dong" in Tibetan, is a container for women to carry brine on their backs in Upper Yanjing Village and Lower Yanjing Village on the east bank of the Lantsang River, which, made of wood or bamboo, is quite primitive. Due to the steep terrain here, this type of brine pail is characteristically deep, thin, and long, as recorded in the old local chronicles: " It's long tube-shaped thingmade of wood, and carried on the back with straps." [1]According to local carpenters, the wood for making brine pails comes from poplar trees. After it is cut down, the tree would be dried in the shade, which usually takes several months. To make a brine pail, the timber is first cut with a saw to the desired length; usually it is cut into large wooden blocks with a large saw, and then into smaller slabs with a knife, etc. These small slabs of wood need to be flattened with a plane, and ultimately made into think planks, each with a smooth surface. When put together, it's inside has to be stuffed with soemthing to hold the shape. The finished barrel approximates a cylinder; with the diameter at the bottom end slightly smaller, it gradually becomes thicker the higher up, which is deliberately designed by the carpenters in order to reduce the burden on the shoulders. The bottom diameter is 20~25 centimeters, whereas the mouth diameter is 22~27 centimeters, about 2 centimeters wider. The pail wall consists of 26 wooden planks (or 15, varying in number), with an outer

1 "Investigation in Yanjing County" by Jin Fei, published in *Political Affairs of the Border Region*, 1931(8).

height of 60 centimeters, a bottom diameter of 25 centimeters, and a wall thickness of 0.8 centimeters. Each plank is 2.5~5 centimeters wide, with a spacing of 6~10 centimeters between the binding iron wires. Tightly bound with iron wires, the wooden planks of the brine pail are closely attached to each other, ensuring that the it is not prone to water leakage. (See Figures 2-14 and 2-15) This type of brine pail is necessarily designed with shoulder straps, which, pronounced "Donglong" in Tibetan, are tied to the lower part of the "Dong" and its upper part close to the carrier's shoulders. Generally woven from fine wool, it is mainly characterized by soft texture, excellent durability, and high tenacity, with a length of about 1.5 meters (depending on the the diameter of the pail) and a width of about 4 centimeters. In addition, it is also necessary to go with a back cushion (pronounced "Tele" in Tibetan). The cushion, whose outer part is handsewn from linen and inner part stuffed with wool or cotton wool, is square in shape, with a side length of about 40 centimeters and a thickness of 3~4 centimeters. The function of the cushion is to avoid wetting clothes, and reduce the friction between the waist and the pail bottom (and its wall), thus alleviating the harm to the body during weight bearing. This type of brine pail, when filled with brine water, weighs about 30 kilograms. But to prevent the brine from spilling out when carrying it their backs, salt makers, instead of filling the pail to the brim, usually fill about 2/3 of its capacity.

Figure 2-14 brine pail

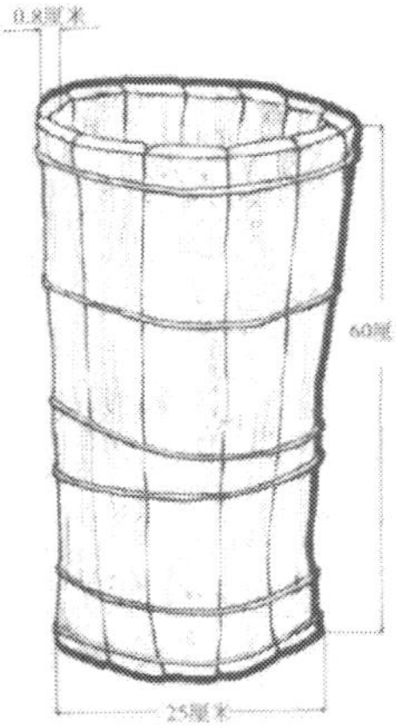

Figure 2-15 the size of a brine pail

Another type is the bark pail, made from birch bark or cherry bark. Round in shape, with a small mouth and a larger bottom, as well as slanted straight wall, it is sewn with woollen cords at the brim and part of the wall, and round the circumference of the bottom. With a mouth diameter of 24 centimeters and a bottom diameter of 41 centimeters, its brim is encircled with birch bark of 6.8 centimeters wide, while its bottom is encircled with birch bark of 5.2 centimeters wide; and the longitudinal birch bark belts on the outer side is 8~19.2 centimeters wide from left to right. Being a container of both brine water and salt, this type of pail is mostly used in the salt pans of Jiada Village, where the terrain is relatively flat, but in Upper Yanjing Village and Lower Yanjing Village it is hardly seen (see Figure 2-16 and Figure 2-17). As it is suitable for carrying brine on the shoulder, a shoulder pole would be needed, too.

What is more, in Jiada Village, there's even a type of brine pail made of galvanized iron sheet, with a diameter of 27 centimeters and a depth of 32 centimeters.

Figure 2-16 bark pail

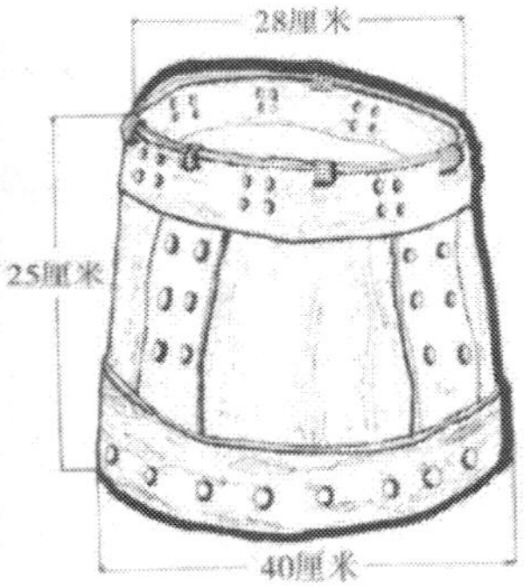

Figure 2-17 the size of a bark pail

The craft of making brine pails has been passed down for about 1,300 years. Before 2007, brine water was mostly carried by manpower, and brine pails had always been an indispensable tool for salt makers in salt production. However, with the popularization of electric power supply and water pumps, fewer and fewer salt makers are using brine pails, and now they have become a new "intangible cultural heritage" project.

3) Shoulder Poles

A shoulder pole, generally made of wood with good toughness, is a type of carrying tool commonly seen in rurual China. Flat in shape, and wavy as a whole, the shoulder pole in the Yangjing area has a straight length of about 142 centimeters. And its middle part is raised upward to form an arc, with a maximum radian of about 8 centimeters. Both end parts are slightly concave downward, with the endpoints slightly upturned and carved with a small groove, each for an iron hook to be fixed with a rope. Such an iron hook is generally 8~12 centimeters long, and the hook head is relatively larger, with a length of about 7.2 centimeters and a width of 1.2~2 centimeters. Each section of rope is 30~40 centimeters long, and can be adjusted according to the height of the carrier. When carrying brine with it, you need to hold the ropes in both hands to keep the bucket balanced.

Figure 2-18 the shoulder pole used in Jiada Village

2. Tools for Repair and Maintaining Salt Pans

In order to carry out sustainable salt production, salt makers have to renovate their salt pans at regular intervals, and even level and smooth them after each salt collection. However, due to the fact that the salt shelves in the Yanjing area are built of timber and mainly supported by structures, heavy tools may not be suitable here, for fear that excessive force will damage the salt pans; and from the safety point of view, there are also hidden dangers, so the tool should be small in size and light in weight. The primary tool for repairing salt pans is a wooden bat, also known as mallet or salt pan pat. Made by cutting a whole piece of Quercus glauca Thunb., it usually comes in the following two shapes:

Type I, the plane of the bat is an arc-sided triangle, with double wings and a single ridge, and the usable surface is curved from the tail end to the tip; the handle is cylindrical. Its total length is about 30.8 centimeters, with a handle length of 10.8 centimeters and diameter of 2.8 centimeters; the body part is 20 centimeters long, 14.4 centimeters wide, and 6 centimeters high (see structural drawing 1 below).

Type II, the plane of the bat is trapezoidal, also with double wings and a single ridge; there are some late-period cutting marks from the ridge to the handle and at the front end of its body part, and the usable surface is quite flat on the whole. Its total length is about 36 centimeters, with a handle length of 13.2 centimeters and diameter of 3.2 centimeters; the body part is 22.8 centimeters long, 16.4 centimeters wide, and 12 centimeters high (see structural drawing 2 below).

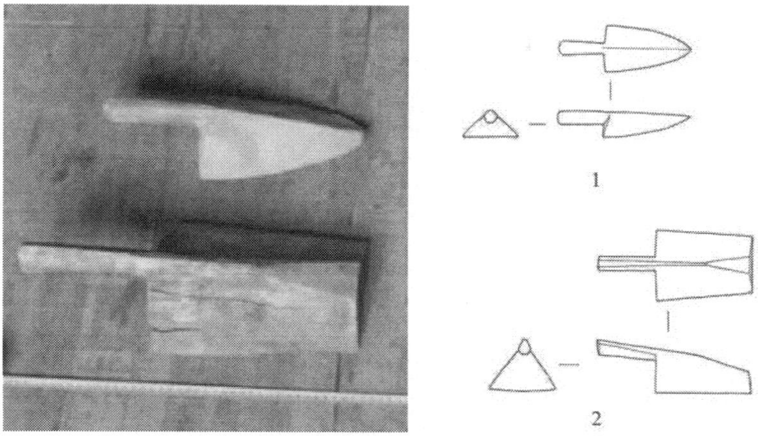

Figure 2-19 a real product photo and structure drawings of wooden bats

In addition, there's another type of bullet-shaped wooden bat, whose body part is 20~30 centimeters long and 14~18 centimeters wide, with a pointed front part; the handle is cylindrical, with a length of 10~12 centimeters and a diameter of 2~3 centimeters. (See Figure 2-20)

Figure 2-20 bullet-shaped wooden bats

3. Salt Collection Tools

Relatively speaking, there are more processes during salt collecttion, so tools may include salt scrapers, salt dustpans, salt baskets, and brooms, etc.

1) Salt Scrapers

The salt scraper is a thin piece of wood or iron. A wooden salt scraping board, in the shape of a trapezoidal blade, when used, the wide (bottom) edge would be in contact with the salt pan, and the relatively narrow (top) edge would be held by one hand. The bottom edge is generally about 40 centimeters long, and the top edge about 30 centimeters long, with a width of 12~14 centimeters and a thickness of about 2 centimeters. As for an iron salt scraper, it is slightly smaller than a wooden salt scraping board, with a thickness of about 1.5 millimeters. And just because the iron sheet is too thin, salt makers' hands might be easily injured, so the side for holding often needs to be rolled up about 1 centimeter inward, forming a circular edge. At present, iron salt scrapers are widely being used in the Yanjing area, while wooden salt scraping boards have almost been phased out.

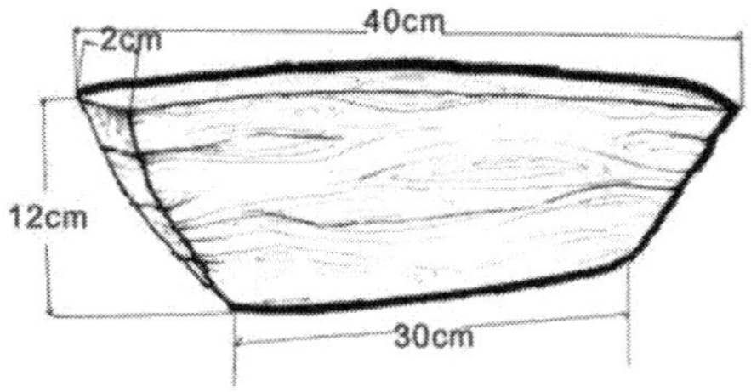

Figure 2-21 the size of a wooden salt scraping board

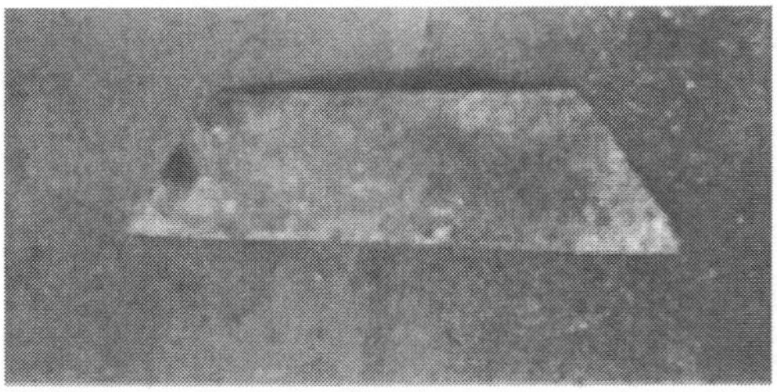

Figure 2-22 an iron salt scraper

2) Salt Dustpans

The salt dustpan is a container for holding salt in the process of collecting it, no different from the dung dustpan used by the Han people in the inland region. Currently, in Yanjing there are two types of salt dustpans: One is woven from thin bamboo strips (see Figure 2-23), and the other made of the waste inner tube of a tyre (see Figure 2-24). These two types are quite similar. The former has a tongue width (the front part) of 38~45 centimeters, a bottom width of about 35 centimeters, a length of 40~45 centimeters, and a height of about 15 centimeters; the latter has a tongue width of 32~40 centimeters, a bottom width of about 40 centimeters, a length of 45~50 centimeters, and a height of about 12 centimeters. The former, with gaps between the bamboo strips, is mainly used for holding the newly collected wet salt from salt pans, so that the water or moisture can be drained; and the latter, as it is made of a wheel tyre, without gaps, might be used to hold dry salt or dirt.

Figure 2-23 a salt dustpan woven from bamboo strips

Figure 2-24 a salt dustpan made of the inner tube of a tyre

3) Salt Baskets

The salt basket, pronounced "Ganba" in Tibetan, is actually a uniquely shaped pack basket woven from bamboo strips (see Figure 2-25). V-shaped, it is large at the top and small at the bottom, with a total length of about 46 centimeters. The upper mouth may be round, oval, or square. The round mouth has a diameter ranging from 46 to 60 centimeters, while the bottom is generally square, with a side length of 12~20 centimeters. The square mouth has a side length ranging from 45 to 55 centimeters, and the bottom shape and its side length are similar to those of the round mouth type.

Figure 2-25 salt baskets

After careful observation, we discovered the mystery of its use. Although salt makers sometimes wait for the salt to dry out before collecting it, most of the time they would collect it earlier than that. They'd gather the dripping wet salt into a dustpan, put it into a basket, and then carry it on the back to a small hole in the salt pan before laying it down there. Due to its pointed bottom, the basket rests firmly on it, and the brine water that is drained out would drip right into the brine storage tank below the salt pan. Similar in function to a salt dustpan, it can also drain the moisture of salt, playing a role in collecting brine water; the craftsmanship is very brilliant, indeed. When dried, the salt would be conveyed to a salt barn.

4) Brooms

In Yanjing, a broom (see Figure 2-26) is usually made by bundling a bunch of plants of Chenopodium album, which mostly grow in the mountains near Juelong Village. The handle is tied twice with thin bamboo strips, iron wires, or strings to fasten the entire broom. The full length is 55~65 centimeters, with the handle length about 25 centimeters. It is mainly used to clean up the remaining salt after scraping it, or sweep the soil layer flat when renovating the salt pans.

Figure 2-26 a broom

This kind of broom is mostly made by residents in Juelung Village. On our way to the mountains in Juelong Village, a villager once told the us that as there are lots of bushes and grasses for making brooms (i.e. the plants of Chenopodium album mentioned earlier) here, most residents are willing to engage in the craft.

The salt making tools in Yanjing are mostly made of local materials and purely handcrafted. Skilled craftsmen would often be adept at making such tools as brine pails, wooden bats, and salt scraping boards, etc. According to some local craftsman, when a tool is to be made, the salt maker need to place an order verbally in advance, informing him of its size and dimensions. And then he will start the making process, and notify the salt maker after it is finished. At this point, the salt maker would bring grain to the craftsman's house for exchange. In Yanjing, many villagers are still able to make wooden bats, salt scrapers, and brooms, but only two or three craftsmen are good at making brine pails. This is bacause their technique requires carpenters to be very patient, meticulous, and precise; once poor workmanship causes a leakage in the brine pail, all work will fail on the eve of complete success. When the wood is ready, he can make one brine pail of this type practically each day. Nowadays, with the development of science and technology, these handmade tools are gradually being substituted or even eliminated. However, as the continuation and inheritance of local ancient skills, they are still the most rich part of the whole history of salt production in Yanjing, embodying the wisdom and sweat of local people.

III. Technological Process

The salt production in Yanjing Township is quite primitive in terms of its production technology. Since ancient times, local Tibetans, Naxi and other ethnic minorities, by giving full play to their ingenuity and talents and cleverly utilizing the local natural conditions and terrain, have created and perfected a complete set of salt-making process suitable for themselves in extremely dangerous environments, and thus brought the manual salt production technology and skills to the extreme. The process can be divided into five steps in detail: drawing brine, conveying brine (including the current pumping brine), evaporating brine in the sun , scraping salt, and collecting salt. To sum up, it mainly includes three steps: conveying brine, evaporating brine, and collecting salt.

1. Conveying Brine

Conveying brine refers to the process in which salt makers convey it to a brine tank for storage, and after being exposed to wind and sun, they then carry the naturally concentrated brine on their backs or shoulders to pour into the salt pans. Before conveying brine, it's necessary to draw it from a brine well, which is considered as the first important step in well salt making, for the quality of brine directly determines the yield and quality of salt production. These two steps are closely connected and thus can be seen as one stage in the salt production process. According to Cui Kexin's descrition, "The first step in salt making is to draw brine water; in Jiada Village, it is drawn from the salt pool and shouldered in two wooden pails to the the "Cayin" (salt pans), while in other villages it is carried on the back in long pails."[1]

Figure 2-27 a salt maker of Jiada Village shouldering brine

Due to the different terrains on both sides of the Lantsang River, the salinity of the brine is higher on the east bank and lower on the west bank, and thus there are also different ways of drawing it. In Jiada Village, as the terrain of the salt pans is relatively flat, people mostly carry brine with a shoulder pole (see Figure 2-27); each pail of it weighs about 20 kilograms. In contrast, for the salt pans of Lower Yanjing Village and Upper Yanjing Village on the east bank, due to their steep terrain and big slopes, people there would carry brine on their backs (see Figure 2-28), which is not only more physically demanding, but also requires certain skills. Someone once described it as follows:

1 Investigation of the Geology and Salt Production in Yanjing County", by Cui Kexin, published in *Xikang Economic Quarterly*, 1944(8).

Figure 2-28 salt makers of Upper Yanjing Village carrying brine on their backs

The carrier descends sideways along a vertical ladder in the brine well, holding onto it with one hand and grasping a pail with the other. When approaching the well water surface, she turns around slightly to face the water and unload the empty pail; standing firmly with her feet apart in an "out-toe" position, which is a little wider than shoulder width, she slowly squats down. With the left hand, she holds the pail along the shoulder straps and moves it to her chest, and with the right hand, she takes out the brine scoop in the pail and places it by her right foot. Then, she holds the pail with both hands, the bottom with the left and the mouth with the right, presses the it horizontally into the water, raise it when more than half filled with both hands, the left side lower than the right, and put it in front of the left toe. Subsequently, she holds the pail brim with the left hand instead and tilt the it slightly towards the well water surface, and with the right hand, the pail is filled to almost 90% of its capcity by scooping (about 20 times); she then gently puts the empty scoop on the water surface in the pail, which not only saves the extra trouble of taking it along, but may also prevent the brine from spilling out when the carrier moves around. After that, the carrier slightly turns left to the pail, puts it to her chest, and hangs the straps between her neck and the waist of the pail. And finally, she stands up slowly, turns right to the ladder, and by lifting the pail, places its bottom on her abdomen for support; holding the pail brim with the right hand and holding on to the ladder with the left, she ascends almost with it carried in the bosom. [1]

1 See "The Study on Salt Culture of Yanjing Nakhi Ethnic Township, Markam County, Tibet", Wu Chengli's Master's Thesis of the School of Sociology and Anthropology, Sun Yat-sen University, 2009, p. 88.

After coming up from the brine well, she switches to carrying the pail on her back, along the steep slope to a brine storage tank. Particularly in Upper Yanjing Village, as the gradient of slopes is nearly 80 degrees, even larger than that of Naxi Village, a carrier can barely walk with the support of her hands against the cliff wall when carrying brine; and some paths are only wide enough for only one person. After reaching the storage tank on the the hillside, she also needs to be skillful when pouring the brine into it. She should never tilt the pail casually, for in that way, not only is the pail, once out of balance, likely to drop into the tank, but also the brine water in it will be splashed out. To find a proper angle, or in the words of Yanjing people, a spot known as "Cangwuzu" (close in pronunciation), the carrier stands there pouring brine out slowly, which is both effort-saving and dexterous.

Due to the steep mountain paths, women's physical strengths wear off more quickly, so they have to work from dawn to dusk. If the brine storage tank is relatively close to the salt well, a salt maker may carry around 400 to 500 pails of brine water each day, while for those farther ones, she is only able to carry around 60 to 100 pails. She has to make a round trip for each pail of brine water, with the closest distance being around 50 meters, and the farthest around 400 meters. So calculated, she may have to walk 40,000~60,000 meters a day at least, and over 100,000 meters at most. The maximum total weight of the brine water she carries each day can reach up to 2,500 kilograms, and even the minimum would be as much as over 500 kilograms[1]. Except when it rains, these women would repeat their monotonous labor every day, making over a hundred round trips on the steep wooden steps. And the pebbles placed on the boundaries of those salt pans are exactly their counters.

The brine is conveyed to the storage tank for two purposes. One is to cool it down. This is because the brine water freshly pumped (or carried) up from the brine well has a relatively high temperature, generally about 40 °C; and if it is poured into the salt pans in a hurry, the quality of the salt thus made will be affected. The other is to increase the concentration of brine. Poured into a public brine storage tank and stored for a few days (preferably 5~6 days), the concentration of the newly

1 "The Study of Mang Kang Tibet Salt Industry" by Tao Hong and Huang Jian, published in *Salt Industry History Research*, 2002 (4).

drawn brine will rise, more or less. Then carried into each household's private brine storage tanks respectively and stored there for a few more days, the brine water will have an even higher concentration. In addition, the firewood removed from the salt pans and the round logs once used for building the salt shelves, if thrown into the brine storage tanks, will absolutely increase the concentration of the brine in them; after all, these logs and wood have already been saturated with brine after a year's leakage and soaking under the salt pans. It is crucial to know the concentration of brine water during the salt making process, because a low concentration will not only prolong the salt collection time but also affect the salt quality. In Yanjing, some salt maker would dip the tip of his finger into a brine storage tank, and then put it into his mouth to judge by taste. Obviously, this method is quite traditional, and thus requires a salt maker to have long-term experience accumulation in order to make accurate judgments.

The brine is thicker from winter to spring, and its quality is the best from March to April. The seepage rate within a brine well varies depending on the location; in some wells, the brine will seep out again just one minute after it is scooped out. Generally speaking, brine can be adequately supplied. But if there's a shortage, salt-maker households will decide the sequence of drawing brine through negotiattion. In the morning, usually salt makers will finish drawing brine from the seven old wells they share, and then begin to draw it from the shallow wells they have dug privately.

For a long time in the past, brine was conveyed all by relying on manpower. In order to relieve the labor intensity of salt makers and improve production efficiency, the Department of Industry of Tibet had tried to draw brine with electricity in the 1980s by investing 200,000 *yuan* in the development of mechanized brine wells, but ultimately terminated due to insufficient electric power and high production costs. However, since 2007, water-raising engines (water pumps) have finally replaced manual brine conveyance. The brine water is first pumped into a large brine storage tank, and then into their respective smaller brine storage tanks by different households. From then on, the historical scene of Naxi and Tibetan women carrying brine on their backs or shoulers had ceased to reappear, and was replaced by electrical wires spread all across in confusion and various pumping equipments within the salt pans. The conversion of carrying brine with pails into pumping it might have been an advance with the times in the local salterns, but other ancient technical processes are still continuing.

2. Evaporating Brine

The salt-making method in Yanjing, with a long history, is quite different from the boiling method in Yunnan and Sichuan provinces; according to "the method of salt making here does not rely on fire...the brine water, poured into the salt pans, and exposed to the sun and wind, will precipitate salt grains when dried..."[1] it's more similar to the solarization method in coastal areas. And the so-called "evaporating salt" is actually "evaporating brine". With its dry climate, strong winds, and high temperatures, Yanjing has favourable conditions for evaporating brine. The essential process of salk making is as follows: the brine water is poured into salt pans, and after being exposed to the scorching sun and blown by winds across the river, it precipitates salt grains through natural evaporation.

The brine is usually poured into salt pans in the afternoon, in the past with wooden pails, but now mostly with water pipes. At this time, if the sediment is filtered with a basket, the salt thus produced would be even purer. The concentrated brine water from the storage tanks are poured into shallow salt pans, with the depth generally maintained at 2~5 centimeters by local salt makers. About the depth of brine water, Naxi women have a simple judgment method; i.e., put a palm flat into the salt pan, and when the palm is completely submerged by the brine (i.e. to its joint). it is the optimum water level. Since the concentration of brine directly determines the yield of salt, it's crucial to be able to make accurate judgments before exposure to sunlight. Unlike in the Yangpu salterns of Hainan Province, where the method used is to cut a section of "Huangyuci" (literally "yellow croaker Tribulus terrestris", a shrub of the Zygophyllaceae family) branch and put it into a brine pool, and if the branch floats on the water surface, then the salt concentration is sufficient; or in such coastal areas as the eastern Zhejiang, where "dried lotus seeds are dropped into the brine water, and as its salinity may be different and the proportion of salt varies, they will float on the water surface due to sufficient concentration and high proportion of salt, or sink to the bottom due to insufficient concentration. When

---

1 See "Journey to the Southwest of Kamu" in *A Compilation of Sichuan-Tibet Travel Records*, p. 444.

the brine is of good quality, the lotus seeds will stand upright just below the water surface." [1]a salt famer in Yanjing would dip her finger tip into the brine tank, and then put it in her mouth to judge by taste. When the brine with a concentration that meets the standard is baked by sunlight, a layer of crystals begins to form on the surface and need to be turned over occasionally. As the salt becomes dry, white grains are gradually precipitated, with a size of about one cubic millimeter. (See Figure 2-29) After the salt has been collected, the salt pans are re-flattened and tamped, and brine water can be poured in to make salt again. Nevertheless, sometimes salt makers also need to check the crystallization of brine in the salt pans. If the crystallized amount is low and the color in the salt pans is not too white, it would be necessary to add brine appropriately once more. It's worth mentioning that when brine water is poured into the salt pans, it must go sideways through the "Qukuang" (pronounced so), for this can filter out the sediment in the brine, and make the salt cleaner. At the same time, it may prevent the pouring brine water from damaging the soil layer in the salt pans, and thus has a protective effect. In addition, the horizontal plane of the salt pan wooden frame must be accurate, otherwise the brine will be lost and result in reduction of output. And due to the gaps between the wooden boards, there are often small crystalline salt columns of varying lengths hanging down, which appear crystal clear and translucent.

filling the salt pan with brine

Figure 2-29 the crystallizing brine in salt pans

1 Hans Ulrich Vogel's "From Boiling to Solarization: A Further Discussion on Chinese Sea Salt-making Technology in the Imperial Era"[A]//*Archaeology of China's Salt Industry (Episode 2): A Comparative Observation from an International Perspective* edited by Li Shuicheng and Lothar von Falkenhausen, Beijing: Science Press, 2010 edition, p. 20.

The cycle of salt making varies depending on the season and weather. In general, "with strong winds and scorching sunshine in spring and summer, on the east bank salt can be swept once a day, while on the west bank once every two days. In autumn and winter, as the sunshine is weak and the wind is cold, on the east bank salt can be swept every three days, while on the west bank it may take four or five days for salt grains to be precipitated." (translator: "Millennium Salt Pans in Tibet: the Ageless Scenery on the Ancient Tea-Horse Road" from www.jxnews.com.cn) From March to June in the Chinese lunar calendar, with fine and warm weather, the salt pans on the east bank generally need 2~3 days due to the high salinity in the brine, while the salt pans of Jiada Village on the west bank may need 4~6 days due to slightly poorer brine quality.

Moreover, salt can only be made for more than half of the year in Yanjing, mainly between October and June of the following year. According to the wind and sunlight intensity, and the impact of rainfall precipitation on salt formation, it can be divided into peak season and off-season. During the peak season, due to strong winds, intense sunlight, and little rainfall, the river water would recede and all brine springs are exposed, which is quite suitable for salt making. Not only are salt grains precipitated quickly, but also they have a whiter color and better quality. The salt made on the west bank is faint red, and due to the peak period starting from March to May each year, which is also the season when peach blossoms bloom on both sides of the Lantsang River, it is called "Peach Blossom Salt" (also known as "Tibetan Red Salt"), while the salt made on the east bank is pure white, thus known as white salt. As for why on two sides of the same river, salt of white and red colors could have been produced, it's said to be related to the rock stratum from which salt brine is produced; the brine on the west bank is produced in laterite, while on the east bank is produced in gray and white rocks. Cui Kexin conducted a more in-depth analysis of this:

Every year from the third to the sixth month of the lunar calendar, with intense sunshine and strong winds, the brine dries rapidly and yields more salt, and is thus called peak season. May and June are the flood season, while July and August are the water recession season. The time of the rise and fall of the river depends on the amount of rainfall. Before the rise, salt makers would hire men to carry brine day and night, and store it in brine pools for salt making in case the brine wells might be submerged. In July and August, when the source of brine water is gradually exhausted, salt makers would demolish the salt pans (and salt shelves), and immerse the mud, sand, beams, and columns in the brine pool, so as to dissolve the salt content in them. After about three days, the salt would be completely dissolved, and the brine is thus suitable for evaporating in the sun. As the yield is not less than that of the peak season, it is considered a peak season, too. In September, October, and November, due to the insufficient sunshine and weak wind force, the yield is reduced to a minimum, accounting for about three tenths of the peak season. And among them, the yield in October or November is even lower than that of September. In December and January of the following year, the yield begins to increase gradually, accounting for about five tenths of the peak season. [1]

From July to September it's the anxious rainy season, making it difficult to produce salt. "...Every summer, when the Lantsang River floods , the well mouths are all submerged. Therefore, salt makers must wait until the river recedes in autumn and winter, and then resume their business..."[2] Due to the influx of rainwater mixed with a large amount of silt, the color of the Lantsang River water changes from green in winter to a muddy red. What is more, the river surface would widen and the water level may rise by nearly ten meters. The turbulent river, like a giant dragon rolling up and down, rushes past through the canyon at great speed. At this time, the brine wells would be filled with rainwater, and the lower salt pans might be submerged by the surging river water; when floods are fierce, even the salt pans erected at higher place are at risk of being washed away. In this case, salt makers can only maintain their production with the brine in the storage tanks. But once it is used up, the production will be unsustainable.

In order to improve salt quality and increase production efficiency, Zhao Erfeng in the Qing Dynasty once tried the boiling method, and after the refining process, it had both unexpectedly fine colour and flavour. But regrettably, "since there was no coal mine outside the Jiayu Pass, if firewood became the only fuel, the cost would almost increase by more than 20 times; thus the old method had to be retained

---

1 "Investigation of the Geology and Salt Production in Yanjing County", by Cui Kexin, published in *Xikang Economic Quarterly*, 1944(8).

2 *Archival Historical Data of Sichuan-Yunnan Border Affairs in the Late Qing Dynasty*, published by Zhonghua Book Company, 1989 edition.

temporarily". [1]In *Local Geographical Records of Yanjing, Batang*, it was also recored that "...(the brine) is only suitable to be evaporated in the sun, insead of being boiled; probably, once the boiling method is adopted, the cost would be too high, affecting its marketing". These salt makers, by giving their subjective initiative into full play, also tried other methods. But due to the lack of local resources or very inconvenient use, they ultimately chose to use the old method of evaporating brine in the sun in salt pans. This also proves from another aspect that the unique salt production method summarized by local salt makers according to the canyon topography and the climatic characteristics such as strong winds and intense sunshine, etc. is the most economical and applicable.

3. Collecting Salt

When the brine in salt pans begins to evaporate, crystallize, and precipitate salt grains, if turned over constantly, it would be dried sooner. But even before the salt is completely dry, or as long as it's not dripping wet when touched by hand, salt makers can start collecting it. Before salt collection, they have to scrape it off first, and these two steps are actually conducted simultaneously. As the final rewarding part of the entire salt making process, it is also the happiest moment for salt makers.

Figure 2-30 scraped salt water being strained off from salt

1 *Archival Historical Data of Sichuan-Yunnan Border Affairs in the Late Qing Dynasty (Middle Volume),* compiled by Sichuan Institute of Ethnic Studies and the editorial group of *Archival Historical Data of Sichuan-Yunnan Border Affairs in the Late Qing Dynasty*, published by Zhonghua Book Company, 1989 edition, p. 446.

With an iron scraper or a thin piece of scraping board, salt makers first scrape off the salt gently from left to right, and then continue to scrape it off from left to right with their backhand. as this cycle repeats, the white crystalline salt grains in the salt pans are scraped into arcs (see Figure 2-30), and all the scraped salt pans look like white chrysanthemums painted on a red canvas. (translator: "Salt Women" coauthored by Zhang Mingfang and Xiao Rui, see http://www.cnwomen.com.cn/2021/11/05/99236535.html.) According to local salt makers, the salt should be scraped off when there's still some moisture in the salt pans, or the white salt can not be separated if it is too dry. And the strength exerted in scraping salt is also very important. With too much strength, the mud will scraped out; with too little strength, the salt cannot be scraped clean. After being dried under the sunshine for more than ten minutes, the moisture in the salt crystals is again reduced. Then the salt maker would gather the salt grains from far to near into heaps with a scraping board and a broom. And finally, she put them into a salt dustpan or a bamboo pack basket (see Figure 2-31). The pack basket is usually placed on a small hole at one of the four corners of a salt pan, its pointed bottom firmly underlaid with a piece of wood. In this way, the residual moisture containing salt can thus leak out through the gaps of the bamboo basket, which not only results in even drier salt, but also allows the water dripping into the brine storage tank to be recycled. After two or three days, when the moisture has almost completely evaporated, the salt is sifted with a salt dustpan, sorted, and then put into salt baskets or bags.

Figure 2-31 women of Lower Yanjing Village collecting salt with a salt dustpan

The salt collecting process requires meticulous accuracy from the perspective of salt-making technology. According to the order in which it is collected, the salt may be divided into different grades due to the varied amount of impurities. Generally speaking, it is divided into three times:

...For the first time, when the brine water is about 70% evaporated, scrape the salt surface with a scraping board, and slightly scrape off the fine and white salt on the surface and put it into a salt bag, which is placed on the boundary between the salt pan and the brine pool, and from which the leaking brine water would directly drip into the pool without loss. For he second time, when the salt is dried, first of all sweep it gently with a broom, and then scrape the salt off with a scraping board, and put it into another bag. For the third time, sweep the remaining salt clean, and with wet sand and mud...much is mixed into the salt. The salt scraped off for the first time has the purest quality, classified as peach blossom salt; the salt for the second time is relatively inferior, known as the second-layer salt; and the salt for the third time is of the poorest quality, known as the third-layer salt. [1]

This shows that in the process of salt collection, it's necessary to first judge the quality of salt precipitated from brine water. If it is of good quality, the salt could be scraped off for three times; other wise, for twice or only once. Salt scraped off for the first time is the surface layer with fine grains, few impurities, and a relatively white color, which, of the best quality, is known as the first-layer salt and mainly for local consumption. Afterwards, the second layer would be scraped off, and as it contains a small amount of soil, the color is noticeably darker: the salt in Jiada Village is reddish, while in Upper Yanjing Village and Lower Yanjing Village are yellowish. The use of second-layer salt depends on its quality. If good enough, it is for human consumption; otherwise, it is fed to livestock or used for Tibetan medicine bath. Finally, salt with more impurities and a darker color is scraped off. Mixed with soil, it's quality is even poorer, and thus all fed to livestock, or poured into brine storage tanks, where the concentration of brine would be increased through its dissolution. At this moment, it is necessary to constantly pat the salt pans with "Lao" (pronounced so, i.e., a wooden bat), not only to beat the remaining salt out of the soil, but also to flatten and solidify the areas that are prone to leakage after being scraped loose, making them as watertight as possible, and thus with the effect of repairing the salt

1 "Investigation of the Geology and Salt Production in Yanjing County", by Cui Kexin, published in *Xikang Economic Quarterly*, 1944(8).

pans. Of course, if the salt quality is poor as a whole, all would be scraped off at once, and there is no such classification as "three layers of salt".

After collecting salt, the salt pans should be swept clean with a small broom, and then flattened and smoothed with a triangular or trapezoidal bat made of Quercus glauca Thunb. Moreover, the dents, if there's any, need to be filled with soil. Finally, the salt pans are patted all over to prevent leakage. Up to this point, the salt production activity is fully completed. Salt makers would sometimes sell their salt to merchants on the day it is produced. They carry it to the loading platform above the salt pans, and there it is taken away by mules specially for salt transportation. If there's no merchant collecting salt on that day, they'd dig a large pit under the salt pans, spread a sheet of plastic, and directly store the salt in it. When the salt needs to be carried away, they'd convey it back home to their salt barns by putting it in sacks or other plastic bags. A salt barn is generally shared by several salt-maker households and merchants. And in oder to prevent salt theft, special marks are imprinted on the surface of the heaps. After the salt in the barns reaches a certain amount, the merchants would transport it to Yanjing Town by a horse caravan, then by motor vehicles or tractors to other regions, and then again by a horse caravan into the mountains, selling to the villagers there. Of course, there are also cases when the salt is directly transpoted to these mountains by a horse caravan.

According to local villagers, with 90 pails of brine water, more than 15 kilograms of salt can be made through solarization; with an area of 7 square meters of salt pans, generally 5~6 kilograms of salt can be made. Thus each salt pan can produce 15~25 kilograms of salt at a time, with a maximum of over 25 kilograms. Due to geographical differences, the yield of a salt pan on the east bank is higher than that on the west bank, and the salt quality is also better. Each salt pan there can generally produce 5 bags of salt in 3 days, with each bag weighing about 50 kilograms. However, the first-layer and second-layer salts for human consumption generally account for only 40% to 50% of the total salt collection. The salt pans of Jiada Village, for geological reasons, have more impurities than those on the east bank. According to historical records, in the late Qing Dynasty, the annual output of salt in Yanjing already reached more than 18,000 *duo*. With each duo equivalent to 70 kilograms, that would be about 1250,000 kilograms, or 1250 tons. According to Dr. Li Hechun's estimation during his survey in 2015, the annual output was around

500,000 kilograms then. [1]

The well salt production process in the Yanjing area, unlike that of the well salt in other regions, has some similarities with the production method of sea salt in ancient China. The salt, poorly colored and coarsely grained, contains a small amount of soil, but has a very strong taste; locals praise the characteristic of it as "very salty". Local Tibetans refer to the salt produced in Yanjing Township as "Tsangpa salt". It is said that salt of different types and effects might be produced according to different needs. Popular among the Tibetan people, it has the following advantages: Firstly, yak butter tea made with it is not only good in color, but also particularly fragrant; thus it has become a must-have for local ethnic minorities when preparing the tea. Secondly, local villagers of each household have a large number of livestock, such as cattle and sheep, etc.; once they're fed with it, the "Tsangpa salt" will not only kill the lice on these animals, but also have a certain effect of fattening them up as well as spurring the female ones into estrus. Moreover, the meat of livestock raised in this way is particularly fresh, tender, and tasty. In addition, "Tsangpa salt", when used for bathing, may play a role in sterilization and cleansing, and has certain therapeutic effects on some skin diseases.

Through field investigations, combined with historical documents, it can be seen that: Due to the complex terrain and landforms, local salt makers, by making use of locally available materials and creating conditions, achieved salt production with simple and primitive tools, forming a unique set of brine mining and evaporating techniques. This salt-making process, easy to learn, does not require any specialized instruction. Salt makers usually take their children with them when going to the salt pans. And these children, by watching what adults do and occasionally participating in it, plus their introducing some common sense and tips, will naturally learn everything over time. According to the local elderly, there were no special taboos or religious customs throughout the whole process of salt production. [2]

---

1 Li Hechun, "Comparative Study on Traditional Salt Technology in the Upriver Basin of the Lancangjiang from Qinghai to Tibet: Based on the Field Investigation of Salt Field in the Nangqian County and Mangkang County", Social Sciences in Yunnan. 2018(06).

2 "Investigation report of Salt Pans in Yanjing, Mangkang County, Qamdo Prefecture, Tibet Autonomous Region", by The Institute of Cultural Relics Protection of Tibet Autonomous Region.

Yanjing salt has always been known as the "work of sunshine and wind". The mining of local saline springs and the germination and development of the salt industry are a living example full of local and ethnic characteristics, reflecting the fearless spirit and ingenuity of ethnic minorities such as the Tibetan and Naxi in their tenacious struggle against nature, which is simply too impressive. The ancient salt production method has been passed down from generation to generation in Yanjing, retaining its original technologies and forms from digging wells in selected locations, building salt pans made of earth and wood structures, evaporating brine, and transporting salt for sale, etc. It occupies an important position in the history of Tibet's salt industry, and has also become an indispensable link in the development chain of well salt production science and technology. As a living fossil of ancient salt production, it is not only of great historical value, but also of important cultural heritage value. The magnificent and spectacular salt pans are the unique scenery on the Ancient Tea-Horse Road, as well as the only swan song in the world. Due to its profound cultural traditions and rich architectural features, the traditional salt making technique of Yanjing in Markam County was included in the second batch of intangible Cultural Heritage List of Tibet Autonomous Region in 2007. What's more, it was included in the second batch of national intangible cultural heritage list in 2008 and was also designated as a national key cultural relic protection unit in 2013.

CHAPTER THREE

Salt administration, Salt Taxation and Its Transportation and Sale

The long history of salt making in Yanjing has contributed to its long-standing and well-established salt production system and salt management system. In the early days, whoever took control of salt resources would possess wealth and the ruling power. For rulers throughout history, salt was the lifeblood of a state and was closely related to the national economy and people's livelihood. Therefore, the issue of salt was always highly valued. In order to keep salt resources firmly under control, they even formulated a series of salt laws, tax laws, as well as related systems and regulations. It can be said that salt affairs were always an official priority in history.

Meanwhile, as an important daily necessity for human beings, salt was also the earliest commodity to enter the market for circulation. Its strong social function could only be achieved through exchange. As a result, the ethnic groups in and around Yanjing, centering on this township, and with the ancient salt routes which took shape in the early period as the passage, gradually formed a fixed salt market area and a non-fixed flow circle.

1. Salt Administration and Salt Taxation

"The ancient Chinese character '鹽' is a hieroglyph consisting of three parts. Its lower part represents the tools, with a court official on the top left and saline water on the upper right. Thus the character itself shows the state's control over salt."[1] From

1 *Salt* authored by Mark Kurlansky, translated by Xia Yeliang and Ding Lingqing, and published by China Machine Press, 2005 edition, p. 9.

the early Spring and Autumn period when the Qi state implemented salt monopoly, throughout dynasties until the period of the Republic of China, salt tax had always been a main source of revenue to maintain the state's rule, and blood for the operation of the national machinery. Within its jurisdiction, no power will easily give up the tax ownership of salt. As the bloodline of a country, sufficient salt taxes may lead to its prosperity and depleted salt taxes may lead to its decline. Thus, the fight for salt is essentially a fight for salt taxes, a fight for sources of revenue.[1]

In ancient China, salt was monopolized, and salt tax became one of the main sources of government finance. Demarcation of salt distribution, as an important guarantee for the realization of monopoly, resulted in a salt production and marketing system under state control through the mode "salt is supplied by certain salt wells and sold in fixed areas".

1) Salt Administration

The political system served as the baton of salt production, transportation and sale. The political environment of Yanjing was always characterized by a very changeable and complex social system full of contention for profit. Legend has it that in those years the Tibetan hero King Gelsall and the Nakhi king Qiangba in Yanjng fought many battles over the local salt resources. Finally, King Gelsall achieved victory and occupied Yanjing. In the late Tubo Dynasty of Tibet, Nakhi prince Youla was made the Nakhi Minister of King Gelsall, and since then salt pans in Yanjing had been administered by Nakhi people. However, in the past three or four hundred years, Yanjing was long under the joint control of secular and theocratic power (political and religious powers combined), and the administration of salt industry by local power and the temple was chiefly through the control of salt tax.

For people in academic circles, their knowledge of Yanjing mainly began in the Ming Dynasty. During the Jiajing period in Ming Dynasty, Chieftain Mu started to expand northward, and his influence reached deep into Yanjing and other areas. Chieftain Mu's traditional mode of administration had a profound impact on the

---

1 "On the Impact and Contribution of China's Salt Tax to Social Development", Shang Jianping, *Chinese Salt Culture* (Series VIII), Chinese Salt Culture Research Center, Sichuan University of Science and Engineering, Sichuan Provincial Key Research Base for Philosophy and Social Sciences.

local society of Yanjing. Regarding the administration of salt pans by Chieftain Mu after he arrived at Yanjing, *Travelogue in Weixi* recorded the situation during the Wanli period: "Whereupon Weixi and Zhongdian, and Batang and Litang which are now under the jurisdiction of Sichuan Province, were all in the control of the Mu clan. He collected taxes, and reported them to the imperial court."[1] Here, "collected taxes" indicates that taxation had become a political and economic measure for the administration of local society. At that time, the west bank of the Lantsang River was administered by someone on behalf of the Nakhi chieftain, but the salt pans on the east bank were administered by the Nakhi people themselves, who even directly participated in salt making. The "garrison soldiers" also assumed the responsibility of collecting taxes. Chieftain Mu mainly occupied the east bank, and due to the danger of crossing the river, it caused great trouble for his administration. Therefore, the most appropriate way was to send officials to supervise salt production and collect taxes on the west bank, and they only needed to transfer salt tax to the east bank when the water level was the lowest in cold winter.

During the Kangxi period in Qing Dynasty, the Batang chieftain in Yanjing followed the original mode of administration and appointed a *xie'ao*[2] to supervise for him. "Yanjing County Chronicles" recorded that "Yanjing was always under the direct jurisdiction of the second highest Batang officer in charge of local armed forces and security, with a *xie'ao* assigned to Yanjing in winter, and to Zongyan in summer", and it went on to state that "There were two *tubaihu*[3]; one was Jicun Qucha, dispatched to stay at Zongya, and the other was Qiongjue, dispatched to stay at Juelung, all holding hereditary positions".[4] In addition, there were two salt bearers serving under the chieftain and the deputy chieftain, who were specifically

1 *Travelogue in Weixi* by Yu Qingyuan, see *Journey to Dali and Other Five Volumes*, The Commercial Press, 1936 edition, p. 1.

2 An important local administrator serving the chieftain and the deputy chieftain under the chieftain system, who was in charge of taxation, errands, and civil affairs, etc.

3 A hereditary military official established in such ethnic minority areas as Tibet, Sichuan, and Gansu during the Qing Dynasty, and this post was often filled by the local chief.

4 "Yanjing County Chronicles: the Chieftain Genealogy", author unknown, published in *Political Affairs of the Border Region*, 1931(6).

responsible for the salt supply in the internal affairs of their bosses. [1] During the reign of Batang chieftain, Yanjing was actually ruled by both monastic and secular powers,and the salt tax there was also collected jointly by the Batang chieftain and the Laweng Temple.

In the thirty-first year of the Guangxu period (1905), the "Batang Incident" shocked the government and the public. In order to seek a new political equilibrium, the Qing government ordered Zhao Erfeng to deal with the crisis with full authority. With his knowledge and experience accumulated over a long period of time in the salt transportation sector[2], Zhao Erfeng started the rectification of the salt industry and carried out reforms in the traditional sense. On the first day of the tenth lunar month in the first year of the Xuantong period (1909), Zhao Erfeng submitted 16 articles of regulations for collecting salt taxes in Yanjing to the department of treasury, which not only detailed the methods and requirements for extracting tax and the intensity of punishment for tax evasions, but also stipulated in explicit terms for the personnel in the salt tax bureau, such as committee members, office clerks proficient in the Manchu language, Tibetan interpreters responsible for going to imperial court or visiting a friendly nation as envoys, watchmen, men doing odd-jobs, and warehouse keepers, as well as the boats (made of animal hides or wood) used for inspection. And the official supervision of Yanjing was also recorded: "Among the border districts where salt is produced, only Yanjing the tiny area whose tax silver has long been collected by the Batang chieftains and Lamas. Now, with the bureaucratization of native officers, personnel should be sent and a bureau should be established to rectify salt affairs, and salt tax is to be collected according to the circumstances. Regardless of the annual revenue, it will be used as the border defence fund. "[3] On November 12 of the same year, Zhao Erfeng, in response to the salt merchant Li Chunde's proposal to turn Yanjing salt affairs into a privately-owned mode, approved by putting it on

1 See *Batang County Annals*, compiled by the Batang County Annals Compilation Committee of Sichuan Province, and published by Sichuan Ethnic Publishing House, 1993 edition, p. 245.

2 See "Biography of Zhao Erfeng" co-authored by Ke Shaomin et al., and published in Volume 469 of *Historical manuscripts of the Qing Dynasty* (the inside-Shanhaiguan-Pass Edition), or No.256 in collected biographies.

3 *Archival Historical Data of Sichuan-Yunnan Border Affairs in the Late Qing Dynasty*, compiled by Sichuan Institute of Ethnic Studies, p. 442.

record, and established a Salt Commerce Bureau. Meanwhile, he issued regulations on salt bureau management, salt prices, salt sales areas, and salt tax collection, etc, normalizing and supervising such aspects as the production and sales of well salt from an official perspective. Afterwards, the regulations were revised to thirty articles, for officials, businessmen, soldiers and civilians to observe. These agreed regulations of the commercial salt bureau have been significantly amended and improved, especially by adding explanations on salt pricing, weighing standards, and channels for purchasing salt among the common people, etc. Moreover, it was but a set of provisional regulations; according to Zhao Erfeng, if there were any matters that had not been mentioned, it could be supplemented or modified at any time. In the second year of the Xuantong period (1910), in order to further regulate the local market and promote the development of the salt industry, he additionally formulated six provisions, i.e., *Flexible Rules and Agreed Salt Prices during the Initial Period of Merchant Supervision in Yanjing*, which ensured smooth progress of the salt industry and in turn ensured the revenue of salt taxes. Judging from Wang Huitong's report on the expenditure of salt taxes in Yanjing, the above policies have already brought initial results to the salt industry. At that time, the total expenditure was 15,701.2519 taels[1] of silver, and the revenue and expenditure surplus of salt tax was 23,153.5106 taels of silver with 97% purity. The reform of the salt industry and the establishment of the commercial salt bureau have achieved the expected results, increasing border defence fund for the bureaucratization of native officers in the Sichuan border area. These measures have played a significant role in promoting the process of bureaucratization of native officers in the Sichuan border area, ending the long-term control of Yanjing jointly by monks (temples) and secular forces (chieftains), and changing the original relationships of administrative subordination. Thus, it can be said that the bureaucratization of native officers in Zhao Erfeng's period was the beginning of institutionalized administration of the salt industry in the history of Yanjing.

From the downfall of the Qing Regime in 1912 to the liberation of Yanjing in 1950, during this short period of more than 30 years, Yanjing had been under the

---

1 A unit of weight, used in the Far East, having various values between one to two and a half ounces. In the Qing Dynasty, 1 tael was equal to 37.30 grams.

alternating jurisdiction of various forces, with intricately intertwined administrative system, and even eventually returned to the Gongga Lama's hands. The continuous changes in the upper system were constantly impacting the entire salt industry system, and ultimately no institutionalized salt industry admistrative system took shape. It was precisely such historical factors that led to the salt industry of Yanjing runnng its own course. However, administrators of any era could not bear to give up the temptation of salt taxes, so they would carry out appropriate reforms to the salt industry from time to time. Salt-maker households, on the other hand, adopted a strategy of hiding when the war broke out and making salt when it was over. Sometimes, under the demagogy of local powers, they participate in some activities, but obviously could not affect the overall situation. These salt-maker households were forced to earn a living, while local forces were driven by salt taxes; the two, in constant interaction, barely pushed the salt industry forward. In the thirtieth year of the Republican period (1941), Cui Kexin described problems of poor salt administration and poor trading system in his *Investigation of the Geology and Salt Production in Yanjing County.*[1] In *The Rectification of Salt Affairs in Yanjing County, Xikang*, it was also recorded: "Since the Republican period, there had been incessant civil strife (in Yanjing). And Tibetans took this opportunity to violate the border, resulting in the slack salt affairs and a shap decline in the county. However, the county magistrates, who served successively there, not only perfunctorily took responsibility for the salt affairs and did not rectify them, but also deceived their superior officers by reporting less salt tax revenue. The administrator who supervised this, due to the long distance, had no choice but to allow them to hide the truth. Thus here, director Cheng Zhongliang of the Financial Coordination Office, in light of this, ordered the county magistrate to rectify the situation effectively." Evidently, the upper level of Xikang were not satisfied with the salt industry administration in Yanjing, and problems with the collection of salt taxes and its production supervision did exist.

After the founding of the People's Republic of China, Yanjing was also caught in the contention for profit and reconciliation between cultures and systems, as well

---

1 "Investigation of the Geology and Salt Production in Yanjing County", by Cui Kexin, published in *Xikang Economic Quarterly*, 1944 (8).

as between central and local authorities, reflecting the impact and confrontation between ethnic culture and external culture in a miniature way. The salt industry operation here began to transform toward the inland way, and in the process of ownership and labor combination conversion, it successively went through stages of "individual ownership--cooperative--collectivization--contracted production to households". In 1965, local salt-maker households were organized to set up a producers' cooperative. Before 1970, salt wells and salt pans were privately owned. From 1970 to 1982, they were owned by people's communes of the day, and due to poor production conditions, the local government had to encourage villagers to engage in salt production in order to develop the local economy. After 1982, free well sinking, free production and free transportation and sale were allowed, and a preferential policy of making salt without paying taxes, uniformly subsidized by the state, was adopted. In 1985, people's communes were abolished in the whole county, and fields and land, as well as salt pans, began to be divided among villagers; thus, a household contract responsibility system for salt production was implemented. Since the 1990s, in order to encourage the salt industry development, the local government even subsidized rice to ensure the basic livelihood of specialized households for salt production. In July 1997, in response to the adverse effects of non-iodized salt on the physical health of local people, the Tibet Autonomous Region implemented the *Regulations on the Supervision of Iodization of Salt to Eliminate Iodine Deficiency Hazards*, in order to ensure people's access to iodized salt they needed normally through legal norms, and also formulated regulations on punishing sellers of non-iodized salt. Since 2008, Tibet has gradually started to distribute iodized salt free of charge. At present, each farmer and herdsman in Yanjing Nakhi Ethnic Township, Markam County may receive 100 kilograms of "Shengkang Iodized Salt" produced in Lhasa, Tibet every year, each package weighing 1 kilogram, and the salt packaging is printed with the words "government subsidies, exclusively for farmers and herdsmen".[1]

"Without salt, no country can be founded, and salt is indispensable for a powerful country", which is a classic precept of national governance throughout

1 Li Hechun. *Impetus and Shackles: Salt and Salt Taxes in the Lantsang River Canyon*. Guangzhou: Sun Yat-sen University Press, 2016: 210.

history. It is precisely because of the implementation of policies in line with the characteristics of the times that the traditional and ancient salt making and operating activities in Yanjing Township have been fully preserved.

2) Salt tax

In Yanjing, by controlling salt taxes, it is possible to have command over the economic lifeblood of local society. Without exception, administrators in different historical periods are interested in salt taxes; whether it was the chieftains, the lamaseries, or Zhao Erfeng and the garrisons stationed here by the central government since the late Qing Dynasty, all raise their revenues through salt taxes. Due the complex political environment in the region where the Yanjing was located, the chieftains and the temple held the local power for as long as 370 years, and jointly collected the salt tax. Therefore, Yanjing was very much affected by a system combining political and religious powers.

As to the salt tax revenue here, except the fact that while recording salt pan supervision during the Wanli period, "collecting taxes here" was mentioned in *Travelogue in Weixi*, an earlier account could be found in "Yanjing County Chronicles": "When Yanjing was not yet conquered (before the bureaucratization of native officers), the salt tax was collected by two or three junior clerks oppointed by the second highest Batang officer, each *duo* (the load carried by an animal) taken away two *pi*..." At that time, the salt tax revenue in Yanjing was mainly controlled in the hands of local officials, headmen, and lamas. In the third year of the Guangxu period in Qing Dynasty (1877), a bureau was established for collecting taxes to replenish border defence fund. The annual output of salt was about 10,000 *duo* then, and each *duo* weighed 60 kilograms, with a tax payment of 1.5 Tibetan silver dollars, which was equivalent to 0.48 tael.[1]

At the end of the Qing Dynasty, there were clear records about the relationship between the taxation in Yanjing and the Laweng Temple on the Lantsang River. When Zhao Erfeng was implementing the bureaucratization of native officers there, "the salt profit in this area has long been forcibly occupied by the temple (Laweng

1 Song Liangxi, Lin Jianyu, Huang Jian, and Cheng Longgang. *A Dictionary of Chinese Salt Industry History*. Shanghai Lexicographical Publishing House, October 2010: p. 108.

Temple)"[1]. According to *Local Geographical Records of Yanjing, Batang*: *Taxation*: "In Yanjing, salt taxes were always collected by a *xie'ao* there appointed by the Batang chieftain, all being highland barley and other miscellaneous grain crops. The chieftain would take away a couple of *pi*, and the Laweng Temple would take away another couple of *pi*. Previously, the food production affairs in Batang were reported to the Sichuan provincial treasurer for record keeping."[2] And it goes on to say: "Since the establishment of the bureau in the thirty-first year of the Guangxu period (1905), the levy was stipulated in explicit terms. Each *duo* weighs 70 kilograms, with a salt tax rate of 3% to 5%, i.e., 1.5 Tibetan silver dollars, or equivalent to 0.525 tael of silver. But since the beginning of its openning in the thirty-third year of the Guangxu period, everything became completely unverifiable. Anyway, from the thirty-third and thirty-fourth year on, salt taxes increased year by year. In the first year of the Xuantong period, tax silver collected through profit tax rate was over 11,300 taels. In October of that year, Zhao, the Minister of Border Affairs, reported to the emperor for the purpose of carrying out experimental regulations. As before, 1.5 Tibetan silver dollars for each *duo* would be taken away..." With regard to the effect of taxation, in just one month from the beginning of November to the end of the thirty-first year of the Guangxu period (1905), the salt tax revenue reached more than 660 taels of silver. To convert the salt tax revenue into the number of *duo*, 1,300 might had been taxed on within one month in the thirty-first year of the Guangxu period (1905).[3] Afterwards, during the administration of the official salt bureau, the salt tax revenue basically remained at around 8,000 to 10,000 taels of silver per year, with the number of *duo* ranging from 16,667 to 20,800. In the first year of the Xuantong period (1909), Zhao Erfeng established a Salt Bureau in Yanjing to administrate salt production, regulate salt prices, and collect salt taxes. For each *duo* of salt,

---

1 *Archival Historical Data of Sichuan-Yunnan Border Affairs in the Late Qing Dynasty (First Volume),* compiled by Sichuan Institute of Ethnic Studies and the editorial group of *Archival Historical Data of Sichuan-Yunnan Border Affairs in the Late Qing Dynasty*, published by Zhonghua Book Company, 1989 edition, p. 103.

2 Duan Pengrui. *Local Geographical Records of Yanjing, Batang*. National Library of China, stereotype edition of the second year of Xuantong (1910), p. 12.

3 Each *duo*, calculated at 1.5 Tibetan silver dollars, was equivalent to 0.48 tael of silver, and the number of *duo* was equal to the amount of salt tax silver divided by 0.48 tael.

calculated at 70 kilograms, 1.5 Tibetan silver dollars would be taken away, which was equivalent to 0.525 taels of silver. According to the statistics of that year, a total of 12,200 taels of tax silver was collected. From lunar January of the second year of the Xuantong period (1910), taxes were collected by the gold standard, i.e., 0.32 tael of gold each Tibetan silver dollars. (Acoording to the local weights and measures, 1 "*ke*" (tranlator: a Tibetan unit of volume or dry measure (holding about 12.5 kilograms of barley).) was equivalent to 1 "*dou*" (tanslator: a unit of dry measure for grain (=1 decalitre).), and 1 "*pi*" was equivalent to 1 "liter". However, among the common people, 20 "*pi*" was converted into 1 "*ke*", while 20 to 24 or 25 "*pi*" used to be converted into 1 "*ke*" by the Batang local authorities.)[1] After Zhao Erfeng's reform of the above stages, the salt industry in Yanjing showed initial results, and salt taxation there was relatively standardized thus gained some effects.

After the fall of the Qing Dynasty, the 13th Dalai Lama returned to Tibet from India and began to implement policies. A few years later, he set up a Salt and Tea Bureau directly under the jurisdiction of bkav-shag (local government of Tibet), with one monk director and one secular director (fourth-rank officials); under it there was one monk offical and one secular official (fifth-rank), plus six staff members. Outposts of the tax office were set up in anterior Tibet and the northern part of Ü-Tsang, Lhoka (Shannan), and Tagong, with 0.6 tael of Tibetan silver taxed on each *duo* by an ox and all turned over to the Tibetan treasury. Except the salt tax in the Qamdo region, which was turned over to the local governor, the other taxes were turned over to the Tibet Treasury.[2]

For more than a decade in the early Republic period, Yanjing was basically under the control of the Gongga Lama. In the thirteenth year of the Republic period (1924), the tax revenue in Yanjing was handed over to the garrison, but the salt tax collection was not ideal. In the sixteenth year of the Republic period (1927), Yanjing Lama Zagong Mise, by colluding from within with Batang bandits from outside, launched attack and burned down the county office; a whole company of the Yanjing garrison was disarmed and could only flee to Yunnan. Under the pretext of rescuing

---

1 Fang Jianchang. "Research on the History of Salt in Tibet". *Salt Industry History Research*, 1995 (1).

2 See Fang Jianchang's "A Brief History of Salt Industry and Salt Administration in Tibet", published in *Journal of Southwest University for Nationalities*, 1993 (1).

Yanjing, the subordinates of Songming Lama collected more than a month's salt tax revenue and left. So, the Gongga Lama had to be asked to assist in the defense and collect the salt tax.[1] Thereafter, the Gonga Lama's power went far beyond collecting taxes. After clashes with Kelsang Zayein in 1932, Yanjing was practically under the control of the Gongka Lama, due to its location on the edge of the area effectively controlled by both Han and Tibetans. Moreover, when Ba'an local armed forces were fighting with the Tibetan army, the Gonga Lama came out to defend and shared the salt tax revenue with the latter.[2] In Cui Kexin's "Investigation of the Geology and Salt Production in Yanjing County", he mentioned that there was a problem of uneven salt taxation in the thirtieth year of the Republic period (1941): The salt taxes on small *duo* and large *duo* were disproportionate, resulting in significant tax losses. Moreover, the tax calculation, based on the *duo*, might easily (cause) the salt merchants to increase the load...reducing the effectiveness (efficiency) of transportation. In this regard, he also put forward suggestions for improving the tax system: The salt tax collection would not distinguish between the sizes of the *duo*, but measure it with *dou*; in this way, the taxation would be fair and equitable, and hence beneficial to both the public and the private.[3] According to statistics from the thirty-sixth year of the Republic period (1947), a total of 725,000 yuan (national currency) was collected in salt tax that year.

Before the liberation of Yanjing, the Batang chieftain sent a *xie'ao* there to collect salt tax, which was paid with highland barley and other miscellaneous grain crops, two *pi* of highland barley (approximately 0.5 kilograms per *pi*) collected for each *duo* of salt. In the early days of liberation, according to *Investigation Materials of Well Salt in Yanjing* compiled in 1954, "regardless of the kind of livestock used for transportation and the amount of load, half a silver dollar is collected for each *duo*, and an additional *dou* (about 6 kilograms) is allowed. Therefore, merchants tried to load the animals as much as possible, each *duo* weighing 75~120 kilograms. From

---

1 See "Yanjing County Chronicles", author unknown, published in *Political Affairs of the Border Region*, 1931(6).

2 See Wang Huailin's *In search of Khamba: A Report from the Hometown of Shangri-La*, published by Sichuan People's Publishing House, 2000 edition, p. 97.

3 "Investigation of the Geology and Salt Production in Yanjing County", by Cui Kexin, published in *Xikang Economic Quarterly*, 1944 (8).

October 1950 to February 1954, a total of 32,374 yuan of salt tax was collected."[1] Starting from 1959, the tax rate was 2 yuan RMB per 50 kilograms; thus by 1979, a total of 548,000 yuan had been collected. In July 1980, salt tax was exempted. After the 1980s, salt pans were distributed to individuals for production. Nowadays, taxes are not paid by salt-maker households, but rather by the merchants who come here to purchase salt. Only when salt makers transport their salt out to sell, they pay the taxes themselves.

When analyzing the salt tax in Yanjing, Fang Jianchang pointed out the importance of four aspects: (1) by increasing border defence fund, (salt tax) drove the implementation of the tax system; (2) by supporting the circulation of Sichuan silver and copper coins in Tibet, it resisted the use of the Indian rupee, and safeguarded China's sovereign position in the financial field; (3) it was conducive to the promotion of standard measuring instruments; (4) it enhanced understanding and unity among the Lhoba, Monba, Mishmi people, Tibetans, and Han people in the Chayu region.[2]

However, the role of salt tax was not always a positive one, for excessive taxation and too harsh treatment of salt makers might dampen their enthusiasm. Therefore, salt law systems were continuously formulated in successive dynasties, so as to regulate the relationship between administrators and salt producers. But more often than not, it was very difficult to achieve a balanced state.

2. Salt Transportation and Sale

According to Thomas Thaler, salt trade was once considered an important way of interpersonal communication, and many experts even regarded the demand for salt

---

1 Quoted from *Investigation on Commerce and Handicraft Industry in Tibet*, co-edited by Institute of Ethnic Studies, Chinese Academy of Social Sciences, and Institute of Social Economics, China Tibetology Research Center, and published by China Tibetology Publishing House, 2000 edition, p. 8.

2 See Ran Guangrong's "A Brief Account of Well Salt Production in Yanjing County, Sichuan at the End of the Qing Dynasty and the Beginning of the Republican Period" collected in *Proceedings of the International Symposium on the History of Chinese Salt Industry*, edited by Peng Zeyi and Wang Renyuan, and published by Sichuan People's Publishing House, 1991 edition, pp. 365-370.

as the root cause of the rise of all trade. Because salt mines are limited, and salt is an indispensable necessity for life, in order to ensure its stable production, it is necessary to have an intensive and sustainable trade network. And it is generally believed that salt trade is actually a common transaction of daily necessities.[1] Although salt making has long been a main livelihood of Yanjing residents, people's intake of it is limited after all. Salt can only fully realize its social function and self-value through exchange and circulation. Therefore, since it was produced, salt has entered the ranks of commodities, and its sales have become the most fundamental way for salt makers to survive.

Yanjing salt has been very popular with a vast number of people in Tibet, and there is a considerable local market. While meeting the needs of local villagers, it has also been sold to Zogang, Zhag'yab, and Zayu in Tibet, Batang and Litang in Sichuan, and Deqin, Zhongdian and other ethnic Tibetan areas in Yunnan.

There are two main forms of trade in Yanjing: "local trade" and "transit trade". The former is mainly operated by Yanjing merchants, while the latter is carried out by some merchants from other places during the transportation of goods via Yanjing. Local merchants generally purchase salt, medicinal herbs, mountain products, and livestock products locally, and transport them to Nandun, Adunzi, Zayu, Zhongdian, Kangting, and other regions, where they may exchange for tea, cloth, Western medicine, and tobacco, etc., and bring them back to the county for sale.

1) Transportation and Sales Routes

The circulation of salt takes place within a certain region and range. After examining the history of Yanjing, the author found that the exchange of salt there formed sales circles in different regions, and with the change of systems, these circles of salt exchange would change accordingly. In general, the core exchange circle is composed of salt makers and their living area; the middle circle is a trade circle formed by the interaction between Yanjing and surrounding ethnic groups and regions, which was realized after a long journey by people, salt and horse caravans;

1 Thomas Thaler's "The Early Salt Industry in Central Europe: Neolithic Salt Production and Trade Patterns", translated by Wen Chenghao and Lin Yongchang; see *Archaeology of China's Salt Industry (Episode 2): A Comparative Observation from an International Perspective* co-authored by Li Shuicheng and Luo Tai, and published by Science Press, 2010 edition, p. 210.

the outermost circle is a vague "imagination", a force that drives the relationship between Yanjing and the national states, and even the world system--invisible as it is, it constantly exercises great influence upon Yanjing.

The factors affectting salt circulation routes and areas are quite complicated, among which geographical environment and the salt industry system are two major ones. From a geographical perspective, due to geographical constraints in eastern Tibet, salt routes mainly consist of the passages opened up on the tablelands at the bottoms of the Lantsang River canyon, and the Nujiang River and Chin-sha River canyons on its east and west sides; of course, the surrounding small valleys are also important channels for salt transportation. Salt makers would first keep their salt in temporary warehouses or at home, and then transport it by mules and horses to distant places for exchange. The transportation and sales system in salt industry became the baton for salt circulation. Since ancient times, the political powers in Yanjing have determined that the sales of salt are mostly limited to their scopes of control. Within these scopes, free trade could be conducted after paying a certain amount of tax; but once beyond them, outposts of the tax office would wait along the way. As far as the trade circles of Yanjing salt are concerned, the northern part is mainly composed of Tibetans; the western part involves Tibetans and such nationalities as Nu, Drung, Monba, Lhoba, and Mishmi people; the southern part is composed of Lisu, Yi, Nakhi, and other nationalities; while the eastern part is mainly composed of Tibetans and Nakhi.

Since the Ming Dynasty, when Chieftain Mu arrived at Yanjing and began large-scale traditional salt making through solarization, the sales of salt here have changed its scopes and routes with the changes taking place in upper political powers and trade relations between ethnic groups. There were two main channels for the transportation and marketing of Yanjing salt in the Ming Dynasty: One was a road leding from Zhaguxi Valley in the north of Yanjing to Batang and Ningjing. In June 2009, a joint archaeological team consisting of the Shaanxi Academy of Archaeology, the Institute of Cultural Relics Protection of Tibet Autonomous Region, and the Archaeology Department of Northwest University discovered the cliffside images in a little Lakang (Buddhist Hall) in the Brag sGo bShad valley of Lower Yanjing

Village, Nakhi Township, Markam County[1], and concluded that it was a cultural relic of the Tubo period. This indicates that during the Tubo period, Zhaguxi Valley was already used as a passage by human beings. In the field investigation, local people told us that this used to be the only way to Batang and Markam long long ago, and afterwards became part of the Ancient Tea-Horse Road. The other, which led south from Yanjing, mainly relied on the geographical advantages of the Lantsang River Canyon, and humans could choose the tableland by the river as a passage. In the late Ming Dynasty, with the opening of the Yunnan-Tibet trade route from Deqin to Yanjing, handmade salt in Yanjing already begun to be sold to Deqin, Weixi, Zhongdian, and Yangla, the northernmost part of Deqin, etc., in Yunnan Province.

Since the Qing Dynasty, with the development of Yunnan-Tibet trade and Sichuan-Tibet trade, especially the further development of the Yunnan-Tibet trade route in the late Ming Dynasty, besides the trunk roads, various trade branch lines, similar to the "capillaries" of a human body, were also joined in. Salt merchants, travelling by climbing up hills and wading through rivers along the Yunnan-Tibet and Tibet-Sichuan Ancient Tea-Horse Roads, sold their goods to Yunnan, Sichuan, and Tibet, etc. Since the Ming and Qing dynasties, Yanjing had long been under the jurisdiction of Chieftain Mingzheng in Batang and the Laweng Temple. Considering the influence of political power on its transportation and sales, salt from Yanjing at that time was sold in large quantities to Batang, Litang, Derong and other places near the east. As to the west and north, according to records of salt in the Qing Dynasty, it was sold to Bitu and Tsarong, etc. Having arrived at Tsarong, one could Keep going in three directions, to Zhayi, Menkong, and Zogang, etc., in the north, to Zayu in the west, or to Gongshan in the south. To the north, due to its proximity to Ningjing County of the time and the Ancient Tea-Horse Road passing through it, salt sales already reached the present-day Markam, Zogang, and the vast area in the southeast of Qamdo. In the sixtieth year of the Kangxi period (1721), Du Changding, an aide to the governor of Yungui (Yunnan and Guizhou provinces), gave a detailed

---

1 See "The First Comprehensive Archaeological Investigation and Record of the Tubo Stone Carving Remains in Eastern Tibet: Two Newly Discovered Tubo Stone Carving Remains in Markam County and Chaya County, Qamdo Prefecture, Tibet" by Xi Lin, Zhang Jianlin, Shar rgan dbang vdus, and Tian Youqian, published in page four of *China Cultural Relics News* on November 13, 2009.

description of the whereabouts of the salt produced in Yanjing, "every year, more than 10,000 *duo* of salt was produced on both sides of the river, and sold to such counties as Ba'an, Lihua, Dege, Derong, Dingxiang, and Daocheng, etc., as well as Tsarong, Mengkong, and Zhongdian in Yunnan Province, and other places", which demonstrates the significant impact of Yanjing on the surrounding areas. Due to the popularity of the local salt, Tibetans from Batang and Litang in Sichuan often drove their horse caravans here to purchase it, too. And this continued until the liberation of Tibet.

At the end of the Qing Dynasty and the beginning of the Republican period, the sales area of Yanjing salt covered several nearby counties (or prefectures), mainly Markam, Qamdo, Zayu, and Zogang, etc. in Tibet, Batang, Lihua, and Derong, etc. in Sichuan, and Gongshan, Deqin, Weixi, Zhongdian and Dali in Yunnan. The vast majority of salt here was locally consumed, while a small portion was sold to neighboring counties. Reaching the Nujiang River in the west and crossing the Chinsha River eastward, it might be described as traversing the three rivers and travelling through the three provinces (or region). In the section "Conditions of the People and Sales" of "Investigation in Yanjing County", the sales area was described as: "It reached Hekou in the east, Zhayi in the west, Jiangka in the north, and Zhongwei to the Yunnan border in the south. Stretching for more than a thousand miles, the sales channels were absolutely plentiful. However, simply because the vast land was sparsely populated, local people's consumption of salt was very little; other than making tea, they fed cattle and horses with it. What is more, the rugged mountain paths made transportation quite difficult. Despite the low salt prices, its sales did not go smoothly[1]. In *Yanjing County Annals*, Liu Zanting stated that "the salt here (salt produced in Yanjing) was sold in different counties in south Kham Tibetan area and the western part of Yunnan".[2] It can be seen that the locality might have been the main sales area, rather than the entire sales scope, and could have been further expanded geographically. Since the Ming and Qing dynasties, the transportation and sales scope of salt industry in regions of northwest Yunnan had basically taken shape.

1 "Investigation in Yanjing County" by Jin Fei, published in *Political Affairs of the Border Region*, 1931(8).

2 Liu Zanting (Qing Dynasty). *Yanjing County Annals* [Z]. Beijing: Library of the Cultural Palace of Nationalities, 1962: 37.

In the early 1950s, with the liberation of Tibet, the roads in Markam County were improved. Since 1959, various sections of highways, such as National Highway 318 and National Highway 214, had been connected one after another. The completion of these road works greatly facilitated the sale of salt. Among them, the Yunnan-Tibet Highway, with a total length of 714 kilometers, was completed and opened to traffic in 1974. Starting from Jinghong, Yunnan, it passes through Markam, Zogang, Qamdo, and Riwoqe in Tibet to Dopuma on the Qinghai-tibet boundary. The section within the Tibetan region is about 803 kilometers long, connecting with the southern line of the Sichuan-Tibet Highway. The Yunnan-Tibet Highway has brought new opportunities for the development of Yanjing, and the traditional life style of the local people has been gradually changed, too. In 2009, a Yanjing archaeological investigation team jointly conducted by the three provinces (region) analyzed that there were three major transportation routes for salt-maker households in Yanjing, Tibet to sell their salt outward: route one being Yanjing--Foshan, Yunnan--Deqin; route two Yanjing--the Brag sGo bShad valley--Xuzhong Township; route three Yanjing--Songmuda power plant valley, which then splits into three: (1) Bida--Zayu County--Tsarong, (2) Duilashan--Zayu County--Tsarong, and (3) Duilashan--Biduo Township, Zogang County.[1] Among them, route two, starting from Yanjing to Xuzhong Township, should have been extended to the areas of Batang and Litang, and even to Yangla in Yunnan and Derong in Sichuan, etc. The sales scope during this period was mainly concentrated in the counties near Markam County, Qamdo Prefecture, such as Zogang and Zayu, etc., as well as Batang, Litang and Derong in Sichuan, and Deqin and Weixi in Yunnan. When the salt output was relatively high, it would also be transported to Yushu, Qinghai and other places for sale. In accordance with his investigation, Cui Kexin noted: "Yanjing salt has a wide market, stretching southward to such counties as Deqin and Weixi in Yunnan, northward to Jiangka, Zhag'yab and until Qamdo, westward to Jigong, Zayu Chayu and Bangda

1 The Institute of Cultural Relics Protection of Tibet Autonomous Region, Shaanxi Academy of Archaeology, and Sichuan Archaeological Research Institute. "Investigation report of Salt Pans in Yanjing, Mangkang County, Qamdo Prefecture, Tibet Autonomous Region" [R]. *Cultural Relics in Southern China*, 2010 (1).

Temple, and eastward to Derong, Dingxiang, Lihua Bai'an and Baiyu."[1] It basically delineated the scope of sale of Yanjing salt. At this time, there were Yanjing salt exchanges in Deqin, Weixi, Gongshan, and Zhongdian in Yunnan Province. In 1978, after the county government established a salt iodization site in Yanjing, the local salt industry began to develop rapidly. The iodized salt was not only sold exclusively in areas plagued with iodine deficiency disorders within the county, but also transported to various parts of Tibet, Sichuan, Yunnan, Qinghai, and other places for sale.[2] However, in recent years, due to the free distribution of iodized salt in Tibet and the harms of non-iodized salt on human health, fewer and fewer people choose to consume Yanjing salt. Its sales scope is becoming smaller and smaller, too, mainly centered around Yanjing, with a radius of 50~100 kilometers, and the sales volume decreases as the distance from Yanjing increases. At present, the sales of Yanjing salt are mainly concentrated in a couple of local villages and villages in Quzika Township near Yanjing, while reaching southwest to Foshan Township in Deqin, east to Xuzhong Township, north to Markam County Town, and west to Bitu Township.

Generally speaking, the sales of Yanjing salt rely on the Ancient Tea-Horse Road, the commercial exchange channels between Yunnan and Tibet, as well as Sichuan and Tibet. Moreover, the salt here did not go through such stages as "circulating salt within demarcation lines" and "selling salt by tickets", etc. in the strict sense, and its sales scope and routes, though existed for sure, were not fixed and unchangeable. This might have something to do with the social history of the region where Yanjing is located, the lack of a systematic salt industry system throughout Tibet over the ages, and the underdeveloped salt industry there.

For a long time, salt makers in Yanjing have formed a mindset centered on local culture, which is based on salt as a necessity of life and goods for exchange. None of the counties around produce salt; within a radius of nearly 200,000 square kilometers, Yanjing has been the only salt-producing area, thus forming a trading circle centered around its salt. Even when transported by horse caravans to distant mountainous areas, Yanjing salt is equally popular there. Before the liberation of

1 "Investigation of the Geology and Salt Production in Yanjing County", by Cui Kexin, published in *Xikang Economic Quarterly*, 1944(8).

2 *Yanjing County Annals*, pp. 366-367.

Yanjing, many merchants went to Yanjing to purchase salt, which was difficult to be obtained within a short period of time; during the Republican period, there was also the phenomena that "salt merchants all brought various daily necessities, such as rice, tea, and clothing, etc., to Yanjing to exchange for salt with salt makers, but due to the large number of pedlars, and the fact that there was no salt for several days after rain, sometimes for tens of days no transaction could be conducted, for it might take even longer"[1]. This scenario is already a thing of the past, and salt makers today would have to find a way to sell their own salt. Although such a transformation seems difficult to endure for salt makers who have experienced the period when merchants and horse caravans came voluntarily to Yanjing for salt, they can only accept reality.

2) Horse Caravans

Located on the Lantsang River banks in the Hengduan Mountains region with high peaks, deep canyons, as well as narrow and precipitous paths, the transportation of Yanjing was extremely inconvenient. The ancient ethnic groups living here were obviously unable to travel long distances to exchange for goods by relying solely on manpower transportation. Hence, horses and mules became special tools for the transportation of local goods, playing an important role in economic and cultural exchanges.

In fact, transportation by horse caravans consisting of "Maguotou" (the leader of a caravan on the Ancient Tea-Horse Road) and "Majiaozi" (the hired horse drivers) developed very early into a major form of Yunnan-Tibet trade. The horse caravans of Bai nationality were esentially responsible for the section of route from Dali to Deqin, extending along "Dali--Lijiang--Shangri-La--Deqin", while the section "Deqin--Markam--Lhasa", owing to the cold climate, is transported by the Tibetan horse caravans instead. Due to the wide scope of exchange outside the Yanjing area, it was mainly achieved through long-distance journeys of the horse caravans in history. For example, on October 1st of the first year of the Xuantong perdiod, Zhao Erfeng, the minister of Sichuan-Yunnan border affairs, reported regulations of the salt taxation in Yanjing for the record: "The salt produced in Yanjing, ...uncivilized

1 "Investigation of the Geology and Salt Production in Yanjing County", by Cui Kexin, published in *Xikang Economic Quarterly*, 1944 (8).

minorities would put it in leather bags, and then transport by horses and mules for sale."[1] In the last century and before, long-distance transportation of finished salt depended chiefly on horses, mules and donkeys. Afterwards, in order to adapt to the narrow paths and steep slopes in western Yunnan, mules produced by crossbreeding horses with donkeys in the horse caravans were used to transport goods instead, for they could move agilely, and had better endurance and were able to carry heavier loads than horses.[2]

The salt produced in Yanjing, whether transferred by salt makers to their homes or points of sale, or transported to various sorrounding areas for exchange, cannot be done without horses and mules. Firstly, to overcome the shortage of manpower was one of the important factors in the formation of horse caravans and their becoming a common means of transportation in the local area. For example, it took 5 to 6 days to travel from here to Deqin, a month to go to Lijiang and get back, or about 8 days to reach Weixi. Without animal labor, it would have been possible. Another important factor was the successful domestication of horses, which gradually became riding tools. For a long time, horses have played an important role in hunting and warfare, but there is still no consensus on whether they were first used for transporting goods or for riding. The Tibetans, as an ancient ethnic group, have a long history of horseback riding. However, it is still unknown whether the horses in the eastern Tibetan region have been fed by local ethnic groups for generations. Yet, Yanjing, close to the northwest region of Yunnan, was also an important channel for the Ancient Tea-Horse Road, and thus the horse caravans played an important role in the long-term trade relations. The vast majority of these caravans entered Tibetan-inhabited areas from such places as Dali, Lijiang, and Zhongdian, etc. It's known that the history of horses in Yunnan is also relatively long. According to some scholar, the emergence of domesticated horses in Yunnan was roughly around the 11th century

1 Li Hechun. "Zhao Erfeng's Reform of Yanjing Salt Industry and Its Significance during Bureaucratization of Chieftains in Chuanbian at the End of Qing Dynasty". *China's Borderland History and Geography Studies*. 2016, 26(02).

2 Yang Xinyu, Zhang Xilu, and Liu Rong. "Economic and Cultural Exchanges Between Tibetan and Bai in the Period of the Republic of China". *Journal of Dali University*. 2015,14(01).

BC.[1] Our understanding of the Qinghai-Tibet Plateau during this period is not yet comprehensive enough. But horse trappings were found in the Sarcophagus tombs in Nagu Touwnship, Deqin County, only 30 kilometers away from Yanjing, and the discovery of these bronze bells may indicate that horses already existed at that time. According to analysis, it shows that horses had already appeared in this area from the Western Zhou Dynasty to the mid Spring and Autumn period.[2]

Duan Pengrui mentioned in his *Local Geographical Records of Yanjing, Batang* that: "The animal husbandry here is far more underdeveloped than that to the north of Batang." It seems that the animal husbandry in Yanjing was not thriving at that time, hence immediately followed by "the total numbers of cattle, and mules and horses in Yanjing were no more than 3,462 and 2,713 respectively", and "such livestock as yaks, dzos, mules, horses, donkeys, and dogs were no different from other places, with Yanjing donkeys of the inferior quality..."[3] Therefore, it's very likely that the horses which made up the caravans were exchanged from Yunnan.

A horse caravan generally consists of 100~200 horses, with at least five people most of the time; a Han Chinese can drive only five horses at a time, while a Tibetan can drive seven. They'd take a short break every fifteen *li* and a long break every thirty *li*. If there was a heavy rain or snow, they might only build a shed for temporary stay, and continue their journey when the weather turned fine again. Sometimes, when a mule got sick on the way, they would not be able to go any further. In such cases, they could only wait for the caravans returning to Yanjing by this path, seek help from them, and go back together. The head mule of a horse caravan was mostly very strong, tied with a large bell. In places where the path was particularly narrow, the clear and loud sound could remind the caravans coming from the other direction, so that people were able to find a proper spot to dodge each other, reducing friction and ensuring everyone's safety. Another version was that the sound

1 See Zhang Zengqi's "Ancient Yunnan Horseback Riding Ethnic Groups and Their Related Issues" in *Collected Works on Yunnan Bronze Culture*, compiled by Yunnan Provincial Museum, and published by Yunnan People's Publishing House, 1991 edition, p. 263.

2 See Wang Mingda and Zhang Xilu's *Horse Caravan Culture*, published by Yunnan People's Publishing House, 1993 edition, p. 9.

3 Duan Pengrui. *Local Geographical Records of Yanjing, Batang*. National Library of China, stereotype edition of the second year of Xuantong (1910), p. 11.

could drive away wild animals; whenever hearing it, they'd be scared and flee at once.

Moreover, while transporting goods on the way, horse caravansmight be robbed by bandits, and this kind of thing happened from time to time. Therefore, the best precaution on the Ancient Tea-Horse Road was to organize horse caravans together, and in most cases, it would be necessary to have more than ten people before setting out on a journey. It's said that the largest caravan group might consist of more than 700 horses and more than 70 people. They were very trustworthy and closely united with each other. Banditry was especially rampant in Benzilan Township, Zhongdian, and they'd often take advantage of favorable terrain to intercept at the front and the back. Wealthy horse drivers were able to purchase firearms to protect themselves, while merchants in poor economic conditions had to rely on those with strong economic strength who could afford to guns, and travel with them, in order to seek protection in these trade journeys. However, this kind of temporary caravan groups also had their own shortcomings. Its main manifestations were as follows: Firstly, it was a long journey to travel by the Ancient Tea-Horse Road at that time, and the horse caravans often had to go through some narrow paths. If there were too many members, it would be difficult to move. Secondly, there would be division of labor among caravan groups throughout the journey; for example, some were responsible for cooking, some for taking care of livestock, and some for security, etc., which might lead to disagreements. As these groups were formed on a temporary basis, and thus difficult to be united, the experienced elderly who often joined in horse caravans would play an important role at this moment. Sometimes, in order to avoid disagreements, people would often take turns doing various tasks. In order to guard against banditry, they'd travel during the day in areas where bandits frequented, and take turns on night shifts to ensure the safety of the entire caravan group. And in some areas where bandits might not be expected, they'd herd all horses and mules by the roadside, and sleep soundly at night.

These horse caravans had to cross many rivers along the routes, and their emergence also made people think about how to replace ziplines and single-log bridges with more solid bridges, and how to replace log rafts and bamboo rafts with larger wooden boats. When the two factors, i.e., horse caravans and salt, came together, an ancient salt transportation routes for horse caravans centered around

Yanjing began to take form, thus becoming the local network of the Ancient Tea-Horse Road. As a kind of ancient routes with strong vitality, they simultaneously embodied properties of both the ancient routes for horse caravans and the salt transportation routes: on the one hand, a network of ancient routes was needed for salt transportation in the perilous mountainous region; on the other hand, horse caravans were able to overcome all kinds of difficulties and obstacles, but required stricter organization to complete the expeditions across the Hengduan Mountains. Countless historical facts have proved that these horse caravans on the Ancient Tea-Horse Road were often faced with extraordinary hardships and dangers, and each of their journeys was one of life and death experience. But meanwhile, it seems that precisely due to the existence of these caravans, the verbal communication along the Ancient Tea-Horse Road was almost barrier-free. These caravans, sharing weal and woe, trudged along the Ancient Tea-Horse Road, have made remarkable contributions to the political, economic, and cultural exchanges between Tibetan and Han people. The ancient routes for horse caravans not only rapidly shortened the transmission time of material and cultural factors, as well as the spatial distance of villages in the region, but also expanded the capacity of civilization dissemination, bringing prosperity to these villages. They might even extend into the Hengduan Mountains region with lofty peaks and high ranges, where people could hardly set foot on.

With the rising and falling whistles and the tinkling of horse bells, a salt-laden horse caravan, whose arrival was always eargerly anticipated, would make the peaceful villages bustling with noise and excitement at once. When arriving there, the caravan would exchange for foodstuffs, such as highland barley, wheat, and corn, and oil occasionally. In order to exchange salt, it was usually necessary to go to a densely populated area. Anyway, those villagers, on seeing the arrival of a horse caravan, would rush about telling the message, and people from each household, carrying bags, basins, and bamboo baskets full of grains and walnuts, etc., or bottles filled with sesame oil, would laugh and talk all the way, and take the initiative to exchange. The exchange site was usually in a village, or in a mountain gully, where more people would gather. If there was no village along the way, the horse caravan had to climb mountains unremittingly, until someone to exchange with might be found. One thing they all knew was that the farther the salt was transported from Yanjing, the higher the price would be. "In order to exchange for more foodstuffs, we are willing to go a longer way, because that way way we may get more grain. For example, a village close to Yanjing can be reached in just one or two days, but the salt carried by 10 mules can only be exchanged for grain carried by 12 mules; however, if we walk for seven or eight days, the salt carried by 10 mules can be exchanged for grain carried by 16 mules. Moreover, horses and mules can graze along the way, saving on fodder."[1] In addition, due to familiarity with the living conditions along the routes for trade, people knew where and what to eat, accumulated a complete set of life experiences in different regions and environments along the way, and learned different survival modes for people, mules, and horses under different circumstances, as well as how to prevent diseases and injuries that might occur easily in different regions. For example, in such high elevational regions as Lhasa, they would neither eat food made of flour nor drink yak meat broth, and when traveling long distances, they'd try to prevent edema in their bodies; the bags for mules and horses to carry goods, woven from wool, were made durable, breathable material that dries easily, and so on.

The Tibetan people's dependence on salt required more horse caravans, and horse caravans of better quality, in addtition to well developed ancient routes for them. The life of horse caravans began to be professionalized, forming their specialized terms as well as business rules, and the ancient routes for them thus become fixed and seasonal. Two major categories, i.e., official caravans and civilian caravans began to take shape. Such ranks as "Daguotou" (the supreme leader of a caravan), "Erguotou" (deputy leader), "Sanguotou" (the thrid most powerful man, responsible for managing the economic accounts and living of a caravan), "Guanshi" (a man in charge of daily affairs), and horse drivers began to appear, too. And then there were caravansaries.[2] Horse caravans was not only a means of goods transport, but also gradually became a messenger of cultural dissemination. The horse caravans from Dali, Yunnan would perform Yunnan Lantern Drama in Tibetan areas, and the horse caravans from Tibet would also play a three-stringed plucked instrument when in Yunnan, attracting many onlookers. Following these caravans, some craftsmen

---

1 Field research data of Yanjing conducted by Li Hechun in November 2013, with Tashi Tsering being the interviewee and Luo Song the Tibetan-Chinese interpreter.

2 Chen Baoya. "The Origination of Tea-Horse Road". *Thinking*, 2004(04).

came to Tibet and became pioneers in spreading new technologies, while the missionaries, with the help of them, entered Tibetan areas to preach their religion.

With the liberation of Tibet, transportation lines began to be built in various regions, especially the section of the National Highway 214 from Yunnan to Yanjing, after its opening to traffic, greatly facilitated the smooth traffic from Yunnan to Yanjing. The horse caravans that used to travel between Yunnan, Tibet, and Sichuan began to decline, and many people had to sell out their horses and engage in other industries. Instead, modern transport vehicles and trucks of all kinds had started their journeys between Sichuan, Yunnan, and Tibet. Today, salt transported from Yanjing to such areas as Foshan Township in Deqin, Mangkang County Town, and Batang is no longer by humans or horses, but can be delivered within a few hours by modern means of transport. In the past, salt used to be transported to Deqin on horseback, and it took 6 to 7 days to travel from Yanjing to Mingyong Village in Deqin; now, the roaring motor vehicles or tractors on the spacious asphalt roads can reach different townships in Deqin County in a short time. The local horse caravans no longer exist, but horses and mules may still be used for transportation in areas inaccessible to vehicles or for short-distance transportation; for example, salt tranported to Zayu and Zogang, etc., still relies on mules and horses. Moreover, Tsarong people still drive their horses here for business to this day; more than 100 people, bringing tsampa and other things, would purchase rice, salt, wheat, and oil, etc.

As the most common livestock in Yanjing, horses and mules, even with the advancement of transportation today, can still be seen on the streets. When asked about the earlier situation in Yanjing, many local elderly and middle-aged people would involuntarily refer to something related to horse caravans. Ellwood once noted, the pastoralism of the horses has influenced the primitive culture of the world much as the invention of the automobile has influenced our modern culture.[1] In fact, for Yanjing salt makers, its impact is even more long-term and far-reaching.

3) Pack-horse Drivers

Pack-horse drivers, known as the men who carry goods, generally refers to the

---

1 Charles A. Ellwood's *Cultural Evolution: A Study of Social Origins and Development*, translated by Zhong Zhaolin, and published by Shanghai Literature and Art Publishing House, 1989 edition, p. 187.

people who carried goods for sale in horse caravans, camel caravans, porters, or a company of travelling merchants. They were the product of ancient commodity trade and cultural exchange mainly relying on manpower and animal power as the means of transportation. In the process of Yanjing salt exchange with the outside world, which was realized through long-distance transportation by humans and horse caravans, pack-horse drivers always played an important role.

Since the Ming and Qing dynasties, Yanjing had long been under the control of the chieftains and lamaseries. For example, Upper Yanjing Village was once the possession of Gangda Temple; the Tibetan merchants there engaged in horse caravan transportation were either lamas in the temple, or their relatives, or local headmen. With the power in their hands, they hired men to help drive the caravan, so as to gain profits. Therefore, the work of driving horses, which required busy traveling outside all the year round, is mostly done by handymen. Through an analysis of two plaints from the Guangxu period, Zhang Xuehui noted that in the late Qing Dynasty, Zhongdian and Deqin, etc. were already important hubs connecting Tibet and Kangting, and pack-horse drivers and mules were playing a key role in the commercial exchanges between them.[1] "In the retail delivery trade of Diqing Tibetan inhabited areas, besides landlords and merchants, pack-horse drivers (handymen) were also indispensable members."[2] They usually farmed the land at home in the spring and summer, and drove horses for their lords in the autumn and winter.

Generally speaking, pack-horse drivers' life of driving horses was very tough, for employers tended to assign them the hardest work, and during the long journeys, they would inevitably suffer hunger and cold, brave the wind and dew, and experience various threats to their lives. As for the safety of the caravan at night, it is basically their responsibility. In Yanjing, the elderly told us that there used to be a set of customs in their families for hiring pack-horse drivers (handymen). If someone came to a household to work as a handyman, the master would first buy him a whole suit of clothes, including a hat, an upper garment, trousers, and boots, etc. If he had to

---

1 See Zhang Xuehui and Wang Yuanjie's "On the Yunnan-Tibet Economy and Trade from Several Archives: And Their Importance to the Study of Social Economy and History of Tibetan Inhabited Areas in Yunnan". *China Tibetology*, 1989(01).

2 Wang Hengjie. *The Social History of Tibetan People in Diqing*. China Tibetology Publishing House, 1995 edition, p. 81.

travel long distance to Lhasa, sometimes one or two more suits would be bought. In addition to ensuring the normal life of a hanyman, before he went on a long journey, the master would send regards by bringing 5 kilograms of rice, and yak butter, tea, and salt to his home. When he returned from a journey, the hanyman would also be paid appropriately, but the amount of wage was mainly based on the wealth and prestige of the master. Generally speaking, wealthy people, caring too much about their reputations, hoped to win the respect of others, and would pay high wages to hanymen if feeling they had fame and prestige.[1]

However, in the vast majority of cases, pack-horse drivers were not treated so well. Ma Zihua, a writer of Bai nationality, once said: “A horse driver, in order to get a little wage from his boss, had to suffer unimaginable hardships. He lived the same life as mules and horses. If there was a caravansary for lodging, he’d sleep on the manger, waiting to feed the animals at midnight. Otherwise, he could only ‘kailiang’ (sleep in the open field) as the bed, with the sky as the curtain, and make a fire pit in the middle, for heating as well as lighting.”[2] “Pack-horse drivers, born poor, had to earn a living through such hard work. However, their interests were often unfairly prejudiced by the masters, “by anually traveling back and forth to Tibet several times, a pack-horse driver may accumulate over two hundred taels of silver in eleven years, but all is stored in his master’s hand. Once the master welsh on his promise, the pack-horse driver will lose every cent, and even cannot pay off debts for he is still in debt to the Tibetan owner of goods. The master...he may not risk himself to go to Tibet in person, but just sit back and reap the benefits”[3].

Thousands of pack-horse drivers, while carrying heavy loads and traveling long distances, tasted the sweets and bitters of life along the way, and exchanged people’s various needs. It is precisely such a primitive means of trade and exchange, as well as every step they have persevered in, that converged into the Ancient Tea-Horse Road which has still been admired and profoundly influential in China to this day. Now, the

1 Field research data of Yanjing conducted by Li Hechun; interview time: October 2013; interpreters: Luo Song and Ngawang Namgyal.

2 Ma Zihua. *Essays on Southern Yunnan*. Yunnan People’s Publishing House, 2002 edition, p.13.

3 See Zhang Xuehui and Wang Yuanjie’s “On the Yunnan-Tibet Economy and Trade from Several Archives: And Their Importance to the Study of Social Economy and History of Tibetan Inhabited Areas in Yunnan”. *China Tibetology*, 1989(01).

scene of horse caravans transporting well salt, especially those distant from modern highways, is vaguely reminiscent of "the Ancient Tea-Horse Road" in the past. And the horse caravan transportation of well salt is actually a microcosm of the Ancient Tea-Horse Road, as well as a "living fossil" of it.

The demand for salt distribution has promoted the development of salt trade; with humans, livestock, and roads as carriers, various ethnic groups continuously interacted with each other due to the demand for salt exchange, facilitating mutual exchanges and communication. To increase opportunities for interaction and exchange what is needed among different ethnic groups is doubtlesslly conducive to promoting positive interplay between each other. According to Thomas Thaler, salt trade was once regarded as one of the most important ways of human communication, and many experts even considered the demand for salt as the root cause of all trade. Due to the limited availability of salt mines and the fact that salt has been an indispensable subsistence for daily life, an intensive and sustainable trade network is necessary to ensure its stable production. It's generally believed that salt trade is actually a common transaction of daily necessities."[1]As a necessity for people's lives, the demand for salt among various ethnic groups in Yanjing has been very strong. Through material exchanges between salt-producing and non-salt-producing areas, there also emerges a phenomenon of communication and integration between different ethnic groups.

In the distant historical context, humans, salt, and horses in Yanjing have constituted a cultural landscape with salt industry culture as the core.

3. Transaction Rules and Means

Since it was produced and entered the ranks of commodities, salt has no longer been an ordinary product, which is inseparable from the fact that salt has always been a necessity for human beings. Only through exchange can salt fully reflect its own value, and even generate values beyond what it should be.

---

1 Thomas Thaler. "The Early Salt Industry in Central Europe: Neolithic Salt Production and Trade Patterns" [M] // Li Shuicheng and Luo Tai. *Archaeology of China's Salt Industry (Episode 2): A Comparative Observation from an International Perspective* [M]. Beijing: Science Press, 2010 edition, p. 210.

Due to the differences in natural conditions and the uneven distribution of resources in Tibet, the self-sufficient production mode could not fully meet people's needs there. In order to regulate supply and demand, a salt-grain exchange mode developed spontaneously between farming and pasturing areas to meet their respective needs. However, due to the profound influence of feudalism for a long time, natural economy occupied the dominant position, and the commodity consciousness of local people in Yanjing was rather backward. In the Qing Dynasty, Xia Hu recorded the exchange of salt when entering the Gongshan area by the Nujiang River: "People in Chang (Gongshan County) all consumed sand-like salt, which was produced in Yanjing County, Xikang Province, and transported for sale by the uncivilized minorities in Tsarong. People generally traded grain for it, rarely buying it with silver coins. Due to the limited production of grain in Tsarong, salt was transported to exchange for it, which was then brought back home for their own consumption."[1] Thus, it can be seen that when salt makers traded with the small retailers, except for a few who used cash for transactions, most of them would exchange other daily necessities for equivalent values through "bartering".

This salt-grain exchange mode, or bartering, having been around since primitive times, occupied a dominant position in Yanjing for a long time, and gradually declined only in the past two decades. Up to now, in Yanjing, salt is still traded in a very primitive bartering mode: salt makers, having collected their salt, wait for traders to purchase it, and transactions are made by fixing prices according to the quality. The purchased salt would then be transported to the township by these salt traders and sold wholesale to horse caravans, and again transported to differnt sales areas. In this process, Yanjing people gradually developed their own set of rules.

1) The Measurement System and Weighing Apparatuses

Throughout history, the measurement system has been of great significance in regulating commodity exchange and maintaining social stability. As a tool for collecting taxes and stabilizing the prices of goods, weighing apparatuses developed rapidly and became complete with the development of national economy and politics.

---

1 Changputong Administrative Councillor's Office. *Changputong Annals* [A], of *Selected Literary and Historical Materials of the Nujiang River*, Volume 18 [C], 1991:14.

The measurement system in Yanjing salt trading in eastern Tibet, emerging from scratch, was caused by its unique economic environment and also related to the historical origins of Tibet. It mainly went through three important stages.

The first was before the bureaucratization of native officers implemented in the Sichuan border by Zhao Erfeng. At that time, it was mostly barter exchange, and weighing apparatuses were rarely used. People often guessed the weight by experience; for example, "a pinch" of salt was pinched up by three fingers in exchange for "a pinch" of tsampa. Hence the description of the situation "uncivilized minorities are born vulgar after all, not knowing what weighing apparatuses are, nor the circulation of money"[1]. Simply because of this, it was often difficult to guarantee that the exchanges ultimately achieved were equivalent. Even if weighing apparatuses were occasionally used, there would still be a phenomenon that "the local minorities' apparatuses, mixed up with the Han people's here and there, were different not only in name from the inland, but also in size, weight, and length"[2]. And in the locality, "the indigenous scales they used, without criterion, made no distinction between *qian* (equivalent to 0.1 tael) and tael; *dou*, known as *ke*, with liter and ge (a unit of dry measure for grain, which is equalant to 1 decilitre) unknown, nor was *dan* (equivalent to 10 *dou*) used here; the so-called *pi*, not uniform in amount, sometimes more than 10 *pi* was equivalent to 1 *ke*, and sometimes more than 20 *pi* equivalent to 1 *ke*. Due to such differences, the collection of coarse food grain was prone to Cheating and deception"[3]. Although Jin Fei once mentioned that during the late Qing and early Republican period, people in the local exchanges of Yanjing used a weighing apparatus, i.e., a kind of indigenous *dou* (possibly Tibetan *dou*), known as "kui", which, "also known as 'mo' (close in pronunciation), was a weighing tool with fixed specification and size: the length, width, and depth were all

---

1 "Investigation in Yanjing County" by Jin Fei, published in *Political Affairs of the Border Region*, 1931(8).

2 *A Memorial to the Throne by Zhao Erfeng from the Sichuan Border*, compiled by Wu Fengpei, and published by Sichuan Ethnic Publishing House, 1984 edition, p. 82.

3 *A Memorial to the Throne by Zhao Erfeng from the Sichuan Border*, compiled by Wu Fengpei, and published by Sichuan Ethnic Publishing House, 1984 edition, p. 82.

20 centimeters, and when filled with salt, each 'mo' weighed about 10 kilograms"[1], the confusion of weighing instruments in the Sichuan border was widespread, indeed.

Obviously, this situation was very unfavorable for the exchange of goods and materials, for it would not only cause many hidden troubles, leading to market confusion and unfair transactions, but also hinder the development of local commodity economy.

In the second stage, Zhao Erfeng, in view of the situation that weighing apparatuses were not uniform in the Sichuan border, vigorously promoted their standardization with the officially designated universal criteria. For this purpose, Zhao Erfeng once requested that "a wooden *dou* be made and handed over to the local official (township commissioner) as an official *dou* of the township, so that people can comply with the manufacturing process on record...Therefore, the official is instructed to immediately make three wooden *dou* in the same size, and transfer them, as the official *dou*, to the officals in Litang, Batang, and Daoba respectively, with common people required to make in accordance with them, and no different sizes allowed"[2]. He also requested that "a *dou* and a steelyard be custom-made, and handed over to local officials as the official *dou* and official steelyard; regardless of or buying or selling, grain and other commodities must always be filled to a level with the mouth of the official *dou*...", and the weighing apparatuses used by common people must "take this as the standard, either more when buying or less when selling is not allowed"[3]. Thus, *dou* and steelyards became two kinds of weighing apparatuses in the process of Zhao Erfeng's pacification of the Sichuan border.

After regaining Yanjing for the government-run mode, Zhao Erfeng stipulated that: "In accordance with Kuping standards, the capacity criterion is set at 15 kilograms per *dou*; as for the length, the offically made ruler would be used as the

1 "The Study on Salt Culture of Yanjing Nakhi Ethnic Township, Markam County, Tibet", Wu Chengli's Master's Thesis of the School of Sociology and Anthropology, Sun Yat-sen University, 2009, p. 98.

2 *On Reproducing Official Dou and Transferring Kuping Official steelyards*, the fourth volume of *Border Affairs Archives*. See *A Historical Review of Sichuan Administration* edited by Kui Yingtao, and published by Bashu Publishing House, 2002 edition, p. 411.

3 *A Memorial to the Throne by Zhao Erfeng from the Sichuan Border*, compiled by Wu Fengpei, and published by Sichuan Ethnic Publishing House, 1984 edition, p. 82.

criterion."[1] In the notice for issuing official *dou* and official steelyards, it was stated that: "According to the Chinese measurement system, *dou* and steelyards were originally used as the criterion. For steelyards, sixteen taels euqals to half a kilogram by Kuping measurement standards, one tael for ten *qian*, one *qian* for ten *fen*, and one *fen* for ten *li*. As for *dou*, ten *shao* (an old unit of capacity, which is equivalent to 1 centilitre) makes one *ge*, ten *ge* makes one liter, ten liters makes one *dou*, and ten *dou* makes one *dan*; these are known to all Han people in the inland, and it is uniform everywhere."[2] Zhao attempted to regulate the measurement system in the Sichuan boder according to the standards of weighing apparatus established in the inland region.

Zhao Erfeng's uniform standards of weighing apparatuses were thus continuously implemented in many parts of the entire Sichuan boder. And on October 26, the thirty-fourth year of the Guangxu period (1908), he reported to the throne that Tachienlu government agency should standardize instruments for measuring length, capacity and weight with uniform criteria. In areas inhabited by ethnic minorities in the Sichuan border, various weighing instruments had not only different names and lengths, but also discrepancies in measurements. To avoid confusion in different trading markets in the future, Zhao Erfeng even formulated a measurement system. For measuring capacity and length, he followed his previous approach, and "...ten steelyards by Kuping standards would be made or purchased by the provincial government for distribution."[3] Each of these official steelyards distributed in different counties near the Sichuan border was marked with a number on the rod for the purpose of inspection and verification, and No. 8 was officially given to Yanjing County. It should be mentioned that during Zhao Erfeng's days, the weighing apparatus used in trading in Yanjing were basically the official steelyard as the uniform criterion. With a uniform measurement mechanism, salt transactions in Yanjing were effectively standardized, laying the foundation for the reform in salt industry.

In "Investigation in Yanjing County", it was noted that: "Since the establishment

---

1 Ibid., p. 83.

2 Ibid., p. 82.

3 Ibid., p. 83.

of the Salt Bureau, the law of Han people was made known here. Each *duo* was set at 60 kilograms, and the price of the well salt was equivalent to two Tibetan silver dollars, with a tax rate of one Tibetan silver dollar."[1] In the attached *Flexible Rules and Agreed Salt Prices during the Initial Period of Merchant Supervision in Yanjing*, salt prices were stated in detail, with dry and wet salt divided into three grades respectively, all using *dou* as the measuring tool.[2] This, to some extent, ensured fairness in the exchange process and taxation, and thus regulated the market. According to *Local Geographical Records of Yanjing, Batang and Yanjing*, 1 "*ke*" in the local measurement system was equivalent to 1 *dou*, and 1 "*pi*" equivalent to 1 liter. Among the people, 20 *pi* equaled to 1 *ke*, while the the Batang government still counted 20 to 24 or 25 *pi* as 1 *ke*. When Duan Pengrui held the position of salt monopoly official in Yanjing County, 1 official *dou* was equivalent to 25 *pi* of the Tibetan measurement system, and 1 official liter equivalent to two and a half *pi*; thus, he required that people should conform to the official standards. In order to accurately weigh salt, official steelyards were also issued to set people's mind right.[3] It must be said that these measures, having achieved some positive results, promoted the development of the local salt industry to a certain extent.

During the Republican period, as the Sichuan border area was involved in the battles of local forces and warlords, not only were there as many as four kinds of currencies in the market, i.e., Indian rupee, Tibetan silver dollar, steel coin, and copper coin, but also weighing apparatuses became non-uniform again.[4] The forms and criteria of instruments for measuring length, capacity and weight used by residents of different regions and counties might differ from place to place, and even within a township, they varied between the east and west neighborhoods, and different industries. As a result, disputes over farmland and goods occurred from time to time, and they even turned into tools of fraudulent practices for corrupt officials

1 "Investigation in Yanjing County" by Jin Fei, published in *Political Affairs of the Border Region*, 1931(8).

2 Ibid.

3 See the author's preface, and such sections as products, taxation, salt pans, measurement system, and the postscript, etc. of *Local Geographical Records of Yanjing, Batang*.

4 See Feng Yunxian's "Actual Investigation of Counties in Xikang: Yanjing County", published in *New Asia*, 1931(05).

and unscrupulous merchants. Before liberation, salt and grain were exchanged on a 1:1 basis, using a container made of wood, narrow on the bottom and wide at the top, which resembled a trapezoid as a whole. Generally, it could hold 6 kilograms of salt. After liberation, Yanjing, increasingly connected to the inland region, gradually adopted a uniform national measurement system, thus entering a new stage of salt industry exchange as well as the third stage of local weighing apparatus development. During Li Hechun's investigation in Yanjing, the major weighing apparatuses for salt exchange there were steelyards and platform scales, with *jin* (half a kilogram) as the primary unit of measure. If the weight of salt was below 50 *jin*, a steelyard would often be used; but if it was above 50 kilograms, a platform scale would be preferred. Meanwhile, at present, there's still a relatively traditional way of exchanging salt and salt making tools in different salterns in Yanjing, and the weighing instrument used in these cases is called "*dou*", known as "*jiabi*" (close in pronunciation) in Tibetan dialect. Also made of wood, it is square in shape, with a large mouth and a small bottom, and trapezoidal on all four sides, which are connected by wedge-shaped mortise and tenon joints. Its mouth edge is about 30 centimeters long, and bottom edge about 24 centimeters; its height is 16 centimeters, and the inside depth 14.5 centimeters; its wall is 1.1 centimeters thick, and bottom 1.5 centimeters thick. (See Figure 3-1) According to local salt makers, each *dou* can hold about 10-12 *jin* of salt, or grain varing from five to six *jin*.[1] *Dou* has been an important measurement tool for salt transactions in the salterns.

Figure 3-1 A wooden *dou*

1 Li Hechun. *Impetus and Shackles: Salt and Salt Taxes in the Lantsang River Canyon*. Guangzhou: Sun Yat-sen University Press, 2016: 73.

It can be seen that in the past a few decades, the measurement system in Yanjing has seemingly gone through periods of confusion, but in fact, it was just an inevitable phenomenon in the process of cultural exchange and integration between Han, Tibetan and other nationalities. Uniquely located in the ethnic corridor, as well as the border area of Tibet, Sichuan and Yunnan provinces (region), on the one hand, Yanjing was hog-tied by the backward productive forces, relations of production, and a system combining political and religious powers, thus inheriting the traditional measurement customs of the Tibetan people; on the other hand, it was deeply influenced by the inland system of weights and measures, thus widely absorbing the measuring instruments, methods, and habits of the Han people. So, to speak, its development trend has been the conformity to the inland measurement system. And such conformity is of great significance to the development of Yanjing, because the standardization of weights and measures is a necessary measure for economic unification, which in turn is the driving force of political unity, national unity and national identity.[1]

2) Varying Salt Prices

As the earliest commodity to enter the market for circulation, the exchange value of salt varies in different historical periods, and may fluctuate with the market conditions. Besides, as a scarce commodity, the accessibility of salt varies from region to region. Therefore, to study the salt prices in Yanjing area, it must be discussed in terms of different historical periods.

Before Zhao Erfeng implemented the bureaucratization of native officers, salt in Yanjing was mainly exchanged through bartering, so its price could only be reflected in the values of the available goods in transactions. About this period, due to limited access to historical literature, the situation was not very clear. Yet it was recorded that: "The goods needed was chiefly salt, often with two or three *pi* of grain in exchange for one *pi* of salt."[2] According to this statement, the exchange ratio between salt and grain was 1:3. However, this was just a normal exchange ratio,

---

1 Chen Gailing and Fan Zhongxiong. "Weights and Measures in Modern Gannan Tibetan Areas from a Multicultural Perspective". Gansu Social Sciences, 2013 (05).

2 *A Memorial to the Throne by Zhao Erfeng from the Sichuan Border*, compiled by Wu Fengpei, and published by Sichuan Ethnic Publishing House, 1984 edition, p. 82.

and sometimes it might even reach 1:10, which means that 100 *jin* of salt could be exchanged for 1,000 *jin* of grain. At the time, when Yanjing people exchanged with the surrounding areas, salt would act as an intermediary, reflecting its value according to the exchange ratio between salt and grain (or other goods). Salt prices also varied depending on how difficult it was obtained and how far it was from the salterns. That is to say, the more difficult salt was obtained, the higher the price would be; the farther away it was from the salterns, or the longer the distance of its transportation was, the higher the price would be, too.

During the period of his administration of Yanjing, Zhao Erfeng, through reforms, laid down various norms and regulations, including those concerned with salt price. In the early days, salt price was not referred to in *Regulations on Salt Taxation in Yanjing* formulated in October of the first year of the Xuantong period (1909), nor was it specified even in the 30 articles of *Agreed Regulations for the Commercial Salt Bureau in Yanjing* formulated in December of the same year, which might have something to do with the fact that Zhao Erfeng, having just taken over Yanjing affairs, still needed to familiarize himself with the local conditions. Yet, these systems preliminarily clarified that salt price shall be negotiated by the Salt Commerce Bureau, salt-maker households, and traders, thus forming a flexible and fluctuating price mechanism. And among them, five articles were related to salt prices:

Article 3: Salt purchased in a monopolistic manner by the designated merchant, its price shall be set by the local officials, on the basis of experiencing and observing the situation nearby, with the merchant, traders, and salt-maker households according to the spring, summer, autumn and winter seasons. Salt may differ in quality, and its price will also vary a lot. This should be made known to everyone through notification in advance, so as to avoid fraudulently buying cheap and selling dear.

Article 4: As to the salt price in all four seasons, local officials must inform the commercial salt bureau to set fairly through consultation with salt-maker households and traders, and announce it in the form of a notice within a limited period of time. In spring, it's on the first day of the first lunar month; in summer, the first day of April; in autumn, the first day of July; in winter, the first day of October.

Article 5: Once salt rice has been set for the season, and announced in the form of a notice by the local government, the commercial salt bureau must purchase it at the fixed price, regardless of the quality of salt produced by the salt-maker households; that is, to offer the current price, without any slight underpayment.

Article 6: If local officials fail to discuss fairly and consult with the commercial salt bureau as well as the majority of merchants and salt-maker households before setting salt price for the season, or their arbitrary pricing disrupt commerce and hinder tax collection, they are bound to be subject to impeachment and conviction.

Article 7: Local officials, after setting the price, and the commercial salt bureau shall each provide a concise report on salt prices of the current season, and submit them respectively to the Minister of Border Affairs for record keeping.[1]

The content of the above five articles thus initially formed a salt pricing mechanism that would change with the seasons. This was because the weather and wind in the four seasons were different, which directly affected salt production, resulting in different yields. And the relation between supply and demand would naturally cause fluctuations in salt prices. To prevent differences in salt pricing among the official salt bureau, the commercial salt bureau, salt merchants, and salt makers, the four groups must set price through negotiation, so as to maintain the normal operation of salt trading market.

In the second year of Xuantong (1910), the local government, by formulating *Flexible Rules and Agreed Salt Prices during the Initial Period of Merchant Supervision in Yanjing*, initially set salt prices and formed a grade system of salt. Hence different grades of salt had different prices. Salt can be divided into dry salt and wet salt, which can again be divided into three grades: upper, middle, and lower. "For the upper-grade salt, one Tibetan silver dollar buys six *dou* of wet salt, or five *dou* of dry salt; for the middle-grade salt, it buys seven *dou* of wet salt, or six *dou* of dry salt; for the lower-grade salt, it buys eight *dou* of wet salt, or seven *dou* of dry salt. The commercial salt bureau shall collect or purchase salt in accordance with the prices set above, uniform in both buying and selling."[2] According to the salt price at

---

1 *Archival Historical Data of Sichuan-Yunnan Border Affairs in the Late Qing Dynasty (Middle Volume),* compiled by Sichuan Institute of Ethnic Studies and the editorial group of *Archival Historical Data of Sichuan-Yunnan Border Affairs in the Late Qing Dynasty*, published by Zhonghua Book Company, 1989 edition, p. 512-515.

2 "Investigation in Yanjing County" by Jin Fei, published in *Political Affairs of the Border Region*, 1931(8).

that time, one Tibetan silver dollar could buy 100 *jin* of upper-grade dry salt. On each *duo* of salt, i.e., 120 *jin*, the government levied a salt tax of 1.5 Tibetan silver dollars (equivalent to 0.48 tael of silver). Taking into account the upper, middle, and lower grades of salt, the tax on a *duo* of salt could buy 140-150 *jin* of salt. Calculated in this way, the price of each *jin* of salt was basically the same as the salt tax.

In the late Qing Dynasty and the early Republican period, as Yanjing was involved in a chaotic contention for profit between the garrison, lamaseries, and Kelsang Zayein, the local salt industry was left completely unattended to, hence the salt price unknown. According to a survey conducted in the thirtieth year of the Republican period (1941), at that time, "as to the price of salt from salt pans, generally during peak months, one steel coin could buy 2.5 *dou* of peach blossom salt, or 3 *dou* of second-layer salt, or 3~4 *dou* of third-layer salt; during slack months, one copper coin could buy 1 *dou* of peach blossom salt, or 1.5 *dou* of second-layer salt, or about 2 *dou* of third-layer salt. If salt maker households needed money urgently, the price would then be squeezed down; but when salt was scarce, the price would rise"[1]. Obviously, salt price at the time was related to its outputs and whether salt makers needed to dispose of it urgently.

Since 1959, taxes in Yanjing were levied in RMB, but documentation on salt prices during this period has not been found. At this point, a tax of 2 *yuan* was levied on 100 *jin* of salt; as a whole, the salt price might have been around 0.15 *yuan* per *jin*. According to a survey conducted in the Mishmi people community in Zayu County in 1976, a medium-sized chicken could be exchanged for one *jin* of salt among them. Extremely short of salt in this area, Mishmi people would often take their own native products, such as coptis they had dug up or other valuable medicinal herbs obtained by chance, to exchange for it with Tibetans in neighboring areas. When salt was expensive, 1 *ke* of grain (such as corn, etc.) could only be exchanged for a small bowl of it, sometimes even couldn't.[2] When trading, it was generally measured in *dou*, and each bag could hold 6~8 *dou* of salt. It was also recorded that when the price was relatively high, 1 *dou* of salt could be exchanged for 1 *dou* of highland barley; when

1 "Investigation of the Geology and Salt Production in Yanjing County", by Cui Kexin, published in *Xikang Economic Quarterly*, 1944 (8).

2 *Survey of the Social History of Mishmi people* (Yunnan People's Publishing House, 1990), pp. 59, 223, 32.

the price was low, as much as 3 dou of salt could be exchanged for 1 dou of it. Salt could also be exchanged for rice, yak butter, and horses, etc. 1~2 *dou* of salt could be exchanged for 1 *jin* of yak butter; 2~3 *dou* of it could be exchanged for 1 *dou* of milk residue. And a bag of salt for a bag of tea.

In recent years, salt has mostly been transacted with currencies, its price fluctuating greatly in different years and months. Moreover, due to the influence of different geological conditions, the red salt produced in salt pans of Jiada Village on the west riverbank contains more impurities than the white salt produced in salt pans of the other villages on the east riverbank, and thus the prices of these two types of salt have always been different. In 2007, the prices of white salt on the east riverbank were as follows: The first-class salt was 100 *yuan* per 100 *jin*, the second-class salt 60~70 *yuan* per 100 *jin*, and the third-class salt 30~40 *yuan* per 100 *jin*. In comparison, salt of the first and second classes produced in Jiada Village was 0.3~0.4 *yuan* lower per *jin*.[1] In 2013, salt prices of each class were rising, with the white salt prices as follows: the first-class salt was 100~120 *yuan* per 100 *jin*, the second-class salt 70~80 *yuan* per 100 *jin*, and third-class salt 40~50 *yuan* per 100 *jin*. And the red salt prices were: the first-class salt was 100 *yuan* per 100 *jin*, second-class salt 60~70 *yuan* per 100 *jin*, and third-class salt 35~45 *yuan* per 100 *jin*. Over the last several years, as the number of salt-maker households has decreased, the salt price has also been increasing. By 2018, the white salt prices were as follows: the first-class salt was 150 *yuan* per 100 *jin*, the second-class salt 95~100 *yuan* per 100 *jin*, and the third-class salt 70~80 *yuan* per 100 *jin*. And the red salt prices were: the first-class salt was 100~120 *yuan* per 100 *jin*, the second-class 70~80 *yuan* per 100 *jin*, and the third-class 40~60 *yuan* per 100 *jin*.

With the advancement of tourism development, the number of tourists visiting Yanjing continues to increase, and the sales methods of salt have also changed. In the past, salt was mostly sold in bags of 100 *jin* by salt makers, but it was not convenient for travel enthusiasts to carry about. Hence, they began to sell it packaged in smaller bags, each weighing 1-2 *jin*. In this way, the salt price is higher than double of that

1 See "The Study on Salt Culture of Yanjing Nakhi Ethnic Township, Markam County, Tibet", Wu Chengli's Master's Thesis of the School of Sociology and Anthropology, Sun Yat-sen University, 2009, p. 100.

in single *jin*. Generally, a 1-kilo-bag of salt can be sold for 2 yuan, while a 2-kilo-bag can be sold for 5 yuan. It is usually packed in white plastic bags without trademarks, but sometimes in colored bags bearing the words "Yanjing salt". These phenomena all demonstrate that local salt makers, no longer selling salt in traditional ways, are striving to increase their income by constantly integrating it into tourism development there.

3) Primitive Transaction Modes

In Yanjing, salt was controlled by local authorities in the early days, so there were few ways to obtain it. Generally speaking, two types of exchange would often be conducted: explicit and implicit. The former refers to the salt-grain exchange with the consent of the local headmen or temples, or that with their direct participation; the latter is a way of doing it secretly in the locality. And the transaction modes of ordinary salt makers, very primitive as it may seem, could be categorized into two types: direct and indirect.

a. Direct Exchange

Direct exchangers could be local consumer groups, as well as salt traders engaging in small-scale businesses. They included both outcomers and local villagers, all of whom valued the profitability of the salt industry. Mostly individual traders specializing in salt sales, they'd collect salt in the salt pans in person, thus mantaining a close relationship with the salt makers.

In different historical periods, various social groups in Yanjing would participate in the sale of salt. During the early days, lamas of Laweng Temple and Gangda Temple here, after joining in one after another, sold salt to the neighboring regions. More tellingly, when salt was sold to such areas as Zayu and Drung River, etc., the lamaseries gained high profits; and they, relying on their religious status and close alliance with the chieftains, even forcibly sold salt to the Drung and Nu ethnic groups.

Salt makers would sell their salt to traders, and sometimes a deal can be made on the same day; if no one come to purchase it that very day, It will be stored in a salt barn, then. In addition to storing salt, a salt barn may also be stacked with production tools, and in some there are even such articles for daily use as stoves, etc. Here, a salt barn is generally shared by several salt-maker households. In order to prevent salt theft, special marks would be imprinted on the surface of the salt heaps with "*buti*"[1]. When selling salt to the traders, except for a small number of transactions conducted in RMB, most salt makers would exchange for other daily goods in the form of "bartering". Instead of using such weighing tools as platform scales and steelyards, they choose a very primitive "*mo*"[2], and each Mo weighs about 10 *jin* when filled with salt.

When the collected salt reaches a certain amount, traders would transport it to Yanjing Town by mules and horses, then by motor vehicles or tractors, and then again by a horse caravan to elsewhere to get the margin in between. Usually, "salt merchants all brought various daily necessities, such as rice, tea, and clothing, etc., to Yanjing to exchange for salt with salt makers"[3]. As a result of the long-term exchange relationship, salt makers and salt traders have become good partners. For salt merchants, the establishment of long-term exchange relationship is conducive to obtaining salt in the hands of fixed partners when salt production is limited but merchants are still too many; for salt makers, only when their salt can be sold smoothly and exchanged for daily necessities, can they maintain their livelihood. Therefore, the relationship between the two is win-win. However, there are also exceptional cases, as recorded by predecessors: "due to the large number of pedlars, and the fact that there was no salt for several days after rain, sometimes for tens of days no transaction could be conducted."[4] Under the circumstances, salt merchants generally have to wait for more than half a month, during which they may need to eat and live together with the salt-maker households, and the latter will provide necessary fodder for their mules and horses for free. On the other hand, these salt merchants, when departing from Yanjing, often leave their belongings with the hosts who have treated them warmly.[5]

---

1 A local tool (pronounced so), whose shape and pattern varies from household to household.

2 It is also known as the weighing tool "*dou*" commonly seen in the past, and a *dou* has a fixed size.

3 "Investigation of the Geology and Salt Production in Yanjing County", by Cui Kexin, published in *Xikang Economic Quarterly*, 1944 (8).

4 Ibid.

5 "The Study on Salt Culture of Yanjing Nakhi Ethnic Township, Markam County, Tibet", Wu Chengli's Master's Thesis of the School of Sociology and Anthropology, Sun Yat-sen University, 2009, p. 95.

When the salt output is high, its exchange will fall into stagnation. At this time, salt makers have to store it in the barn, waiting for the arrival of new merchants. And in times of a severe shortage of food, their only solution is to reduce the amount of goods that salt can be exchanged for. Most of time, these Yanjing salt makers, in order to exchange for more grain and yak butter, will take the initiative to carry their salt on horseback to various regions; and it may take more than ten days of travel to reach the parts where transportation is inconvenient and salt is urgently needed. In this way, they could get more grain or yak butter with the salt they had laboriously carried there than in the Yanjing area. And in their opinion, it can also save livestock feed during the journey. Therefore, the quick-minded salt makers, who dare to take risks, would overcome geographical constraints and transport their salt to further places for exchange.

They sell salt mainly in exchange for grain, in addition to tea, medicinal herbs, oil, cloth, and other daily necessities. Cheng Fengxiang once recorded: "In the Chayu (Zayu) area, there are still local traders from Menkong who have transported well salt here in exchange for coptis."[1] This example shows that salt was often exchanged for important goods. Moreover, there is now a relatively common form of direct exchange in Yanjing, i.e., the exchange of salt and salt-making tools. Such a traditional and ancient way of exchange still exists tenaciously in the current era of money transaction.

Some common tools are naturally indispensable to salt making, and the existence of this demand has formed the tradition of exchanging tools with salt. In the past, a newly made brine pail could be exchanged for 124 *jin* of salt. A brine scoop can be exchanged for more than one and a half "*jiabi*" of salt, which is estimated to be about 9 *jin*. The amount of salt a broom can be exchanged for depends mainly on its size; a small one can be exchanged for five or six *jin* of salt, while a larger one for seven or eight *jin* of salt. According to our observation in the investigation, a salt maker, having selected two wooden bats, filled the seller's bag with salt measured by "*jiabi*"--12 "*jiabi*" of salt in total! In this way, a wooden bat is equivalent to 6 "*jiabi*"

1 *Archival Historical Data of Sichuan-Yunnan Border Affairs in the Late Qing Dynasty (Middle Volume)*, compiled by Sichuan Institute of Ethnic Studies and the editorial group of *Archival Historical Data of Sichuan-Yunnan Border Affairs in the Late Qing Dynasty*, published by Zhonghua Book Company, 1989 edition, p. 606.

of salt; that is, each wooden bat can be exchanged for 72 *jin* of salt. (See Figure 3-2, Figure 3-3) However, most of the time, the transaction is conducted through people's estimation, with fair dealing mainly depending on a person's moral quality. For those honest traders, they would fill up their "*jiabi*" with salt, lest the craftsmen suffer losses. But of course, there are also some more cunning guys, who deliberately smooth out the mouth of a "*jiabi*", so that a little less salt will be given each time. Villagers who do this would often become the subject of after-dinner gossip.

Figure 3-2 salt makers selecting salt-making tools

Figure 3-3 the exchange of salt and salt tools

b. Indirect Exchange

Where there is salt, there will be salt merchants. Due to its special value, salt became one of the earliest universal equivalents. Some people, seeing this, begin to specialize in salt trading. They hoard it when the price is low and sell it out when the price is high; or they may purchase it from one place and transport to another for sales, so as to be more profitable. And this has been a common occurance in Yanjing. For example, "in the urban and southeastern areas of Batang County, as well as the Kangning Temple area, there were some mule caravans specially devoted to salt trading. They purchased salt cheaply from Yanjing, and by transporting it to Batang, sold it at a higher price to the sedentary merchants there. The latter would then sell it at an even higher price to customers, or again transport it to other parts of the south

Kham Tibetan area for sale"[1]. These salt traders, different from salt consumer groups and small-scale salt vendors, serve as intermediaries between salt-maker households and the consumer groups.

The Salt Affairs Bureau established in Yanjing in the late Qing Dynasty was only an institution for administering salt affairs, not one for salt sales. And there's no salt specialty store in Yanjing. The salt transaction is entirely conducted by individuals, almost in a state of full production and full sales. Most of the salt is sold locally, either to local shops or local merchants engaged in salt business; a small portion is sold to other regions, mainly in Tibetan inhabited areas, and transported by horse caravans. Nowadays, the prosperous scene of horse caravans coming to the salterns for salt may not be seen again, and the exchange of salt is mainly carried out at several salt collection spots. Moreover, there aren't many long-term salt purchasers on Puding Street, only three, and all distributed on the street in front of the police station.

The first collection spot is 10 meters right from the main entrance of the police station. Only the shop door is visible, with the inside completely dark. The house of this salt collection spot, as a legacy of the previous supply and marketing cooperative, has been rented by the trader. The signboard above the door reads in large Chinese characters "Salt Collection Spot", with some Tibetan characters[2] at the top and such chinese characters below: "Main business: white salt, red salt, big tea", as well as the shopkeeper's contact number (see Figure 3-4).

---

1 Tashi Langjia: "The Ancient Salt Road in Batang", published in *A New Collection of Batang Literature and Art*, issue 3, and quoted from Ran Guangrong's "A Brief Account of Well Salt Production in Yanjing County, Sichuan Province during the Late Qing Dynasty and the Early Republican Period"; see *Proceedings of the International Symposium on the History of China's Salt Industry* co-edited by Peng Zeyi and Wang Renyuan, and published by Sichuan People's Publishing House, 1991 edition, p. 357. It is now difficult to find the source of this article cited by Fang Jianchang. *A New Collection of Batang Literature and Art* is suspected to be *Local History of Batang*.

2 In Yanjing, each store name needs to be printed in both Chinese and Tibetan, which is a uniform requirement of the local industrial and commercial office.

Figure 3-4 a salt collection spot

The second shop is located next to the first salt collection spot, and the owner has also rented the house of the original supply and marketing cooperative. The signboard of the store still reads "Yanjing Supply and Marketing Cooerative" in Tibetan and Chinese respectively, and below the Chinese characters is printed with: "Main business: salt, general merchandise, subsidiary food, wine, Tibetan costume" and the owner's contact number (see Figure 3-5). The entrance to the store is slightly larger, with plastic pots and brooms of different sizes hanging on both sides. Moreover, on the ground on either side of the doorway, there are salt baskets, bamboo baskets, plastic buckets, mops and other sundry goods, too. During our investigation in Yanjing, the owner shared her experience with us.

In 1964, when I was 16 years old, there was much financial difficulty in the family because I had six sisters, so I came to work in the supply and marketing cooperative through personal connections. At that time, Yanjing Supply and Marketing Cooperative dealt in locally produced white salt and red salt. Two employees would go to the salterns at any time to purchase salt, and conveyed it back for sale; they may purchase as much as 240 tons annually. The current situation is that there's a difference of 5 cents between the purchase price and sales price for each *jin* of salt. That is, selling 100 *jin* makes a profit of 5 *yuan*, selling 1,000 *jin* makes a profit of 50 *yuan*, and selling 1 ton makes a profit of 100 *yuan*. The annual profit of the supply and marketing cooperative (referring to the two salt collection spots that has now rented the house of the original supply and marketing cooperative) is 24,000 *yuan*, with an average profit of 12,000 *yuan* per person (Jo-nga Lhamo and Yang Pei). March, April and May of the solar calendar each year is a perfect season for collecting salt, when the weather is warm and fine, the brine is of good quality, and the salt making cycle becomes short. Once the salt pans are filled with brine water, salt grains will be precipitated in just two or three days. Afterwards, as the supply and marketing cooperative underwent structural reform, my colleague Yang Pei and I, who had been working together for many years, bought this house and dealt in various general merchandise here, selling white salt and red salt.[1]

Figure 3-5 Yanjing Supply and Marketing Cooperative

The third salt merchant on the Yanjing Street is Deng Deng, a villager from Lower Yanjing Village (the father of Tashi Yangzong mentioned in this book). His shop is opposite to the second shop mentioned above, and the house was built by his own family. The signboard of the store reads "Deng Deng General Store", still in Tibetan and Chinese characters respectively, with such words below the Chinese: "Main business: salt, tea, subsidiary food, grain, highland barley wine, wine" and the owner's contact number (see Figure 3-6). According to the shop owner, he has been selling salt in Yanjing for more than 20 years. Ten years ago, there were a large number of people making salt here, and the amount of salt bought at a time was also relatively large, mostly as much as a few hundred *jin*--some would even buy more than one thousand *jin*! But now, the number of people buying local salt is gradually decreasing, with an amount ranging from more than ten *jin* to forty or fifty *jin* each time. At present, as salt is hard to sell here, and a lot of salt is being made in their own salt pans, the main purpose of his shop is to help his daughter sell out her salt smoothly.

1 Li Hechun. *Impetus and Shackles: Salt and Salt Taxes in the Lantsang River Canyon*. Guangzhou: Sun Yat-sen University Press, 2016: 81.

Figure 3-6 Deng Deng General Store

In addition to the above-mentioned three on the Yanjing Street, there are also one or two small traders in Upper Yanjing Village who collect salt. But these salt collection spots are often locked, without much business. In Jiada Village, there are almost no fixed salt collection spots. Hence the salt makers here, when having made more than ten bags of salt, will transport by tractors to Yanjing Street and sell it to the three salt traders mentioned above. In recent years, salt sales have dwindled a lot, for few people would take the initiative to buy salt from these salterns. So local people have to find ways to take advantage of tourism development to sell salt to tourists. For example, in order to better promote their salt in areas such as Zogang, the Jiada villagers on the west bank of the Lantsang River have come up with the saying "Fed with white salt, the livestock will be plagued by lice, and not easily fattened up; on the contrary, fed with red salt, not only will it kill the lice on them, but also have a certain effect of fattening them up". As a result, in Zogang and other areas, the red salt produced in Jiada Village is much more popular than the white salt produced in Upper and Lower Yanjing villages on the east bank. Even until now, a small number of people in Zogang, mainly the elderly, still hold this view. So, if people selling white salt go to Zogang for business, they may dig up some red clay, crush it, and then mix it with water and salt moderately to form "red salt" to the buyers' liking.

c. Transaction Rules

In the long-term process of salt-grain exchange, salt dealers and salt makers have formed their own set of exchange rules, which is to both ensure the interests of all parties and guarantee the long-term exchange itself.

According to local residents of Yanjing, a long time ago, people here, with weak commodity awareness, were reluctant to engage in face-to-face transactions. At that time, some people brought salt (produced in Yanjing) from Tsarong to exchange for grain with the locals; the way they took was to put the things they needed in a prominent position on the roadside, and secretly find a secluded spot, watching quietly and from afar. He who came in exchange for grain knew, of course, that he could take it away by simply leaving his goods where it had been placed. After this person already went away, the former party would then come out to pick up what they had exchanged. As for those who came to exchange things, in fact, the tribal headman would always send someone to keep watch on them. Once they took the other party's goods without putting their own in the same place in a timely manner, or it was found that the values of the goods exchanged differed greatly, the headman would order the watcher to take action against them. Thus, both buyers and sellers are invisibly adhering to the exchange rules.

When driving horse caravans from a place far away to Yanjing for business, salt traders naturally needed to prepare grain, yak butter, meat, or other goods to exchange with local salt makers. In general, each household who came here for exchange would become exchange partners with a fixed owner, and this relationship might have been maintained for generations. As a result, there was a great deal of trust between them, and the whole exchange process would also be very pleasant. The exchange was generally carried out in a salt maker's home, during which, the salt owner could not touch the salt, nor could the grain owner touch the grain; instead, a trusted villager would be asked to act as an intermediary and organize the exchange activity. Both parties placed their own goods or salt in the house, and after negotiating the exchange ratio, completed the exchange with the direct participation of the intermediary, so as to ensure fairness and make the transaction go smoothly. This was also a measure taken when it was difficult to ensure full amount exchange or equality due to the lack of standard weighing instruments. Even if salt owners and grain owners (salt traders) had established a long-term trusting exchange relationship, such a principle was still retained. During the exchange process, sometimes there might be a shortage of exchangeable goods in the hands of a salt trader. However, he still wanted to exchange for more salt, because it was not easy to make the long journey to Yanjing every time, after all. At this point, he would need to find a trusted intermediary to ensure that the amount of goods that were not brought this time would be replenished next time.

In recent years, salt collected from the salterns of Upper Yanjing Village was mainly stored in a little shed on the upper side of the Catholic Church near National Highway 214. When salt traders who specialized in collecting salt came, it would be directly sold to them. Through on-the-spot observation, we found that those "salt traders" were inextricably linked to the salt makers, either as relatives or friends. In general, people would not bargain; they might conduct transactions when everything was not far off, for everyone was not used to haggling over the price. There were two specific transaction methods: "*kaniu*" and "*kajiu*". The former refers to being level with the brim, which means that the amount of grain or salt measured is exactly at the same level with brime of the basin. I have seen them measure it this way: fill a basin with grain or salt, and wipe across the brim with the right elbow, smoothing it flat; the movement was extremely simple and neat. And the latter refers to overbrimming. If the grain brought along by villagers is characterized by full kernels and sufficient quantity, the amount of salt exchanged will be higher than the basin brim, and the increased height depends on the generosity of the other party. During the transaction, both parties might even talk and joke with each other; the scene was full of harmony and joy. This form of transaction conducted in the light of the principle of trustworthiness, everyone is very at ease, and not too fussy about anything. Sometimes, salt makers would directly give salt to the traders as a present; salt making seems to be just a way of life for them, rather than a means of profit.

Today, such salt-grain exchanges have become quite rare, for salt makers can obtain grain from their relatives and only need to give them salt in return when it is collected. Actually, this is still salt-grain exchange, but it is not an immediate exchange; instead, the amount of grain or salt supplied by the other party is recorded first, and products of equal values will be given in return next time. In other words, the exchange still exists, only without intermediaries, and it has become a kind of reciprocal gift behavior with a human touch between relatives or acquaintances.

The salt-grain exchange, with multifaceted impact on society, is always conducted within traditional regions. Some herdsmen, on coming to farming areas, may choose to establish a landlord-tenant relationship with the local farmer households. This relationship often resulted in successive generations of close family relationships, with great affection and hospitality, and through the landlords' help, salt and livestock products are sold to other farmers. If they have not been completely exchanged, these goods can also be stored with the landlord and sold again next year. Sometimes, the herdsmen may need relatively more grain, but the salt they bring along is insufficient; in this case, they can also take away the grain first and replenish it by transporting salt here the next year. In this kind of interaction, people gradually deepen their cultural connections and psychological communication with each other. What is more noteworthy is that during the salt-grain exchange period, temple fairs are often held in farming areas, during which religious, sports and various cultural performances are organized, enabling farmers and herdsmen from different regions to carry out cultural exchanges, and promoting cultural identity between different regions of the Tibetan Plateau.[1]

d. Monetary Function

The local people of Yanjing have long been backward in commodity awareness. Before the emergence of currency, they basically relied on bartering one thing for another, with certain items of value in their own right that did not wear out easily acting as general equivalents. In the process of Yanjing investigation, our vision gradually formed such a concern, that is, salt had become the general equivalent of exchange within and outside the village, and even in the bartering achieved through long journeys. Salt, traded as a commodity with other tribes, actually functioned as currency for them.

For example, in the bartering within the area, salt has become a measure of the value of any exchangeable object; it is common to exchange a "*jiabi*" of salt for several "*jiabi*" of highland barley or wheat, or a spoonful of salt for several spoonful of oil, etc. In the process of forming an exchange chain, salt, as a general equivalent, is shown in a clear view. For example, if merchants from outside want to trade salt, the first thing they need to do is to exchange grain for salt in Yanjing. When they transport it to other regions, salt will be exchanged for more grain, and thus is undoubtedly playing an intermediary role. Figure 3-7 the exchange circle and goods for bartering.

When bartering, if it is exchanging salt with oil, a container called "scoop" would be used. A scoop full of salt can be generally exchanged for a scoop of oil,

---

1 Li Jianshang. "Salt-Grain Exchange and Its Impact on Tibetan Society". *Tibet Studies*, 1994 (01).

which is basically a one for one exchange. If the exchanged goods were buckwheat flour and corn, then a wooden container (referred to as "wooden *dou*" earlier), also known as "*jiabi*" in Tibetan, would be used. In this case, one *jiabi* of salt could be exchanged for eight or nine *jiabi* of buckwheat flour or corn. When exchanging for rice, it is usually conducted in a one for one manner, too; i.e., one *jiabi* of salt for one *jiabi* of rice. As for those who lived at higher altitudes, the road conditions were generally poor, so the phenomenon of salt shortage was very common, and the exchange could be one for two.

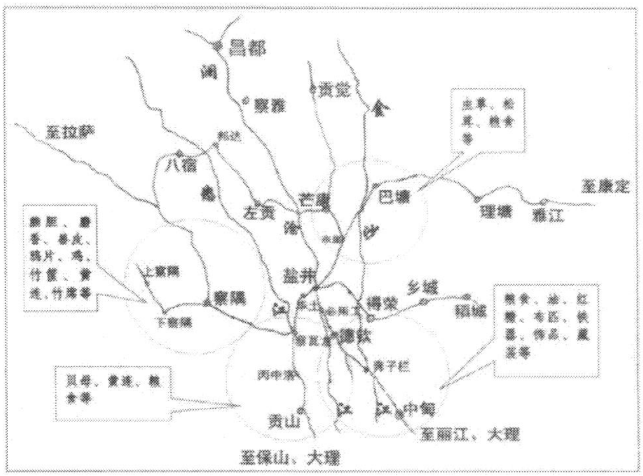

Figure 3-7 the exchange circle and goods for bartering

To obtain higher profits through salt trading was quite universal in the area of the three parallel rivers, and under the political and religious system, it became even more prominent. As a daily necessity of the Drung people in the western part of Yanjing, salt, which was also a scarce commodity, had always been occupied and controlled by the upper classes. The exchange scope of Yanjing salt might have been extended to the Gongshan area by the Nujiang River, which was administered by chieftains and Tibetan lords before liberation; when they sent men to collect tributes there, salt would be lent to the Drung people, too. For example, when entering the Drung River area, they'd tranport salt by horses, and the local Drung people might be forced to buy it. But many families were not able to provide exchangeable goods on the spot, so they had to accept the salt first and make a supplementary payment when the other party came next time, thus forming a debt relationship.[1] In this relationship, the monetary function of salt can be clearly seen. Moreover, salt also has the function of storing value. For example, in Tsarong, the headman tend to lend salt to the local people, "in spring, he lent four handleless cups of salt (about more than 100 grams for each), and in autumn he'd collect six kinds of goods in return: an animal hide, one *jin* of coptis, one *jin* of beeswax, a dustpan, a bamboo tube, and a box made of bamboo strips, etc."[2].

With the development of commodity economy, salt exchange gradually moved towards marketization, and the way of exchange was no longer the traditional bartering, but the continuous implementation of monetary transactions. However, the salt-grain exchange in the Yanjing area, as a product of history, was to solve the problem of grain sources for people in pastoral areas, and salt source and part of the livestock products for people in farming areas under the condition of low productivity level. After all, bartering was a backward mode of circulation. With the development of socialist construction in Tibet, the drawbacks of salt-grain exchange began to appear, and thus it was bound to die out and become history. Through salt trade and exchange, communitions among people along the trade routes increased a lot, with different ethnic groups intermingling with each other, and ties and cultural identity between different regions have also been greatly strengthened.

---

1 See *Essentials of Drung River Culture: A Study on the Socail Changes of Qiu People and Their Neighboring Peoples* co-authored by Zhang Jinfu and Luo Bo, and published by Sun Yat-sen University Press, 2013 edition, p. 148.

2 Yang Yuxiang's *Ethnic Groups under the Snow Line of Bershulla Ridge*, published by Yunnan University Press, 2002 edition, p. 61.

CHAPTER FOUR

A Distinctive Cultural System Centered around Salt

Salk making is not only a material production activity related to salt among the people, but also a traditional folk activity. The salt industry, besides promoting the development of regional economy and social changes, also reflected the rise and decline of regional kingship and ethnic groups, and profoundly affected the ideology and folk customs of the local people. The Yanjing area in Tibet has long been a blessed place where diverse cultures converge and develop in harmony, with Tibetans, Nakhi, Han and other ethnic groups, as well as Tibetan Buddhism, Catholicism, Dongba religion and primitive religion coexisting here. And with salt as the medium, a relatively unique "salt culture" system, centering around the production, circulation and exchange of salt and other related activities, gradually took shape.

Professor Zeng Fanying, a renowned salt culture expert, once pointed out: Salt is both a cultural phenomenon and a cultural carrier; the object of salt culture research includes all material, spiritual, and institutional cultures related to salt.[1] Yanjing salt culture is very rich in content and variety. In addition to such material culture as salt production tools and technologies mentioned earlier, there is also institutional culture, such as salt administration and salt taxation, etc., and during this process, spiritual culture such as Yanjing stories, salt industry beliefs, and taboos has also come into being. The spiritual culture in Yanjing salt culture, which includes not only the social and cultural psychology, but also the social fashion and ethos related to Yanjing Salt,

---

1 Zeng Fanying. "Connotation and Characteristics of Salt Culture" [J]. *Journal of Sichuan University of Science & Engineering (Social Sciences Edition)*, 2006 (1): 47.

is one that exists in the forms of consciousness, concept and psychology, etc. The spiritual qualities, life attitudes, and ways of life required in salt industry activities have deeply permeated the hearts of the local people.

1. Cultural Integration among Ethnic Groups

Well salt culture, as a product of the collision and integration of salt production and sales, scientific and technological development, and immigrant cultures, is characterized by diversity, openness, inclusiveness, complexity, and contradictoriness, etc. The ancient ethnic groups in Yanjing, after thousands of years of development, has now formed a community dominated by the Nakhi and Tibetan people, and scattered with Han, Bai, Lisu, Pumi and other ethnic groups. These ethnic groups, living in the canyon and affected by natural factors, are closely related to each other, forming a symbiotic phenomenon. On the basis of blood relationships, professional relationships, languages, religious beliefs, marriages, and other aspects, they have in turn formed the inner circle of traditional salt industry culture in Yanjing.

Due to its abundant salt mines and important role in Yunnan-Tibet transportation, Yanjing, undergoing various cultural changes for more than a thousand years, has become the forefront of the integration and conflict between Nakhi and Tibetan cultures. Although it is a region where the Tibetan and Naxi people live in compact communities, these two ethnic groups are "a bit of me in you and a bit of you in me". Ethnic boundaries are blurred in Yanjing society. They share a common way of life, a common living area, and even a common historical memory, but it is hard to interpret the community of residents living here with the term "nationality". Thanks to the multi-ethnic settlement, local ethnic compositions and origins experienced a long-term integration and assimilation process, too.

1) The Tibetan Nationality

"Zang" is a Chinese term, and they call themselves "*fan*" (Tibetan pronunciation "bod-pa"). According to the Tibetan terms for people living in different regions, people in Yanjing Township, Markam County, eastern Tibet call themselves "Khamba". As Markam is a Zang-inhabited county, where the Tibetan nationality accounts for over 90% of the total population, in Yanjing it is also largest ethnic group.

Judging from the excavated Karuo ruins in Qamdo and the cultural relics unearthed from the Sarcophagus tombs in such Kham Tibetan areas as Batang in 1978, original inhabitants had already settled down here, living and multiplying before the Warring States period.

During the pre-Qin period, the Di and Qiang tribes in the Yellow River basin migrated southward along the Dadu River Valley, Yalong River Valley, and Jinsha River Valley of the Hengduan Mountains on the Qinghai Tibet Plateau. One of them, after arriving in the areas of Batang and Markam, lived on animal husbandry and hunting, and established the "White Wolf Kingdom" in Batang during the Western Han Dynasty, with jurisdiction over Batang and Markam, etc. Hence, in early Chinese historical materials, the Markam area was referred to as "Rong" or "Xiqiang".

During the Yongping period in Eastern Han Dynasty (58-75 AD), Yizhou prefectural governor, Zhu Fu of Liang State, disseminated the grace of the Han Empire to the White Wolf Kingdom. As a result, Tang Zui, the White Wolf King, offered to pay Emperor Ming "a variety of tributes and bow at his feet", and the land of Batang and Markam also submitted to the authority the Han Empire with the White Wolf Kingdom. In the Sui Dynasty, Tubo emerged in the present-day Lhoka region of Tibet and gradually expanded eastward. In the second year of the Qianfeng period in Tang Dynasty (667 AD), Songtsen Gampo, the 32nd btsan-po of Tubo, dispatched troops to conquer the White Wolf Kingdom. From then on, Markam was under Tubo's rule for 210 years. While ruled by Tubo, "thanks to the promotion of Buddhism and the establishment of temples, Tibetan Buddhism became deeply rooted in the hearts of the people along with political forces. And over hundreds of years, Tubo gradually merged with Qiang tribes in Kham Tibetan areas into one ethnic group".[1] "The Tibetan nationality, having gradually taken form and developed in the course of thousands of years of history, was the result of the mutual natural assimilation and common development of the Tubo and various ethnic groups living

---

1 See the ancient Tibetan historical accounts of past events *Dunhuang Version: Historical Documents of the Tubo Kingdom.*

in the western frontier of the motherland, mainly through long-term intertwined coexistence with the ancient Qiang tribes"[1]. Therefore, the formation of the Tibetan nationality in Yanjing undoubtedly resulted from the integration of Tubo people, Markam indigenous people, as well as the White Wolf Qiang tribe.

In the fourth year of the Qianfu period in Tang Dynasty (877 AD), after the Tubo regime was destroyed by the great uprising of slaves and civilians, the society, over 400 years, was characterized by "dispersed ethnic groups everywhere, with thousands of large clans and hundreds of small ones not reunified any more...each has its own chieftain for divided rule"[2]. From the fourth year of the Zhengde period in Ming Dynasty (1509) to the Qing Dynasty and the Republican period, with the reign over Markam of Chieftain Mu of the Nakhi ethnic group in Lijiang, Yunnan, and the entry of the Qing and Sichuan armies into Tibet, some Nakhi people who remained in Yanjing, and the Han officials and soldiers of the Qing Dynasty and the Republican period, began to intermarry with local Tibetan women and reproduce their descendants. This integrated not only the elements of the Nakhi ethnic group, but also those of the Han nationality, into the Tibetan nationality taking form in Markam.

The Tibetan nationality in Yanjing speak the Kham dialect, and its written language take the form of Tibetan characters. Characterized by strong pronunciation airflow and lowered intonation, in verbal communication, there are often many borrowed words from Chinese, such as *dianhua* (telephone), *dianshi* (television), maoxian (wool), and *baicai* (cabbage), etc., and loanwords (mainly English and Hindi), such as "grasse" for glass, "bugirin" for flashlight, and "brzeg" for machine gun, etc. Besides, there is still a local dialect called "Mage" in Rumei Township and Cuowa Township, where the local people can neither understand nor speak Tibetan, so this local dialect has become the only way of communication.

The Tibetan nationality in Yanjing have their own unique food structure and eating habits, among which yak butter, tea, tsampa, beef and mutton are known as the "four treasures" of Tibetan diet, in addition to highland barley wine and various dairy products. As an important Tibetan food, tsampa is very simple to make: stir fry

---

1 See *History of Ming Dynasty: Records of Tibet.*

2 *The Additional Tibetan History* compiled by Liu Liqian, Xihua University Press, 1945.

highland barley and grind it into flour, and it's done. There are various ways to eat, but the most common is to mix tea juice and yak butter with tsampa and milk residue thoroughly by hand in a small bowl, and knead it into small balls. Yak butter tea has been an indispensable beverage for the Tibetan nationality in Tibet. To make it, yak butter, salt, and tea are essential. Yak butter is the cream extracted from cow or sheep milk, with the best being a golden type extracted from yak milk in summer, whereas a different type extracted from sheep milk being pure white. When drinking tea, the Tibetan nationality are particular about the order of the older and the younger, as well as hosts and guests. Guests are not expected to drink too hurriedly or quickly. Generally speaking, three bowls would be the most auspicious.

The costume of the Tibetan nationality in Yanjing, regardless of gender, has been preserved in its entirety to this day. People living in different parts wear different apparel and accessories. Men's costume is robust and bold; while women's is elegant and unrestrained, particularly with jewelry, gold, and jade as accessories, forming a unique style for plateau women. The basic features of the Tibetan costume are long sleeves, wide waist, long skirts, and long boots. Moreover, the Tibetan nationality, regardless of gender, are very particular about ornaments, the texture of which includes silver, gold, pearls, agate, jade, emerald, coral, and amber, etc., and they are widely used in head wears, hair accessories, earrings, necklaces, waist accessories, and rings. Tibetan people attach great importance to "Khata" and regard it as the most precious gift. "Khata" is a snow-white fabric, generally about twenty to thirty centimeters wide and one to two meters long, woven from yarn or silk. Whenever there is a festive event, or a visitor from afar, or a courtesy call to the elderly, or a farewell before a long journey, it would be presented as a gesture of respect.

The fact that the Tibetan nationality has always been the main ethnic group in Yanjing did not change before, and will not change in the future.

2) The Nakhi Nationality

The Nakhi nationality, known as "Moso", are called "Jiang" by the Tibetan nationality in Markam. Local old people also affirmed that in ancient times, the true Nakhi ethnic group were known as "gongbu". In Tibet, "gongbu" refers to parts of the Nyingchi region and its local residents, who migrated to "Moso" areas and gradually merged into a group.

In the fourth year of the Zhengde period in Ming Dynasty (1509), most of the Markam area was occupied by Chieftain Mu of the Nakhi ethnic group in Lijiang, Yunnan Province, and the "Chawadamizong" was established with Yanjing as the center for governance. Chieftain Mu, by adopting the immigration policy of "slaughtering the local residents and relocating Moso soldiers for garrison"[1], forcibly moved a large number of Nakhi people from Lijiang to Markam over time. Living in Yanjing, most of these Nakhi people were serfs of Chieftain Mu during peacetime, who had to cultivate the land and build houses. But during wartime, they would serve as his soldiers, garrisoning the fortress and dashing about in a battlefield. At the end of the twelfth year of the Chongzhen period in Ming Dynasty (1639), Güshi Khan, chieftain of the khoshut tribe of Mongols, attacked and occupied Markam, destroying the 130-year rule of Chieftain Mu there. At that time, some of the Nakhi ethnic group fled back to Lijiang, while others remained in Markam, settling down and living together with the local Tibetans.

After the Ming and Qing dynasties and the Republican period, the Nakhi ethnic group in Yanjing had already integrated with the Tibetans. But they, who had previously moved to Yanjing as "victors", are gradually blurring their identity today. Except that the Yanjing Nakhi nationality close to Lijiang still preserved some customs that are different from both the Nakhi ethnic group in Lijiang and the Tibetans in Yanjing, such customs as clothing, food, shelter, transportation, marriage, and funeral, etc. here have quietly changed. So to speak, except for their ethnic attributes, everything else is no different from that of Tibetans.

They usually speak fluent Tibetan with a Nakhi accent. As to Nakhi language, only some Nakhi elders are still able to speak; young people no longer understand it except for a little Nakhi colloquialism. They live in Tibetan-style blockhouses, with the interior furnishings basically the same as those of the Tibetans. These Flat-roofed houses consist of a first floor and a ground floor in addition to an attic with opening: on the ground floor are stables for cattle, sheep, and horses; on the first floor are bedrooms and the kitchen; on the roof is grain to be dried in the sun and stacked. Their staple food includes noodles, tsampa, corn, pork and mutton, and their favorite beverages are yak butter tea, highland barley wine, and Tibetan liquor, etc. They

1 See *Additional Manuscript of Yunnan Annals*.

celebrate Tibetan festivals and believe in Tibetan Buddhism. While their marriage customs are basically the same as those of the local Tibetans, their funeral customs may include water burial, cremation, and celestial burial, etc. The choice of funeral type is determined by the divination of the wizard.

The Nakhi nationality in Yanjing generally wear Tibetan costume, and there are also a few young men and women who choose to wear Han clothing. Traditional Nakhi costume with a lot of silver ornaments is rarely seen here, but according to local Nakhi women, their earrings still retain the traditional Nakhi style. Young women like to wear a silver bar-shaped ornament--about seven or eight centimeters long and wound with silver wire--diagonally on the right side of the bun; whereas older women like to wrap a red kerchief about 1.5 meters long around their heads. "Usually, Tibetan women braid their hair with colorful strings, and to wrap their heads with a long fuchsia kerchief is typical of Nakhih costume."[1] Men on festivals or festive days like to wear a small waistcoat hemmed with gold thread.

Yanjing Nakhi nationality celebrate the spring festival by the Tibetan calendar. Before the New Year, every household would clean their houses, wash their clothes, and prepare such food as "yuzedu" (steamed buns in Nakhi language) and cooked pig heads, etc. On New Year's Eve, a whole family get together and offer "yuzedu" respectfully to the stove, water vat, and the deceased; children who have set up separate families would bring a portion of pork, a "yuzedu", and a bottle of Tibetan liquor to their parents and relatives, and kowtow to their parents and the stove. On New Year's Day, the whole family enjoyed themselves at home, accepting New Year's wishes from the elders, and they must eat longevity noodles at noon. The second day of the New year is Labor Day, when men would go to cut firewood in the mountains, known as gathering "yongxin" (auspicious firewood), while women would go to the salterns to collect salt, known as "blessing salt"; the laborers' constant laughter and singing filled the air. The third day is for sacrificing to the mountain god. By riding horses and carrying guns, men would offer sacrifices to the sacred mountain "Naiqiao", and the sacrificial activities are presided over by the village chief or a man of prestige in the whole village (before democratic reforms, they used to be arranged by the local headman or a rich and influential family).

1 "Salt Makers by Lantsang River", author unknown, *Gansu farmers Newspaper*, August 2, 2005.

After offering sacrifices to the mountain god, people would not only sing and dance on their way back, but also engage in a shooting competition, the first prize winner of which might be rewarded with a bar of tea and a Khata. The rest of the villagers would cook “qing” (congee made from crushed wheat by adding yak butter, meat, milk residue, and beans) in a spacious area at the foot of the mountain, waiting for the return of worshippers. As soon as they come back, people start drinking and eating. When everyone has dined and wined to satiety, all the men and women in the village would dance Xianzi and Guozhuang around the bonfire. With merry singing and wild dancing all night, they celebrate last year’s harvest and make wishes for the coming year. The fourth day is the guest day; each household would first bring 10 kilograms of grain for feeding New Year pigs to their relatives receiving the guests. When a guest comes, an elder of the host’s family should greet him (her) at the door with spices in hand; on entering the house, the guest should go to the kitchen to pick up a cypress branch and stand listening to the elderly’s New Year wishes; after that, the guest puts the cypress branches into the stove for simmering mulberry and sits down one by one, waiting for the elderly of the host’s family to express their gratitude before drinking and eating. In the evening, the guests return home and select a strong pig, which is black all over with four white hooves, as the New Year pig. They propose a toast to it and feed twice as much fodder; meanwhile, they simmer mulberry with cypress branches, wishing the New Year pig thick fat and more meat, thereby bringing good luck to the whole family. Moreover, all the family members should go to the slaughterhouse and plant three trees (in a row), two of which are pine trees representing the parents of the pig, and one a cypress tree representing the piglet, to signify the prosperity of the livestock and good luck in the coming year. The Yanjing Nakhi nationality mainly engage in collective activities during the Spring Festival. The food and New Year pigs needed for these activities are provided by each household in the village in turn, while the required highland barley liquor is brewed with pooled grain from these households or brought along by them.[1]

After hundreds of years, Nakhi people nowadays wear Tibetan costume, speak Tibetan language, live in Tibetan-style houses, spin prayer wheels, and go to temples

1 See *Yanjing County Annals*, p. 266.

asking for their newborn children's Tibetan names from the living Buddha; their way of life has essentially been Tibetanized.[1] However, their kinship terms (father is called "Awu", mother "Amei", father's elder brother "Abudi", father's younger brother "Abuji", mother's brothers "Agu", and father's sisters "Ani", etc.) still retain the characteristics of those in ancient Nakhi language, which are very similar to the kinship terms still exist in such regions as Tacheng mountainous area in Lijiang.[2] The Nakhi culture is still being passed down here, but only faintly discernible.

3) The Han and Other Nationalities

In addition to Nakhi and Tibetan nationalities, which make up the vast majority of the population, the great family of nationalities in Yanjing also includes Han, Bai, Lisu, and Pumi, etc. These nationalities, by contributing their own strength to the social development of Yanjing, have played a significant role in promoting regional economic and cultural prosperity.

a. The Han Nationality

Before the Ming and Qing dynasties, there were few records of the Han nationality's social activities in Yanjing. In the fifty-eighth year of the Kangxi period in Qing Dynasty (1719), when Yue Zhongqi led his troops on a westward expedition to Tibet, some Shaanxi travelling traders migrated with the army from Kangting to Jiangka and Yanjing for business. By this time, the Han nationality first registered for permanent residence in Yanjing. Later, the Qing government established garrison officers, grain officers, and the Yanjing Salt Bureau, and stationed troops in Markam; the number of Han people coming to Yanjing gradually increased. At the end of the Qing Dynasty, Zhao Erfeng, by implementing the bureaucratization of native officers in Markam, and recruiting reclamators and craftsmen from the inland, in addition to the consequent arrival of civil and military officials and soldiers, as well as merchants from Sichuan, Yunnan, and Shaanxi, led to a sharp increase in the number of Han immigrants to Yanjing. During the Republican period, a total of 467 Han households and 700 people came to Markam for business or work. Among them, there were 384

---

1 Ran Jiayan and Wu Hongjin: "More than A Thousand Nakhi People Are Tibetanized" [N], *Chongqing Economic Times*, June 18, 2006.

2 Yan Xiaohua. "Investigation on the Religions and Current Situation of the Yanjing Villages in Tibet Border Region". *China Tibetology*. 2009(04).

households and 576 people in Yanjing.[1]

The majority of Han nationality, who remained in Yanjing through the ages, not only intermarried with local Tibetan women and reproduced their descendants, but also introduced a fine variety of grain crops such as wheat and corn, as well as various vegetables; meanwhile, they also brought in such manual work as salt making, hairdressing, sewing, tanning, brewing, woodcarving, and silver censer making, etc. All these had a profound impact on the ethnic economy of Yanjing.

After liberation, group after group of government cadres, people's teachers, technicians, construction workers, and soldiers of the People's Liberation Army were dispatched by the Party and government to support the construction of Tibet in Markam. Henceforth, many cadres, workers, and veterans of the Han nationality have settled down in Yanjing and lived in harmony with the Tibetan people for a long time, contributing to the prosperity and development of Yanjing.

b. Other Nationalities

- The Bai Nationality

The Bai nationality, calling themselves "Baizi", "Baini", and "Baihuo", collectively known as "Bai people", belong to a branch of the Di and Qiang tribes. They are a minority ethnic group with a long history and culture in the southwestern border of China, mostly distributed in Yunnan, Guizhou, Hunan, and other provinces, of which Yunnan Province has the largest population, mainly living in compact communities in Dali Bai Autonomous Prefecture. The population of Bai nationality in Yanjing is relatively small. As for the specific time when they started their activities here, according to the available records, it was probably in the late Qing Dynasty: "Zhao Erfeng's policies attracted a large number of merchants to and from the Kham Tibetan region and its surrounding areas. In the early Republican period, there were already tens of thousands of nonlocal merchants in the Kham Tibetan areas, including a large number of Yunnanese, mainly Bai and Nakhi ethnic groups from western Yunnan."[2]

There are many traditional festivals of the Bai nationality. The "Third Month

1 See *Markam County Annals*, p. 267.

2 Yang Xinyu, Zhang Xilu, and Liu Rong. "Economic and Cultural Exchanges Between Tibetan and Bai in the Period of the Republic of China". Journal of Dali University, 2015, 14 (01).

Fair", which dates back to 1,300 years ago in Tang dynasty, has been their largest annual festival, now known as the "Third Month Fair Ethnic Festival". Every year, from March 15th of the lunar calendar, people of local and nearby ethnic groups gather in Dali, Yunnan Province to exchange materials and hold such activities as horse racing, archery, and dragon dancing, etc. for five to seven days. In addition, there are such ethnic festivals as the "Torch Festival" (also known as the Star Return Festival) and the "Butterfly Festival", too. The Torch Festival is held every year on the 25th day of the sixth month of the lunar calendar. On that day, men and women, old and young, would gather to worship their ancestors. Through activities such as worshipping torches, igniting torches, holding torches, and jumping torches, they wish that the harvests of the five grains were plentiful and the domestic animals are all thriving.

The Bai nationality's costume, after a long historical development process, has formed its own ethnic characteristics. The Bai people adore white, and their costume styles are slightly different from place to place, with white being considered noble. Men often wear a kind of white Chinese-style jacket with buttons down the front, with a black collared gown--or several leather or satin collared gowns--over it, commonly known as "three drops of water"; on the waist they'd wear a leather belt or embroidered bellyband, going with blue or black long trousers. Women's costume varies from place to place, too. They mostly wear a white jacket and a red waistcoat, or a light blue jacket with a black velvet collared gown over it; on the waist they'd wear a short embroidered apron, going with blue loose trousers and a pair of embroidered "hundred-knot shoes". Unmarried women would coil their single braid on top of the head, and tie a bright red string around a white kerchief; while in contrast, the red and white complement each other very well. With the short embroidered apron, they appear to be even more vivid-colored, elegant and in good taste. But married women would change to buns. Bai women have the custom of wearing earrings and bracelets. And the headwear, with wind, flowers, snow, and moon as its theme.

The Bai people refer to women, regardless of age, as "Jinhua" (golden flower), and men as "Apeng Ge" (brother Apeng).

- The Lisu Nationality

During Ming and Qing dynasties, they were known as "Lixie" or "Li Su", and they called themselves "Lisu". The name was first seen in writings during the Tang Dynasty, when Fan Chuo referred to them as "Li Su" in *History of Nanzhao* and believed them to be a constituent part of the "Wuman"[1] people at that time. Except for slightly different characters in historical records, the name has been used for over a thousand years to this day. This ethnic group bears a close relation to the Yi and Nakhi nationalities in terms of ethnic origin. They first lived in the Chin-sha River basin at the border between Sichuan and Yunnan. "In the mid-16th century, due to the protracted war between the Mu clan, prefectural magistrate of Lijiang, and the ruling power in Tibet, a large number of Lisu people, unable to tolerate the oppression and exploitation of Chieftain Mu, moved into the the Nujiang River region under the leadership of their tribal chief "Tubiba". At the end of the 17th century, some Lisu people entered such mountainous areas as Zhina, Zhanxi and Sudian in Yingjiang County border along the migration route 'the Nujiang River--Pianma--Guyong, Tengchong', and then moved to all directions in Yingjiang, forming the characteristics of large dispersion and small concentration..."[2] It might be inferred that the first group of Lisu people arrived at Yanjing no earlier than this period.

Lisu people's houses, mainly made of bamboo, wood and grass, are of two types: the wooden fork house and the bucket house. Their major festivals include Kuoshi Festival (December 20, equivalent to the Spring Festival of the Han nationality), Torch Festival (June), Harvest Festival (October), Mid-Autumn Festival, Bathing Festival, and Knifeladder-Climbing Festival, etc., among which the Knifeladder-Climbing Festival, held every year on the eighth day of the second lunar month, especially tugs at one's heartstrings. In the center of the gathering venue, there are two or three 20-meter-long erected thick wooden poles, on which more than 30 long knives have been inserted, with the blades facing upward. Some men in their prime, dressed in red cloth, with their heads wrapped in red, and barefooted, would jump onto the knifeladder and climb upward on the blades after dancing and singing around. The first to the top needs to do very difficult handstands and set off firecrackers. Thousands of spectators, while looking up, burst into cheers from time

1 An ancient ethnic name, for citizens of the ancient Nanzhao Kingdom and one of the ancestors of the Dai ethnic group, who originated from the Di and Qiang tribes.

2 See the introduction of Yingjiang on the official website of Yingjiang County People's Government.

to time. It is said that the knifeladder was set up to instill in the Lisu people the spirit of "daring to climb the mountain of knives and venture into the sea of fire".

The Lisu people are very fond of singing in antiphonal style, and there's a saying that goes, "You cannot live without salt, and you cannot be sensible without singing old songs". Folk songs have almost become the "second language" of the Lisu people. Whether in various productive activities or weddings and funerals, they would sing songs; even when engaged in a lawsuit or mediating a dispute, they'd settle it by the way of singing. It is said that there are more than 10 types of folk songs, which are plain and touching, rich in tunes, and mostly improvised in the form of narrative poems. Lisu people's dance forms are also full of variety. In addition to the heel dance, the three-string dance, and the *lusheng* (a reed-pipe wind instrument) dance, etc., there are other dance types reflecting their daily labor and life, the process of their fighting against nature, and their open-minded spiritual outlook, such as the monkeys' finger-guessing dance, the pigeon drinking water dance, and the bird king dance, etc. that imitate animal movements and habits, as well as the wedding dance and the laundry dance, etc. that depict daily life. And their traditional instruments include pipa, Jew's harp, four strings, and *lusheng*.

The traditional costume of Lisu women consists of a long gown or short skirt made of cloth, with a red belt or apron around the waist, and white cloth leggings tied around the legs; the style is very simple. They wear ornaments made of stringed red beads and shells on their heads, and a necklace made of stringed colorful beads around their necks. Men generally wear a Chinese-style collarless short gown with buttons down the front, and knee-length trousers, with a floral cloth bag of the Lisu style slung diagonally. When herding cattle and sheep in the mountains, they like to wear a waistcoat made of sheepskin; when hunting or working outside, they'd carry a Lisu ethnic machete and a quiver on the back.

The Lisu people are mainly engaged in agriculture, with rice, corn, buckwheat, and millet as their staple foods. Both men and women enjoy drinking alcohol, and are good at it; they often entertain guests with watery wine. Historically, the development of Lisu society had been relatively slow, and until the democratic reforms in the 1950s, it still retained some remnants of the primitive clan system.

- The Pumi Nationality

The term "Pumi", a phonetic translation of Primi language, originally meant

"Bai people". They use it to call themselves, while Tibetan people call them "Ba" or "Bo". According to the Chinese historical records, ethnic tales and ethnological data, the Pumi nationality, which might have originated from the nomadic Di and Qiang tribes in ancient China, is one of the Chinese ethnic groups with a long history and ancient culture. So far, few detailed historical records of the Pumi people's social activities in Yanjing could be found, but it is estimated that they migrated from Yunnan.

The major religious belief of Pumi people is the primitive religion, with a small number believing in Taoism or Tibetan Buddhism. And offering sacrifices to mountain gods is one of the most solemn ritual activities for them. Most Pumi festivals are closely related to ritual activities. During these festivals, people would often get together, singing and dancing. In addition to celebrating the traditional Spring Festival, they also have the Dragon Boat Festival, the Snow Threshold Mountain Tour Festival, and the coming-of-age ceremony, etc., which have recorded the cultural development over thousands of years. Among them, the Snow Threshold Mountain Tour Festival on the fifth day of the fifth lunar month every year is a unique one for the Pumi people. During the festival, women would go up the mountains to gather medicinal herbs, hoping that the family will be healthy in the following days. There are also be some related festival activities, such as wrestling and archery competitions, as well as love song duets, and musical accompaniment, etc., showing the splendid cultural traits of the Pumi ethnic group.

As to the costume, the Pumi people prefer white, which is a symbol of purity, nobility, and auspiciousness in their hearts. According to the ancient Pumi customs, children, regardless of gender, would wear a linen long gown which buttons on the right before the age of thirteen. Girls wear a braid in front of their hair ornaments, with red and green beads tied to it, while boys wear a braid on the front, left, and right sides of their heads respectively, without any bead. After the coming-of-age ceremony at thirteen, they would change to adult attire. The costume of Pumi men is basically the same in various regions, i.e., a short linen garment and loose long trousers, with a long gown over them, and a belt tied around the waist. Pumi women often wear long hair and are considered beautiful for their thick braids. They like to add yak tail hairs and silk threads into a braid, coil it on the top of the head, and wrap it in black cloth. Take the Pumi women in Ninglang region for example, they tend to wear a pleated long skirt that has been cinched around the waist, with a red thread usually embroidered horizontally across its middle part. They say this is their ancestors' migration route; after death, people need to follow it to find their own destination, otherwise they will not be able to return to the hometown. Regardless of the clothes style, both men and women would add a sheepskin waistcoat or cape at last.

The ancestors of Pumi people mainly relied on hunting and animal husbandry for a living, and only in the late Ming and early Qing dynasties, they gradually began a farming lifestyle, working at sunrise and resting at sunset. So, the diet of this ethnic group is characterized by meat. Living in the plateau region for all these years, their staple foods include corn, wheat, and highland barley, etc.

4) The Ethnic Integration

Yanjing, located at the border of three provinces (and region) and on either side of the Lantsang River, one of the "three parallel rivers", has experienced the rule of various regimes and ethnic interaction, exchange and integration in different historical periods, gradually forming a pattern of multi-ethnic and multi-cultural gathering in this magical land. These multi-ethnic groups have maintained their own cultural traditions (religions, productions, architectures, diets, and languages, etc.), while at the same time developing and undergoing some changes. Relying on their own initiative, they have actively adapted to social and cultural progress, thus presenting the harmonious coexistence of multi-culture and multi-ethnic groups.[1]

a. The Borrowing and Blending between the Tibetan, Han, and Nakhi Nationalities

The mutual trade that began in the Tang Dynasty led to the development of the Markam region. With the increase of trade activities on the Ancient Tea-Horse Road, Nandun, Markam became an important trading spot and trade was booming here. About Nandun's location at the intersection of the Sichuan-Tibet and Yunnan-Tibet Ancient Tea-Horse Roads, Liu Zanting, in his *Ningjing County Annals in the Republican Period*, recorded that "the merchants have flocked from all parts of the

---

1 Han Yan. "Presented in *Weixi Jianwen Ji*: Muliti-ethnic Integration and Development in Northwest Yunnan". *Journal of Ethnology*, 2022,13(09).

country" to Nandun, revealing the diversity and integration of Han and Tibetan cultures. For example, in Nandun "there are Guan Yu Temples, referred to as Han Temples by the natives". And the cultural integration was also reflected in the increase of intermarriage phenomena: "As Nandun has long been a trading spot, the natives learned Han etiquette unexpectedly. Afterwards, some Han people settled down here or married into and live with the brides' families, and was thus half-assimilated." During the Tang Dynasty, sculptures of Maha Vairocana, Avalokitesvara Bodhisattva, Ksitigarbha Bodhisattva, Lao Tzu, and Chuang Tzu, as well as numerous statues of Songtsen Gampo, Princess Wencheng, Bhrikuti Devi, and blon-stong-btsan were produced in Markam, showing the features of multicultural exchanges and integration between Han and Tibetan people.[1]

In the fourth year of the Zhengde period in Ming Dynasty (1509), Chieftain Mu of the Nakhi ethnic group in Lijiang, Yunnan, occupied most parts of Markam and forcibly relocated a large number of Nakhi people from Lijiang to Markam, where they took root and multiplied since then. After migrating to Yanjing, they on the one hand became salt makers and salt traders for generations over the long period of time; on the other hand, after years of fierce battles and coexistence, most of them were either Tibetanized as members of the Tibetan ethnic group, or lived together with the Tibetan people, or retained their own characteristics, forming Nakhi Villages in the Tibetan region. Although local Tibetans were politically dominated by the Nakhi ethnic group, their culture exerted a great influence on the latter's economic and cultural life to varying degrees. For example, the Nakhi people learned to grow highland barley, eat tsampa, drink yak butter tea, wear Tibetan costume, and speak Tibetan language; their living habits have completely become Tibetanized. At the same time, they also introduced some advanced planting techniques to the Tibetans, exerting an influence in return. In daily life, the Naxi people make pipa meat with their own wisdom, as was recorded in Yu Qingyuan's *Travelogue in Weixi* during the Qianlong period in Qing Dynasty: "People become rich by raising many horses, cattle, sheep, and pipa pigs, and the chieftain raises even a lot more. In winter, pigs are butchered, their bones and hooves removed, and shaped like a pipa, hence the

1 Zhao Guodong and Liu Hua. "Tea Distribution, Culture and Life in Mangkang, Tibet". *The Border Economy and Culture*, 2022(08).

name." Making pipa meat has been a tradition of the Nakhi ethnic group. Under their influence, almost every Tibetan family in this area would make pipa meat every year.

During the Kangxi period in Qing Dynasty, Tuowengbu and his younger brother Tashi Tsering in Batang were made chieftain and deputy chieftain respectively by the central government, with jurisdiction over the present-day Batang, Derong, as well as Yanjing, Mangling, Zhubalong, Xisonggong, and Kuobu in Markam, and Zhongdian and Deqin in Yunnan. From then on, the religious culture, folklore culture, and folk art culture of Batang had a significant impact on Markam, especially the Yanjing area. The most salient facts were that the "Mao noodles" in Batang gradually developed into "Jiajia noodles" (or "adding noodles", which would be explained in the ensuing chapters) in Yanjing, and the development of Markam Xianzi Dance also absorbed the essence of Batang Xianzi. In the late Qing Dynasty, the central government appointed Zhao Erfeng as Minister of Sichuan-Yunnan Border Affairs. While carrying out the bureaucratization of native officers in Yanjing and Ningjng, Zhao also established a reclamation office, renovated transportation, set up Mandarin Language schools, improved medical and healthcare conditions, and encouraged intermarriage between Tibetan and Han people. Moreover, he forcibly popularized Han agricultural techniques, medical techniques, hygienic knowledge, wedding and funeral customs, and clan names. Objectively, all these measures not only facilitated the exchanges between Han and Tibetan cultures, but also greatly promoted the local economic and social development. And because of the salt business, the Nakhi ethnic group in Yanjing often interacted with inland people, and thus the Han nationality also had a certain influence on their customs, which was already the case in the late Qing Dynasty: "The customs in Yanjing are the same as those of the Batang natives, but just because it's close to the Yunnan border and people here interact with the inland people, everything is gradually imbued with the Han style."[1]

During the Republican period, a county primary school was established in Yanjing, too, teaching such courses as Mandarin Chinese, Tibetan language, arithmetic, music, physical education, and art, etc. In this period, the influx of Han,

---

1 *Local Geographical Records of Yanjing, Batang*, No. 15 of the Chinese Ethnic History and Geography Data Series, compiled by the Library of Central Academy of National Minorities, 1979.

Nakhi, and other ethnic groups increased a lot compared with the past. They were mainly engaged in handicrafts, commerce, and other walks of life according to their own characteristics, which had a certain effect on promoting the social and economic development of Tibetan region. Since then, Yanjing became the only Nakhi village community in Tibet, as well as the gathering place for a small number of Nakhi people outside Yunnan. The local Nakhi people formed and preserved some customs and habits that were different from both the Nakhi ethnic group in Lijiang and the surrounding Tibetans. Moreover, in this small township where various ethnic groups lived together, intermarriage between them became very common, and families composed of Nakhi and Tibetan nationalities were everywhere. Intermarriage between ethnic groups has been the most direct and thorough form of ethnic integration and exchanges, as well as a direct reflection and manifestation of harmonious ethnic relations.

Historically, the Nakhi and Tibetan people in Yanjing experienced fierce military conflicts, but now the relationship between the two ethnic groups is very harmonious. In productive activities, both ethnic groups have salt makers, and are engaged in animal husbandry and farming; while in emerging commercial and transportation industries they cooperate with each other. As to language, the Nakhi people are fluent in both Nakhi and Tibetan, and most of the local Tibetans also speak Nakhi. In the Yanjing Nakhi dialect, Tibetan characters make up nearly one-third, and nearly two-thirds are related to Lijiang. Both ethnic groups fully respect each other's beliefs and traditions. The Nakhi people here, in their long-term interaction with the Tibetans, have become an ethnic group more inclined to Tibetan cultural identity. It is this harmonious atmosphere that has created a healthy and stable society, fostering an open cultural mindset.

At the same time, the Tibetans, while maintaining their own traditions, have been gradually influenced by different cultures, too. For example, although the Tibetans still keep the tradition of "covering houses with soil and preferring storied houses", with most of their dwellings being Tibetan style blockhouses, yet due to the influence of Han culture, they have gradually changed the architectural structure and style of their dwellings. Some villagers, by adding a protruding and roofed balcony to the second floor of the earth houses, not only have facilitated their daily lives but also changed the Tibetan habit of spending their time inside the blockhouses. Many

Tibetans have even built their courtyards with cement pouring method, and planted flowers and fruit trees, greatly expanding their living space. In recent years, with the help of the national targeted poverty alleviation policies, many Tibetans have built separate one-storey cement houses, and installed solar energy equipment donated by the state for daily washing. All these indicate that the cultural customs of local Tibetan people have undergone significant changes with the support of the state and society.[1]

b. The Borrowing and Blending between Other Nationalities

- Between Pumi and Tibetan People

The Pumi ethnic group has a close kinship with the Tibetan people. According to an ancient tale of the Pumi ethnic group, the Tibetan, Pumi, and Nakhi people are brothers born to the same mother, with "gami" (Tibetan) being the eldest, "paimi" (Pumi) the second, and "nami" (Nakhi) the youngest. And there's another tale: the relationship between "xifan" (Tubo people) and Tibetans is that of maternal uncle and nephew, Tibetans being the uncle; due to the disharmony between the two, "xifan" moved southward, but the time is unknown.

In fact, Pumi and Tibetan people not only share a common ethnic origin, but have also maintained a natural sense of mutual respect and trust from ancient times to the present. And this trust and closeness have been shaped by history. In ancient times, the Pumi ancestors lived together with the Kham Tibetans in the Mount Gongga snowy plateau, with similar lifestyles, economic types, and cultural beliefs. Due to the influence of geographical environment and times, the Pumi people's dwellings, costume, and customs were mostly the same as the Tibetans in addition to their very similar politics, economy, and religious beliefs. Since the end of Ming Dynasty and the beginning of Qing Dynasty, an organizational form "combining political and religious powers" similar to that in Tibet and other regions has emerged, fully identifying with Tibet politically and economically. Though retaining its own customs and habits, the ethnic group is greatly influenced by Tibetan culture. In view of this, Mr. Ren Naiqiang said: "During the Ming and Qing dynasties, the yellow

---

1 Han Yan. "Presented in *Weixi Jianwen Ji*: Muliti-ethnic Integration and Development in Northwest Yunnan". *Journal of Ethnology*, 2022,13(09).

sect of Lamaism[1] was introduced into this area, and many of the natives was thus Tibetanized. However, a small part of them, under the protection of chieftain Guabie (and others), still retained the Qiang customs, who are now known as the "Pumi ethnic group".[2] It can be seen that such long-term friendly exchanges and mutual influence between these Pumi people and the Tibetans have gradually formed the similarity in cultural and psychological qualities.

- Between Lisu, Han, Tibetan, and Nakhi people

Tracing back to its origin, the Lisu ethnic group took form through the integration of the ancient northwest Qiang tribe moving southward along the three parallel rivers with the local indigenous people. In the mid-8th century AD, they were ruled by three major forces, i.e., the Tang Empire, the Nanzhao Kingdom, and the Tubo kingdom forces; during the Yuan and Ming dynasties, they were under the rule of Chieftain Mu of the Nakhi ethnic group in Lijiang and slave owners of the Yi ethnic group on the east bank of Chin-sha River respectively. Under the domination and exploitation of other ethnic groups for long, the Lisu people, who could not tolerate the heavy burden of military services, fought many times but all ended in failure. In order to survive, they had to constantly migrate deep into mountains and forests. It is no exaggeration to say that the history of the Lisu people is, to some extent, one made up of continuous migration due to discrimination from other ethnic groups.

As the Lisu people used to migrate constantly to avoid wars and natural disasters in different historical periods, traveling between the mountain peaks and rarely settling down for a long time, they were also known as "the ethnic group chasing the sun". Later, the Lisu people on the Chin-sha River banks continually migrated to the high mountain and canyon areas near the Lantsang River, the Nujiang River, and the Drung River. And those who lived in the river canyon areas and mixed with the Tibetans, were mostly Tibetanized under the influence of the latter's social and cultural customs. Some Lisu people even accepted Tibetan cultural customs, especially their belief in Tibetan Buddhism, and followed their Tibetan singing and

---

1 The Gelug Sect of Tibetan Buddhism. The monks wear yellow hats, hence the name.

2 See Ren Naiqiang's *Exploring the Origin and Development of Qiang Nationality*, Chongqing Publishing House, 1984 edition, p. 106.

dancing culture, costume, and folk houses, forming the local "Tibetan Lisu" culture. As Tibetans and the Lisu people here have been living together and mixing with each other for a long time, the Tibetan Lisu people, while maintaining part of the Lisu ethnic traditions, have absorbed some Tibetan cultural traditions, too, showing a certain degree of hybridity in their culture. What is more, the ritual activities of the Tibetan Lisu people have been influenced by the Nakhi ethnic culture. Meanwhile, local Tibetans have also been unceasingly accepting the Lisu and Han cultures, thus gradually forming a pattern of multi-cultural symbiosis.[1] This is of enlightening significance for us to understand the emergence and development of ethnic cultures, indeed.

- Between Bai and Tibetan People

The Markam Tibetans and the Bai ethnic group in western Yunnan, living in proximity, with continuous economic and cultural exchanges, have exerted mutual influence and integrated with each other, forming a unique economic and cultural belt.

After the outbreak of the Opium War in 1840, Western imperialist countries constantly intervened in Tibetan affairs. The Qing government, aware of the important interdependence between Sichuan, Yunnan and Tibet in the Kham Tibetan areas, appointed Zhao Erfeng as Minister of Border Affairs of Sichuan and Yunnan in 1906. And he, by putting forward "Three Strategies for Pacifying the Kham areas" and "Six Measures for the Governance of Sichuan Border", vigorously developed commerce in the region, "expanding the Yunnan-Kham and Sichuan-Kham commercial exchange channels, and dispatching escort troops to protect trade caravans of different provinces traveling to and from the Kham areas"[2]. As a result, Zhao Erfeng's policies attracted a large number of merchants to travel to and from the Kham Tibetan areas and its surrounding regions. In the early Republican period, there were tens of thousands of nonlocal merchants here, including a large number of Bai merchants from Yunnan. Lots of Bai merchants entered Xikang, which resulted in the prosperity of economy along the routes and the promotion of economic and

1 Han Yan. "Presented in *Weixi Jianwen Ji*: Muliti-ethnic Integration and Development in Northwest Yunnan". *Journal of Ethnology*, 2022,13(09).

2 Zhou Zhisheng. "Yunnan Merchants and Modern Yunnan-Tibet Trade Exchanges" [J]. *Tibet Studies*, 2003(1): 78-81.

cultural exchanges between the Bai and Tibetan people. The exuberant prosperity of commerce in southern Xikang from the late Qing Dynasty to the early Republican period largely relied on the support of Bai and Nakhi merchants from Yunnan (as evidenced by the Yunnan Bridge spanning the Zhaqu River in Qamdo to this day).

During the early and middle years of the Republican period, the major channels of economic and cultural exchanges between the Tibetan and Bai people were trade and caravan culture. Since there was a great difference between Tibetan and Bai cultures, Bai architects and ethnic craftsmen, in order to gain a foothold in the vast Tibetan market, had to familiarize themselves with the unique Tibetan culture. These smart architects and ethnic craftsmen took every opportunity to learn from Tibetans, and by giving full play to their wisdom and talents, incorporated some ideas from Bai culture into Tibetan culture. Thus, they won the trust of the Tibetan people and monks, and become messengers of economic and cultural exchanges between the Tibetan and Bai ethnic groups. Due to the impact of extensive economic and cultural exchanges between the Tibetan and Bai people during the Republican period, there have always been some informal religious activities and a few forms of communication, such as intermarriage, between the Tibetan and Bai ethnic groups.[1]

As can be seen, the ethnic composition of Yanjing resulted from the integration of natives and outcomers through the ages. Various ethnic cultures, historical backgrounds, and customs gathered and collided in such a small area. Although there might have been misunderstandings, estrangements, and even conflicts, over time, through adaptation and dialogue, they have long achieved mutual respect, seeking common ground while reserving differences, and cooperating with each other here, thus making it a blessed place where diverse cultures converge and coexist harmoniously. Many daily necessaries and customs are widely exchanged and disseminated among various ethnic groups; some of them are even integrated into the daily lives of Han and other ethnic minorities, such as drinking yak butter tea, drinking highland barley liquor, dancing Tibetan dances, singing Tibetan songs, living in Tibetan houses, and speaking Tibetan language, etc. Moreover, some families are enjoying the combination of Chinese and Tibetan cultures in their daily

1 Yang Xinyu, Zhang Xilu, and Liu Rong. "Economic and Cultural Exchanges Between Tibetan and Bai in the Period of the Republic of China". *Journal of Dali University*, 2015, 14 (01).

lives. They celebrate the Tibetan New Year, as well as the Spring Festival and the Mid-Autumn Festival; they speak Chinese, as well as Tibetan; they enshrine and worship statues of Buddha, while pasting Spring Festival couplets; they eat rice and vegetables, as well as tsampa and beef; they wear Tibetan costume, as well as Han costume and western-style clothes; they believe in Buddhism with great compassion and wisdom, while admiring heroes who can both compose and perform military exercises. This interplay of diverse cultures is a true portrayal of the integration of different ethnic groups in Markam.[1] Today, the Nakhi, Tibetan, and other ethnic groups, living in harmony in Yanjing, are actively adapting to globalization and the rapid development and cultural changes of modern society.

2. Harmonious Coexistence of Different Religions

In Yanjing, the harsh natural environment filled the locals with fantasies and dependence on the divine power, and they expressed their gratitude and earnestly hopes to gods with a tenacious will to survive and pious rituals. Therefore, the Yanjing culture exhibits a strong and mysterious religious charm, and its colorful figurative interpretations are mostly closely related to ethnic religions. Whether it is the Tibetan, Nakhi or other ethnic groups living here, they regard religion as an indispensable part of their spirit and life. The religious beliefs of the villagers are very unique, and in the long-term practice of living together, the indigenous communities in Yanjing, with their intelligence and wisdom as well as inclusive mentality, have jointly created a regional religious culture with prominent characteristics, presenting a situation of diverse coexistence in varied colorful splendor.

As the only area in Tibetan where Christianity has been successfully disseminated, there are both Buddhist temples and churches in Yanjing Township. Most Tibetans believe in Catholicism, while Nakhi people believe more in Tibetan Buddhism, in addition to a few people believing in ancient Bon religion and Dongba religion, etc. Nowadays, Tibetan Buddhism, Catholicism, Bon religion, Nakhi Dongba religion, Lisu primitive religion, and Taoism, etc. have been brought together

1 *Yanjing County Annals*, pp. 356-357.

in Yanjing, which is also a major feature of the religious culture here.

1) Major Religion Types

a. Bon Religion

The Bon religion, as the most primitive local religious belief of the Tibetan people, was once the "national religion" of the Tubo Kingdom. In ancient Tibet, "Bon" was a common name for the believers of Bonpo, which predated Buddhism. In terms of content, Bon religion is an animistic belief, whose worshiped objects include heaven, earth, sun, moon, stars, thunder and lightning, hail, mountains and rivers, earth and stone, grass and trees, birds and beasts, and other natural things. It lays emphasis on such activities as sacrifices, Lamaist devil dance, divination, and averting a misfortune or disaster by prayers. Bon religion can be said to be a regional form of animism in Tibet. According to historical records, the Yanjing villages used to be under the jurisdiction of the Kham areas, and the long-dwelling people here were mainly Tibetans. Before the Naxi ethnic group, the bureaucratization of native officers in the Qing Dynasty, and Catholicism entered Yanjing, the Tibetans here believed in Bon religion or Tibetan Buddhism. At that time, the Kham Tibetan areas, including the Yanjing villages, was the activity center of Bon religion.[1]

However, as Buddhism gained dominance in Tibetan areas, the influence of Bon religion in Yanjing was gradually weakened. Its doctrines incorporated a large amount of Buddhist content, and its cultural traits and thinking structures of the primitive state were mostly lost, while the Bonpo cultural factors in Dongba religion still retained its original style and features. From the perspective of ethnic development, Dongba religion of the Nakhi ethnic group is closely related to the primitive Bonpo religion of Tibetan people. The Tibetan and Nakhi ancestors have the same ethnic origin relationship--ancient Qiang nomads. Judging from the worldviews, icons, religious concepts, and forms of religious activities of the Dongba and ancient Bonpo religions, their religious forms are quite similar, and as primitive religious forms, they might have been different branches of the same source. Among the Nakhi people, ancient sound and meaning of this belief are still preserved, which

1 Yan Xiaohua. "Investigation on the Religions and Current Situation of the Yanjing Villages in Tibet Border Region". *China Tibetology*. 2009(04).

may indicate the relationship between the Dongba culture and the primitive religion worshipping light.[1]

Before the peaceful liberation of Tibet, there were many Bon believers in Yanjing, Markam and other areas. By continuously absorbing nutrients from Buddhism, Bon religion enriched itself and made much progress, which is also a factor that cannot be ignored in its preservation to this day.

b. Dongba Religion

Dongba religion was originally an ancient religion prevalent among the Nakhi people in Yunnan; the priests (master interpreter and chanter of the scriptures) who presided over religious activities were known as "Dongba", hence the name. In Tibetan language, the term "Dongba" (or "Dunba") means "one who points the way" or "wise man". Dongba religion, as a polytheism, includes ancestor worship, ghost-god worship, and nature worship, etc.; its ritual activities include offering sacrifice to Heaven, funeral ceremony, exorcism, warding off calamity, and divination, etc. It was a combination of ancient Tibetan Bonpo religion and the primitive beliefs of the Nakhi people, and was later widely inherited by the Nakhi ethnic group. After Songtsen Gampo's reign, Bonpo believers in Tibet were expelled to the remote border areas, mainly going and seeking refuge with various indigenous residents in the Han-Tibetan corridor, especially the Moso or Nakhi people.[2]

After the Nakhi people moved into Yanjing, they also introduced their own ethnic belief, Dongba religion, to this area. As is well known, Dongba religion, originating from primitive witchcraft, has the characteristics of primitive witchcraft and religion at the same time. Dongba priests, as religious professionals, enjoy a very high status in the Nakhi society, and are regarded as the medium between man, gods, and ghosts. They can interact with gods, talk to ghosts, welcome good fortune and expel the evil spirit, eliminate disasters among the people, and bring peace and happiness to the world. Just as Tibetan Buddhism is based on *Tripitaka*, Dongba religion is centered on the Dongba sutras. The Nakhi ethnic group's belief in

1 Tao Zhanqi. "A Study of the Relationship between Dongba Religion and Ancient Bonpo Religion from the Perspective of Mythological Classics and Cultural Factors Involved". Tibet Studies, 1998(04).

2 Rolf Alfred Stein. *La civilisation tibétaine*, compiled and printed by the Chinese Literature Editing Office of the Tibetan Academy of Social Sciences, p. 253.

Dongba religion is an important symbol of inheriting Nakhi culture and uniting the hearts of the people, too. When Nakhi people began to settle down in Yanjing, they were first faced the strong impact of Tibetans, especially the economic exploitation of Tibetan chieftains and the cultural influence of Tibetan Buddhism. While living and multiplying in Yanjing for generations, some Nakhi people have been gradually assimilated by the Tibetan culture dominated by Tibetan Buddhism and the living customs of Khamba Tibetans, but some are still clinging tenaciously to their own religious beliefs.[1]

Nowadays, most Nakhi people in Yanjing have become Buddhists, and the consciousness of nature worship with a hint of wizardry is fading away. During the Cultural Revolution, Dongba scriptures and dharma-vessels were destroyed, and Dongba rituals were banned, resulting in its cultural loss. Nevertheless, in some specific places and periods of production and life, there are still animistic Dongba characteristics remaining. The relatively well-preserved religious ritual is the annual Spring Festival activity "offering sacrifice to Heaven". Known as "Meibi", it is one of the major traditional sacrificial activities for the Nakhi people, and plays an important role in maintaining their psychological quality and ethnic culture. The Nakhi people claim to be the people of heaven, so worshipping heaven has been the core of their culture.

In addition, although many previous investigators said that there was no longer any ritual custom of Nakhi Dongba religion in Yanjing, according to Dr. Li Hechun's field research, the ritual of offering sacrifice to Heaven, which is held on the fifth and eighth days of the first lunar month every year, has been retained in the village, and there are groups of people specially devoted to this. At present, there are still three Dongba priests in Lower Yanjing Village. Without scriptures, they could only recite them orally. Sometimes, Dongba believers also worship in the temples of the yellow sect of Lamaism. Before liberation, Tibetan headmen had forbidden the locals to practice Dongba religion. Today, the elderly in the village not only believe that the custom of offering sacrifice to heaven should be passed down, but also advocate speaking Nakhi dialect. Up to now, the Nakhi people in Lower Yanjing Village

---

1 Yan Xiaohua. "Investigation on the Religions and Current Situation of the Yanjing Villages in Tibet Border Region". *China Tibetology*. 2009(04).

are still offering sacrifice to their ethnic protector god "Sanduo", who has been worshiped by the whole Nakhi ethnic group.

c. Tibetan Buddhism

Since Buddhism was introduced into Tibet from India and China respectively in the 7th century, after hundreds of years of struggle with Bon religion, Buddhism, by absorbing some excellent elements of Bon, developed into a new localized type in Tibet in the second half of the 10th century, i.e., Tibetan Buddhism, which thus became the common religion of Tibetan people. Its core ideology was to persuade people to be prodigal of benefactions, unite for the good, abstain from killing and eliminate conflicts, and live in harmony. After the middle of the 11th century, various sects began to take form within Tibetan Buddhism. Among them, many of the Kagyu, Nyingma, and Gelug founders were Tibetans in the Kham areas. Especially the Nyingma sect, which was very typical of Esoteric Buddhism, always maintained a thriving and enduring influence in the Kham areas, with comparatively more temples and a large number of followers.

Most Tibetans in the Yanjing villages, while practicing Tibetan Buddhism, mainly revered the Gelug sect. This sect, which came into being in the early 15th century, was founded by Tsongkhapa (originally known as Lobsang Drakpa), a famous religious reformer in the history of Tibetan Buddhism. In the seventh year of the Yongle period in Ming Dynasty (1409), Tsongkhapa, after initiating the Monlam Prayer Festival (also known as Great Dharma Assembly for Prayer) at Jokhang Temple in Lhasa, built the Ganden Monastry in the northeast of Lhasa, officially marking the formation of this sect. Due to its rigorous canons, strict and impartial commandments, and complete doctrines, hence the name "Gelug" (i.e., good rules or good laws). The Laweng Temple and Gangda Temple nearby were the main venues for its activities. After the rise of Tibetan Buddhism in the Yanjing villages, although there were still some Bon believers for a period of time, yet its form and doctrines had gradually incorporated lots of Buddhist content. From the Ming and Qing dynasties to the Republican period, monks of Laweng Temple and Gangda Temple, as well as Tibetans and chieftains who believed in Buddhism, in order to defend their local religions, were clashing with Western missionaries throughout the years.[1]

---

1 Ibid.

Today, there are a large number of Buddhists living in Lower Yanjing Village. As a matter of fact, in Yanjing cultural system with various cultures interwoven and superimposed, Buddhist culture is still the core and main part. The vast majority of Nakhi people in Yanjing believe in Tibetan Buddhism, especially those who are living in a compact community in Lower Yanjing Village and engaging in salt pans. It has been exerting an influence on their value judgments and basic ethics.

d. Catholicism

So far, Yanjing has been the only area in Tibet with a Catholic church and lots of believers. As the most eye-catching building in Upper Yanjing Village, the Catholic church located here, which is not very tall, makes this ancient village even more mysterious.

In the mid-19th century, many missionaries from western countries attempted to spread Catholicism in Tibet, but all failed, with Yanjing being the only exception. At that time, foreign Catholic priests, dressed up as traders, entered Tibet from India and came along the Yunnan-Tibet Ancient Tea-Horse Road. After several twists and turns, they successfully brought the western Catholic church to Yanjing, where Tibetan Buddhism and Dongba religion were at the core of people's spiritual beliefs. According to the official version, this only Catholic church in Tibet was built by two French priests (one by the Chinese name of Deng Deliang) in 1855, with a history of 166 years. But there are also different versions among the people. Some say that Catholicism was introduced to Tibet in 1878 by a French missionary surnamed Bai, who went to Batang to join another priest preaching there in 1880, was later expelled from there in 1887, and had to return to Yanjing via Tachienlu (present-day Kangting) before building the Catholic church. Another version, in Liu Jun's paper "An Overview and Analysis of Foreign Churches in the Kham Areas" published in *Tibet Studies* in 1991, states that "the church in Yanjing was founded by a French priest, Chinese name Lv Siduo, in the 1860s". Therefore, among so many versions of folk tales, it is difficult to determine which might be true today, but the distinctive features of the church are true, indeed. Located next to the Yunnan-Tibet Highway, the church, whose full name is "Blessed Church of the Virgin Mary", and covers an area of more than 6,000 square meters, including 15 rooms, is a typical Tibetan-style wooden structure building. If it weren't for the big cross on the roof, no one would have associated it with a church.

From the introduction of Catholicism to Yanjing till the liberation of this area in 1950, as many as 17 missionaries came one after another to preach in this Catholic church, including French, German, and Swiss, as well as those sent by the Church from Kangting and Batang in Sichuan, and Weixi, Deqin, and other places in Yunnan. After many years of operation by these missionaries, some Tibetan and Nakhi salt makers already converted to Catholicism in the late Qing Dynasty. According to *Local Geographical Records of Yanjing, Batang*: "Today, there are only about 70 households in the vicinity of Yanjing, with the Catholics accounting for two in ten." That is, twenty percent of the inhabitants, mainly including Tibetans living in Upper Yanjing Village and a small part of Nakhi people, already believed in Catholicism. According to statistics, there were 342 Catholics in Yanjing in the 1860s.[1]

At present, there are several priests, friars, and nuns, etc. in the church. The Tibetan versions of *Bible* and *Biography of the Virgin Mary* here should have been the only fully preserved Tibetan Catholic scriptures in Tibet. The great festival is Christmas in December every year. Beforehand, the congregation would donate to the Catholic Church according to their own economic conditions. During Christmas, the church would offer a feast of delicious food and local specialty--wine without distiller's yeast--to the congregation who come to celebrate the festival. And in the evening, the congregation, dressed up in festive attire, would sing and dance until dawn.

These Catholics usually have Catholic names, such as Paul, Paulus, Yanne, Dolce, and Maria, etc., and many of them, after being given a Catholic name, will not choose an additional Tibetan name, and will be buried in full accordance with Catholic rituals after death. In general, they follow the tradition of their fathers, praying with both hands on the cross on their chests before eating and drinking in the early morning and at dusk, and the scriptures they recite are mostly translated and published Tibetan prayer scriptures. Their chanting, similar to the tune of the local Guozhuang Dance, is very distinctive. Some Catholics, though completely illiterate, are able to recite large sections of scriptures, and the God accompanying them throughout their lives is not Śākyamuni, but Jesus.

1 Ibid.

2) The Coexistence of Various Religions

In Yanjing, the Nakhi people, instead of practicing their own Dongba religion, have followed Tibetan Buddhism, while the Tibetans believe in Catholicism. Most of the Catholics here are Tibetans, especially in Upper Yanjing Village; thus they, with obvious Tibetan features, reveal the profound influence of traditional Tibetan Buddhism. Up to now, nearly 70% of the residents in Upper Yanjing Village are Catholics, the rest being Tibetan Buddhists; whereas people in Lower Yanjing Village all believe in Tibetan Buddhism. And there are the declining Dongba religion and Bon religion, too. Diverse cultures coexist harmoniously in this canyon area of Hengduan Mountains, making the ancient township, where the most original salt-making techniques are still preserved, a little bit more mysterious.

a. Catholicism Coexisting with Tibetan Buddhism

During the Xianfeng period in Qing Dynasty, French Catholicism was introduced to Markam, and in the process of its development, though mutually exclusive with different sects of Tibetan Buddhist for various reasons, the characteristics of coexistence and common development were very prominent.

First of all, the Yanjing Church is a combination of Western architecture with Tibetan art, fully reflecting its way of preaching in accordance with local customs since the introduction of Catholicism. Its exterior is typical of Tibetan residential architecture, while the interior structure is typical of a Gothic vault. The church decoration also shows many traces of Tibetan culture, such as the refreshing Tibetan Thangka pictures of gods, and the Khatas around the pictures of these gods, etc., the biblical frescoes on the ceiling still retaining the Catholic style. The façade wall in the church is hung with a huge portrait of Jesus and various holy images, under which there's a cross and holy water, as well as offerings such as lit candles in pairs, similar to Tibetan Buddhist temples or scripture halls in form and content. The holy water in the church, called "Equ" or "Qudanba" by Catholics, is completely in accordance with the religious terminology of Tibetan Buddhism. And the Catholic "holy end" (or "qukang") set up by the villagers at home is likewise no different from the "Buddhist End" in Tibetan Buddhism, borrowing the same term.[1] Even the

---

1 KalSang Gyal and Deji droma. "The History and Current Situation of Multi-religion Coexistence in Tibetan Areas", *China Tibetology*, 2008(02).

traditional Tibetan Khata is offered in front of the statue of Virgin Mary.

In terms of religious activities, the Catholics in Upper Yanjing Village, like other Tibetan people, regard the Losar Tibetan New Year's Day as the beginning of a year, though Christmas is still a major festival. In daily life, the priests and nuns, no different from ordinary Tibetans, would wear Tibetan costume, recite the Tibetan version of the Bible, and sing hymns in Tibetan language. During traditional Catholic festivals, Yanjing Church may invite Catholics of nearby churches and leaders of the township government, or even the abbot of Gangda Temple and Buddhists from the village, to attend the rituals. Firstly, the priest would preside over a mass to the Catholics, preaching and reviewing the life over the past year; then he would receive the holy communion for them, and all Catholics and invitees dine in the courtyard of the church; finally, under the careful organization of the township government leaders, as well as priests and nuns of the church, everyone engages in the elegant Guozhuang Dance and the bold and unrestrained Xianzi Dance, all through the night. The Gangda Temple of Tibetan Buddhism may also invite priests and Catholics to watch the "Lamaist devil dance" of Tibetan Buddhism performed by monks in the temple and celebrate the Buddhist festival during the annual "Festival of Lamaist devil dance".[1] Although Upper Yanjing and Lower Yanjing are two separated villages, and their religious beliefs are clearly differentiated, there is no gap in the hearts of Tibetan people. In the words of Father Lu Rendi, "Yanjing Catholic Church, with a history of over a hundred years, is currently the only Catholic building in Tibet that has been preserved and in continuous use. Its existence proves that different religions can coexist harmoniously in Tibet. "[2]

Not only are some religious activities of the Catholicism strongly colored by Tibetan Buddhism, but also in ordinary believers' homes, "Buddha images and crosses are often enshrined and worshiped in a mixed manner", and even the miraculous scene of Jesus and Sakyamuni's images coexisting one room might be seen, reflecting the interpenetration between the two. In Yanjing today, there are both Catholics and Tibetan Buddhists within a lot of families, and they respect

1 Paul and Ze Yong. "A Brief History of Catholicism in Yanjing". *Tibet Studies*, 2000(3).

2 Zhang Jiawei, Huang Shengang, and Ye Hui. "Visiting the Only Catholic Church in Tibet: Coexistence of Jesus and Sakyamuni's Images" [N], *People's Daily Overseas Edition*, January 29, 2008.

each other's beliefs. Meanwhile, there are also cases of religious conversion when a woman marries into the bridegroom's family or a man marries into and lives with the bride's family. Intermarriage between families of different faiths would be possible here, and conversion to the religion of the other family is not uncommon, either.[1] Many local young men and women may believe in Buddhism or Catholicism respectively when falling in love with each other, and their families will not prevent them from getting married. After marriage, the couple still believe in their respective religions, and thus images of Jesus and Sakyamuni can coexist in the same room. As to what newborn babies will believe in, it depends mostly on their own wishes.[2] The beliefs within family members may also vary. Today, in Upper Yanjing Village, it has been a common phenomenon that members of a family either believe in God or practice Buddhism. Portraits of Chairman Mao, Sakyamuni, and Jesus hung in the same oratory inside the house, each prays according to his/her own faith, and they have never had a dispute over their different beliefs. There are even couples who, because they believe in Catholicism and Tibetan Buddhism respectively, go out together whenever the husband attend church. The wife would then move while chanting the Buddhist scriptures and rotating a prayer wheel in the pagoda at the end of the village, and after church the husband would return home with her.[3] Cases like this are quite common in the Yanjing villages. However, considering the unity of time in the production, living, and religious activities of all family members, a whole family may choose to believe in the same religion, too.

Of course, Catholicism, while bringing many exotic customs to the Yanjing villages, has also presented a unique interest and charm in the long-term integration with Tibetan culture. For more than 20 years, unlike the clashes of blood and fire occurred here in recent historical periods, the two religions have long buried the hatchet. The Nakhi people who believe in Buddhism and the Tibetans who believe in Catholicism are at peace, understanding and respecting each other's beliefs; the Nakhi people and Tibetans of the same Buddhist faith appear to be even closer and inseparable. The long-term ethnic integration has made the Nakhi way of life

---

1 Dorjee Tsering. "The Present and Past of Tibetan Catholicism" [J]. *Chinese Folklore*, 1999 (3).

2 Wu Yongxian. "Entering the Only Catholic Church in Tibet" [N]. *Faith Weekly*, December 18, 2006.

3 Bian Ji. "Yanjing: Ancient Salt Mining and Catholic Faith". *China's Tibet*, 2008 (3).

compatible and coexisting with local Tibetan customs. Thus, it can be said that the customs of the Tibetan and Nakhi ethnic groups are intermingled here in Yanjing.

In Upper Yanjing Village, the names of Catholics, similar to those of Europeans, are mostly given by the priests; after death, they would be buried in Catholic rituals, but their way of life and customs are no different from those of Buddhist Tibetans. One may well say that the 530-odd Catholics in Upper Yanjing Village are a typical combination of Chinese and Western elements. They celebrate the Losar Tibetan New Year, as well as Christmas and Easter. Every Christmas, believers will hold a carnival. Five or six stoves will be set up in the church courtyard; the villager with recognized culinary skills will serve as the chef, and each household will send helpers. Hundreds of believers will gather on a patch of open ground for dinner, and in the evening, they will sing and dance merrily. All of these expenses come from the Mass offering, which is kept by a dedicated person, and the details of which will be disclosed after the event. The Buddhas and Christ's halos of wisdom and compassion have been integrated in this small township in the remote Hengduan Mountainous Region, educating people and benefiting the world, and thus becoming a unique cultural highlight of Tibet.

b. Mixed Relationship between Buddhism and Bon Religion

After Bon religion entered the Kham areas, on the one hand, it actively relied on the local political force--the chieftains, and on the other hand, it constantly absorbed nutrients conducive to its own development from Buddhist theories. For example, the Bon canons might be divided into "nine yàna"[1], i.e., nine parts. The first four parts, known as "causal yàna", mainly focus on the teachings and rituals of Bon religion, such as divination, praying to the spirits to bring evil upon someone, lauding, praying, and exorcism, etc. These parts have mostly preserved the basic content of Bon religion in its entirety. The latter five parts, known as "resultant yàna", specifically talk about how to transcend the suffering of Samsāra; the contents of these parts and the scriptures that give an account of these contents are mainly derived from Buddhism, whose canons have already been changed into Bon sutras.

From the perspective of the gods enshrined in Bon religion, a considerable

1 In Buddhism, "yàna" refers to the teachings that can carry sentient beings to the other shore of liberation.

number of them have come from Buddhism, and many portraits or statues of gods are also modeled after the images of Buddhist deities. The triad of the Buddha, the dharma, and the sangha enshrined and worshiped in Bon religion have been learned from Buddhism, too. Moreover, the living Buddha system, as one of the basic features of Tibetan Buddhism, was introduced into Bon religion by some Bon temples afterwards. Hence, just because there are so many Buddhist traces in Bon religion, lots of people would mistake it for a sect of Tibetan Buddhism. In fact, Bon believers never identify themselves with Buddhists. Although they are learning from Buddhism, they do not copy it, but selectively transform and absorb Buddhist doctrines according to the interests of their own religion. Therefore, in addition to some similar or identical features with Tibetan Buddhism, there are also many different or even fundamentally opposite contents in Bon religion.

Compared with Tibetan Buddhism, it reveals obvious traits of the primitive religion. For example, among the gods enshrined and worshiped in Dengqing Temple in Dege County, besides Tonpa Shenrab, Tara Buddha, and Padmasambhava, there are also various gods of nature representing the incarnation of such natural spirits as heaven, earth, mountains, rocks, and trees, etc. The mountains and lakes in Tibetan areas, often revered as sacred mountains and sacred waters, are worshipped by many people. This is clearly a reflection of nature worship in the primitive religion. Moreover, such wizardry as divination, praying to the spirits to bring evil upon someone, exorcism and evocation, and treating an illness with Lamaist devil dance, etc. are not only important religious activities for Bon monks, but also indispensable elements in their secular lives.

The symbol of Bon religion, "卍", as opposed to the "卐" of Tibetan Buddhism, is known as "Yongzhong". Signifying Bon religion, it has the meaning of auspiciousness, holiness, eternity, and firmness, etc., as well as the function of expelling evil spirits and accepting good fortune. In Bon temples, spiritual pagodas, scriptures, and believers' residences as well as on their clothes and hats, etc., the symbol "卍" can be seen everywhere. The symbol "卐" in Buddhism, known as Swastika, is the opposite of the Bon symbol, but has the same implication and social functions.[1]

1 Lin Junhua. "Bonpo, the History of An Ancient Religion in the Kham Areas". Tibet Tour, 2002 (04).

The direction in which Bon believers move while chanting scriptures and rotating a prayer wheel is completely opposite to that of the Tibetan Buddhist sects. In Bon religion, believers are required to walk chanting scriptures counterclockwise, rotating the prayer wheel with the right hand and turning the prayer cylinder with the left hand. But in Tibetan Buddhism it's just the opposite. It is said that neither Bon believers nor any sect of Tibetan Buddhists could reverse the rotation direction of the prayer wheel, otherwise "even ten more rotations will not work".

The bells used by Bon Lamas when practicing magic are also somewhat different from those of the Tibetan Buddhist sects. These Bon bells, flat-shaped and large-mouthed, are swung with the mouth upward when in use; whereas the bells of different Tibetan Buddhist sects, being deep inside, are swung with the mouth downward.

In addition, Bon temples are closely related to farming and animal husbandry activities. There are very few monks living in Bon temples. Ordinarily, the vast majority of them would stay at their respective homes, doing housework during the day and reciting scriptures in the morning and evening, while leaving only several to watch over the temples. Only when religious activities are conducted will these monks gather in the temples; after the event, everything returns to the way it was. This feature of incorporating religious practices into farming or animal husbandry activities is also unique to all sects of Tibetan Buddhism.

From the above-mentioned characteristics, it can be seen that the current Bon religion is different from the early primitive Bon religion, let alone Tibetan Buddhism, but a fusion of the two.

c. The Relationship between Bon and Primitive Religions

Bon religion, having developed from the inherent primitive religion in the Tibetan region, is closely related to a "indigenous" and ancient primitive religious belief widely spread among people in the Kham areas.

From the perspective of the content of belief, this primitive religion, like Bon religion, worships all natural things, ancestors, and the spirits. In the Kham areas, mountain peaks, rivers, and lakes that have been regarded as sacred mountains, sacred waters, and sacred lakes by the people can be found everywhere, and each county, each township, and even each village has sacred mountains, sacred waters, and sacred lakes of its own. These sacred mountains, sacred waters, and sacred lakes are either the dwellings or the incarnations of gods, who have been dominating the production and living, as well as the childbirth, old age, illness, and burial arrangements of the human world. Only by bowing in worship, can people obtain the protection of gods; any act of irreverence will surely be punished by them. In addition to the worship of mountains and waters, the primitive religion also attaches great importance to the worship and belief of heaven, earth, trees, fire, animals, ancestors, heroes, and spirits.

In the way of communication and exchanges between gods and human beings, Bon religion also shares a lot of similarities with the primitive religion. In Bon religion, it is the monks who go between gods and humans and are responsible for their communication; whereas in the primitive religion, there are also such exceptional talents, known as "wizards". These wizards, going back and forth between gods and humans through their wizardry, may help people to pray for blessings from gods, remove impending ill fortune, expel the evil spirit, perform the divination, pray to the spirits to bring evil upon someone, and so on. The role of wizards is basically the same as that of Bon monks. And now, some have even introduced canons of the Nyingma sect of Tibetan Buddhism and Bon religion while practicing witchcraft.

As mentioned above, there is not much difference between Bon religion and the inherent primitive religion in the Kham areas. However, Bon, as an artificially founded religion, is still obviously distinct from the primitive religion. In terms of sacrifice, the primitive religion is more original, economical, and practical than Bon religion. Take the ways of worshipping and offering sacrifices to sacred mountains, sacred waters and sacred lakes for example, they'd hang "Nongda" printed on cloth or scatter "Nongda" printed on paper in sacred mountains or beside sacred waters; they'd heap up "Mani stone mounds"[1] or build "Laze"[2]; they'd walk around

1 Mani stone mounds, also known as "sacred mounds", are altars made of stones and slabs, on which there are mostly the six syllable Mantra, a mind which perceives both past and future, statues of gods, and various auspicious patterns engraved. They are masterpieces of Tibetan folk artists and can be seen almost everywhere in the mountains, intersections, lakesides, and riverbanks of Tibet.

2 A "Laze" is a pile of stones, earth, and rocks, inserted with long bamboo poles, long arrows, long sticks, spears, and also tied with prayer flags, in the Tibetan mountain passes, hillsides, the main peaks, and the boundaries, etc. It is popular among Tibetans, Mongolians, Yugur people, Mosuo people, and some Nakhi areas.

the mountains, visit the lakes, "simmer mulberry", and offer various foods on the traditional days of sacrifice, and in some pastoral areas there's even blood sacrifice to the mountain gods preserved; they'd set free domestic or wild animals in sacred mountains, sacred waters, and sacred lakes; they'd stack white stones above the roof or door frame to symbolize the worship of mountain gods; they'd constantly recite the blessing scriptures whenever passing by sacred mountains, sacred waters, and sacred lakes. There are also significant differences between wizards of the primitive religion and Bon monks. This is mainly manifested in the fact that their wizardry has been generally passed down through inheritance or apprenticeship. They do not have fixed religious venues (temples), nor are they separated from productive labor. Leading a completely secular life, their distinctiveness from ordinary people would be shown only when invited to pray for blessings, remove impending ill fortune, expel the evil spirit, perform the divination, and pray to the spirits to bring evil upon someone, etc. Due to the different cultural traditions in different regions, the address terms of wizards vary a lot, and their wizardry activities also vary.

Of course, this phenomenon of integration resulting from coexistence could often be found among other religions in the Yanjing area. But considering space limitation, I will not list them all. Despite their widely varying doctrines and conceptual definitions, these religions have very similar fundamental purposes, all explaining the meaning of world existence and life existence, and on this basis, finding spiritual sustenance for people. Thus, it leaves room for a wide range of developments, integrations, and possible goals to be achieved. Religious integration is a manifestation and result of the inclusiveness of Yanjing culture.

From the perspective of multiculturalism as the joint construction of social spiritual wealth by mankind, both the Yanjing Catholic Church and the ancient Dongba culture in Yanjing have played their own roles of different cultural features or religious identities in the historical and cultural arena of Tibetan areas. This not only reflects the historical opportunity for the coexistence of multiple religions in individual Tibetan areas in history, but also shows the current situation of diverse and rich foreign cultures in the real society here. The Yanjing villages, a typical compact community where Chinese and Western cultures coexist harmoniously, has aroused the keen interest and extensive attention from the academic circle as well as Chinese and foreign tourists, and to study its economic and cultural impact is of great significance for analyzing how to promote the harmonious development of Tibetan areas.

3. The Unique Charm of Salt Folk Customs

Folk customs, as a kind of historical reality, is the inheritable life culture of an ethnic group. In real life, it presents as a set of unwritten and stylized rules, or a cultural psychological pattern, which is prevalent in and unique to the society of a minority nationality. Having undergoing a repeated accumulation process, once formed, it would develop according to its regular formula, like rays, and be inadvertently embedded in daily food, clothing, shelter, and transportation, as well as utensils, ideas, words and deeds.

The long history of salt industry development, while producing rich material wealth for Yanjing, has also created diversified spiritual wealth and accumulated charming customs with distinctive local characteristics. The formation of customs in various ethnic groups, such as literature and art, language, beliefs, apparel and accessories, diet, entertainment, festivals, etiquette, weddings and funerals, production, and daily life, etc., closely related to salt, and based on its discovery and development, all bears its obvious mark. The salt industry customs, as a salt culture spirit deeply rooted in the land of Yanjing and the hearts of Yanjing people, still retain the unique local values and ways of thinking.

1) Tales and Legends

The folk tales and popular legends circulating in the Yanjing area are numerous and rich in content, mainly including *The Story of Agutomba*, *The Legend of Dameiyong*, *The Legend of King Gelsall*, *The Legend of Princess Wencheng*, as well as various mythological stories and fairy tales of nature. These tales and legends, with rich imagination and vivid and tortuous plots, warmly eulogize Markam people's industriousness, wisdom and ethnic sentiments with a clearcut stand on what to love and what to hate, and lash out at some ugly behaviors in society, reflecting the common yearning of the ethnic minorities for good fortune and a better life.

a. The Legend of Dameiyong

Located on the Lantsang River bank, Dameiyong Snow Mountain, as the main peak of Tenasserim Chain, is the highest mountain in Markam County, with an altitude of 6,434 meters, and also one of the famous snow mountains in eastern Tibet.

Legend has it that a long time ago, Dameyong, the third daughter of Meri Snow Mountain God in Deqin, Yunnan, descended to the world without permission and came to the Quzika area when the local epidemic disease was spreading. Seeing the tragic scene of the field strewn with corpses and people living on the edge of starvation, pity welled up in Da Meiyong's heart. So, she decided to cure the epidemic disease for the people and benefit all sentient beings by making holes to draw water and regulating the temperature with tears. And since then, there has been 108 unique springs. When the steaming hot spring water gushed out of the ground magically, people cheered and ran towards them with joy, drinking while bathing, and all diseases were gone immediately. From then on, the local people regarded these hot springs as sacred, calling them manna from heaven. And it was said that drinking the spring water would eliminate disasters and dispel diseases, and bathing in it would purify the soul and prolong life.

The beautiful and benevolent Fairy Dameiyong, who had violated the rules of heaven by secretly participating in the trivial affairs of the world, was degraded to the mortal world by the gods. And thus, she incarnated into a majestic snow-capped mountain, standing in the rear of Quzika, and forever protecting the people of this area. In order to commemorate the grace of Fairy Dameiyong, the locals not only named this snow mountain "Dameiyong Snow Mountain", but also honored its peak as "Goddess Peak", worshipping for generations.

In addition, there's a lake called Lake Tsonavam behind the sacred mountain Dameiyong. Legend has it that there was a large black python in the lake, which was man-eating monster. There was once an eminent monk dedicated to subduing ghosts trying to hunt down this large venomous snake and chasing it all the way from YamdrokTso to here. But Goddess Dameiyong, by hiding it in this lake, saved the python's life. Since then, it has thus stopped eating human flesh, and the area has also become peaceful.

Throughout the Tibetan and Kham areas, sacred mountain worship is a widespread phenomenon of belief, and the female image of the ancestor god (Dameron, the daughter of the mountain god) suggests that the Yanjing people are highly likely to have experienced a matriarchal clan social stage as an independent group. According to local accounts, they are the descendants of the daughter of Snow Mountain, not the legend of the macaque commonly passed down by Tibetans, or the legend of six ancestors. Yanjing people, by associating these deified images and cultural elements from outsiders with the local salt culture, achieved their own cultural construction.

b. Legends about the Brine

How exactly did the brine come into being in Yanjing? For a long time, there have been multiple versions among the ordinary people in the Yanjing area, including legends about the Golden Rooster and the Silver Hen, as well as a legends about phoenixes. There are at least three legends involving the Golden Rooster and the Silver Hen.

One legend is as follows: A long time ago, people in the area around Yanjing (now Yanjing and Quzika Township) suffered from severe rheumatism and were relatively short, which seriously affected their productive labor, and they once lived a hard life. One day, Dameiyong, the third daughter of Kawagarbo (also known as Meri Snow Mountain), witnessed the suffering of the people living here on her way west to Lhasa for worship. Kindhearted and merciful, she felt obliged to do something good for the local people, but was troubled by not knowing how to help them. Later, Sakyamuni reminded her that the only thing that could help the impoverished people was her dowry, so she left the only dowry her parents gave her, i.e., a golden rooster and a silver hen, in Yanjing: the rooster on the west bank and the hen on the east bank of Lantsang River. Dameiyong told the elders of the two villages on both banks to be sure to take a look at what was happening at the bank the next morning. Then villagers of Jiada Village on the west bank found a rooster constantly circling around a piece of land, with brine bubbling up all over the valley. And people on the east bank also found a hen cackling while laying eggs in the soil. As she had to protect her eggs, the brine was hidden deep, and thus deep wells needed to be dug in order to draw it. Moreover, the brine of Jiada Village on the west bank was conjured by the Golden Rooster, so salt made there has been reddish, while the brine of Lower Yanjing Village and Upper Yanjing Village on the east bank was conjured by the Silver Hen, so salt made there has been white.

Another legend is: Long ago, a pair of magical rooster and hen lived in the lower reaches of the Lantsang River in Yunnan Province. One day, a menstruating woman in Yanjing, who was greedy for the hot water in the brine wells (the brine in these wells was hot water, but its temperature varied from well to well in different locations), washed her pants stained with menstrual blood in one of them, polluting the brine--the source of the salt industry. And this greatly angered the rooster and the hen, so the two traveled a long distance to Yanjing to guard them. The silvery-white hen landed on the east bank of the Lantsang River in Yanjing, causing the salt produced there to be white; the rooster with a big red comb landed on the west bank, causing the salt produced in Jiada Village there to be reddish.

And another legend: In Zayu, Tibet, there lived a pair of rooster and hen lived long long ago. And thanks to their divine blessing, the local people were all well-fed and well-clothed, and they could harvest corn once every seven days. But one day, a little child, being too mischievous, wiped his buttocks with leftover tsampa, and such an act of wasting food directly angered the rooster and the hen. So, in a fit of temper, this pair of divine chickens flew to Yanjing, the hen landing on the west bank of the Lantsang River, while the rooster landing on its east bank. The rooster was bold, and would continue to live in his own nest even if the river water rose above its top. Therefore, the brine on the east bank was seemingly not only hidden at the bottom of the river, but also concentrated in one or two spots, producing white salt. And the hen, relatively timid, was afraid of the river water. Whenever it rose a little, she would walk higher and move her nest from place to place, for fear of dirtying her red dress. As a result, not only has the salt produced on the west bank been reddish, but also the brine will move towards the shore with the rise of the river.

In addition, there's a legend about the Golden Phoenix and the Silver Phoenix.

In the past, Yanjing, located at the border of Sichuan, Yunnan, and Tibet provinces (region), was plagued by turmoil and chaos of war all year round, and the indigenous people were living in extreme misery. Seeing the people of Yanjing suffering so much, kind-hearted Dameiyong deeply sympathized with them. So, she reported the difficulties of Yanjing people to Avalokitesvara Bodhisattva, Protector of the snowland. The latter opened her third discerning eye and took a look: not only was the Yanjing area plagued by war and chaos, but also the natural conditions there were very harsh, with people living in extreme hardship. Immediately, two red and white "A" (Amitabha) characters flew out from her chest, and soon turned into a red and a white phoenixes, both making for Yanjing, and landing on the Lantsang

River banks before incarnating into brine wells respectively. Since then, brine water has been drawn by the local indigenous people, and salt made by evaporating it in the sun, which was even more valuable than gold, has become the main source of livelihood for the local populace. And this place is also known as "Chakaluo" (Tibetan, which means a place for salt production; "Cha" means "salt") and "Yanjing" (Chinese name) for this reason.

Now, people in Yanjing often say that of the two phoenixes, the red one was male and the white one was female. The female phoenix, in order to protect their nest and eggs, would often be submerged in the river water every year during the summer floods, and so were the brine wells; the male phoenix, who needn't protect the nest, could naturally move back as the river rose, and the brine wells were not flooded, either. The white phoenix stayed on the east bank of the Lantsang River, so salt produced there has been white; the red phoenix stayed on the west bank, so salt produced there has been reddish.

The religious use of chickens by the Kham people can be found in many historical materials. The rooster, the hen, as well as the male and female phoenixes, which have been deified successively as the others outside of humanity, are undoubtedly the embodiment of some primitive nature worship and animist concepts reflecting the consistency between Yanjing people and the stage of clan society at that time.

c. "The War between Jiang and Ling" and Other Legends Related to King Gesar

As the main figure in the Tibetan folk hero epic, King Gesar's position in the hearts of Tibetan people is roughly equivalent to that of Genghis Khan in the hearts of Mongolian people. Up to now, his tales have also been widely spread in the Yanjing people's residential area.

According to legend, King Gesar was the youngest of the three sons of the White Brahma King in the Upper Realm. At that time, demons ran amok, injuring and devouring people in the world of man. Avalokitesvara Bodhisattva, after discussing with the king, proposed that some heavenly gods be assigned to descend to the world for the purpose of subduing them. Hence the prince and his two siblings resolutely shouldered this important task. He was born the son of the abandoned wife of a small tribal chief. When still in his mother's womb, King Gesar was slandered by his uncle. Later, framed up by this uncle, he had to leave his hometown and led a wandering life with his mother, relying on each other for survival. From childhood to adolescence, he almost lived in the impoverished mountain areas all the time, digging for rats from underground and hunting wild animals for food, yet he had great ambitions. At the age of fifteen, he married Aburig mo. After that, with his divine power, he contended for the throne with a number of master-hands in the horse race meeting, and finally won the game and became king in victory. With the honorific title "Gesar", he was the ruler of the blackheaded Tibetans, too. In the year following his reign as king, he began his campaign to conquer north and south, engaging in desperate battles with various evil forces. He'd either command a huge army against his enemy, charging and fighting, or defeat it simply with tactics. Through constant metamorphoses, he successively subdued the cannibal demons, and expelled the invaders who had been plundering the people. Then, by subduing 18 large tribes, 7 medium clans, and 4 small clans, he conquered more than 150 tribes of all sizes, and finally helped the territory of the Ling Kingdom be reunified, winning freedom, peace and happiness for these tribes. When all enemies were defeated and all the evils were removed from the people, this super omnipotent hero got everything settled in the three realms before returning to the Kingdom of Heaven.

As for "The War between Jiang and Ling", legend has it that King Gesar, the Tibetan ethnic hero, once fought over the salt in Yanjing against Qiangba, the local Naxi King. As a result, King Gesar defeated Qiangba, occupied the salt pans, and captured the Naxi King's son Yilatogyu (also known as Youla). In the late Tubo Dynasty, Naxi Prince Yilatogyu surrendered to King Gesar and became his Naxi Minister, managing the salt pans there. And King Gesar granted him the salt wells and salt pans as a reward. It is said that the oldest and most primitive salt production method was also passed down from that time.

Similar to the above, there are still many more vivid and beautiful historical legends that have survived. As legends, they, lying between faith and fantasy, were created in the consciousness of common recognition by people of Yanjing and other Tibetan areas living in different times. Although their roots were in ancient times, they continue to flourish and grow. They have been passed down from generation to generation in Yanjing, all reflecting a strong "brine" flavor and highlighting the unique charm of the local well salt culture.

2) Local Festivals

Local festivals are a unique form of community activity. In the process of development and evolution of the salt industry in Yanjing, the operators and producers, either to promote the development of salt industry productivity, or to safeguard their personal interests, or to pray for the protection of gods, or to integrate groups, or to vent their emotions, or to realize their own aesthetic pursuits, gradually formed local festivals related to salt making or hosted by industry groups.

a. The New Year Festival

The masses in Yanjing are accustomed to celebrate the New Year with the "Spring Festival", which falls on the first day of the first lunar month every year (some also celebrate the Tibetan new year in addition to the Spring Festival). No matter according to which calendar is the new year festival celebrated, every household demands a family reunion. Those who are away from home must hurry back before the New Year, and girls who have got married must return to their parents' home before this day, too.

On New Year's Eve, every family has to offer sacrifices in front of a niche for Buddha, and light yak butter lamps day and night. They'd tie a Khata around the main pillar of the living room, and some may even decorate auspicious patterns with white powder on the beams and columns in the room. What's more, all the family members are busy preparing the New Year's Eve dinner. When eating the New Year's Eve dinner in the afternoon, they must drink "Getu", i.e., a Nine Treasure

Congee (made of 9 ingredients, including flour, meat, milk residue, and ginseng fruit, etc.), and they must all be stuffed, for legend has it that the King of Hell will weigh everyone on this day. In the evening, the whole family enjoy themselves to the fullest, until late at night. After midnight, housewives from different families, by carrying buckets to the riverside, scramble for the first golden and silver water of the New Year, which is called "Sercho Encho" in Tibetan. At dawn the next day, the man of each house goes to the nearest mountain top and their own mulberry stoves to burn incense and offer sacrifices, praying for a bumper harvest and good fortune for the whole family in the coming year. On the first day of the new year, people are dressed in festive attire; children go to the elders' houses to pay New Year's greetings, while adults gather at home and cannot go out dropping around. Starting from the second day, people exchange New Year's greetings and wish each other "tashi delek" (good luck and happiness) in the New Year. The third day, the tenth day, and the fifteenth day are days of worshipping Buddha and offering sacrifice to the mountain gods, when men from each household must decorate their horse gear, bring food as well as pine and cypress branches for offering sacrifice to the mountain gods, and ride to the sacred mountain in splendid attire, so as to burn incense, simmer mulberry, and pray, wishing the New Year free from disasters and diseases, and everyone healthy. The rest of the family, also wearing the most ornate clothes and various expensive accessories, will bring Khata, Tibetan incense, yak butter, cypress branches, and small money to the lamaseries, simmering mulberry, rotating prayer wheels, and worshipping Buddha. They will offer Khata and small money to the Buddha, and add yak butter to the lamps, wishing that the harvests of the five grains are plentiful, the domestic animals are all thriving, and everyone has good luck in the new year.

During the New Year Fetival, urban and rural masses must also dance Xianzi and Guozhuang, and hold various folk recreational activities, which will not end until the 15th day of this month.[1]

b. Labor Day

The Labor Day is mainly a festival of the Naxi ethnic group in Yanjing, which falls on the second day of the Losar Tibetan New Year.

Early that morning, men will go up the mountains to cut firewood, known as

1 See *Markam County Annals*, p. 283.

"Yongxin" (blessed firewood). And women will collect salt from the salt pans. When the salt has been collected, these women will sprinkle some on their way home. After returning home, the first handful of salt is offered to the Kitchen God, and the second is put into a salt jar. The rest is given to relatives and friends, as well as families without salt pans, hence called "lucky and precious salt", so as to achieve a bumper output next year, without disasters or diseases. Moreover, they may sing incessantly in the midst of the labor: "There's a tree with three branches. One branch reaches the cloudy sky, wishing the vast happiness as high as the heaven; another sticks into a huge rock, wishing the human body harder than the stone; and another stretches out to the river, wishing the life longer than the water..." Yanjing people treat the protector god in their minds with great caution.

c. "Napa" Festival

"Napa", which means "to butcher pigs for offering sacrifices to ancestors", is not only a ritual of worshiping heaven and gods, but also a traditional ritual of the Naxi people in Yanjing with the longest history, the grandest scale and the richest cultural connotation. This festival, unique to the Naxi nationality, is a combination of Dongba culture and Tibetan culture.

Li Jing of the Yuan Dynasty recorded in his *Brief Annals of Yunnan*: "Moso people (Naxi people), extremely solemn and neat, climb up mountains to offer sacrifice to Heaven. There are often more than a hundred men and women, holding hands with each other and singing and dancing by going round in circles for joy."[1] According to the traditional custom, every family in the village takes turns to hold this sacrificial activity. When a certain day from the sixth to the eleventh of the first lunar month every year has been fixed as the "Napa" Festival by a wizard through divination, the family responsible for hosting the sacrificial activity (as it is held once a year, all families take turns to be the host) will get busy, not only preparing sacrificial supplies, but also choosing a pure black boar to be used as a sacrificial offering. The evening before the festival, men of each family must bring half a kilogram of wheat, half a kilogram of rice, and a bottle of wine, and by riding a large horse, gather in the host's home for dinner. After dark, the ritual of "Beizeng" is held,

---

1 "Dongba Sacrificial Scripture of the Naxi Nationality: *Ancient Songs of Offering Sacrifices to Heaven*" [EB/OL], China Ethnic Literature Network (http://iel.org.cn).

i.e., to ride a horse while shouting "Wozengda" and run through the whole village, and then return to their homes respectively.

The next day, all people, regardless of age and sex, get up at about 4 a.m., and after freshening up, hurry to the sacrificial site one after another. The hosting family also needs to feed the sacrificial fat pig with wine, making it dead drunk, and then carry it to the sacrificial site for butchering, burn off its hair over fire, dismember its body, and cut off its haslet. It is said that an old Dongba priest of the Naxi nationality can estimate the disaster, harvest, death and other signs of the coming year from the shape and condition of the pig's splenic organ and scapula. If everything is normal, it proves that there will be favorable climatic weathers and a bumper grain harvest in this year, and they will make offerings to their ancestors, showing gratitude for the blessings; if there's a disease in the haslet, it proves that disaster may strike, and they will also offer sacrifices to their ancestors, praying for exorcism and disaster prevention, and all living beings blessed with good luck and peace. When the ritual of offering sacrifice to Heaven begins, a Naxi elderly will pray to heaven by reciting the Dongba scripture. In such a formal ritual, there should be a Dongba[1] reciting the *Chongbantu* (also known as *Human Migration*)[2]. But since no one in the Yanjing area can recite it any more, the priest will usually sing a section of prayer-like scripture. These are all the Naxi people's pursuit and happiness standards for health, longevity, harvest, harmony, and protection from heaven.

Afterwards, all households gather at the host's house to celebrate the festival, singing and dancing. This festival generally lasts two or three days, and next year it will be another family' turn to serve as the host.[3]

d. Marriage Festival

In Jiada Village of the Yanjing area, there's a unique and peculiar festival, i.e., a collective fake wedding ceremony that is held to revere and please the mountain gods

---

1 Name of the traditional religious clergy of the Naxi nationality, which is translated as "wise man". They are not only the highest intellectuals of the nationality, but also the main inheritors of Dongba culture.

2 *Chongbantu* gives an account of the origin of heaven, earth, and all things, reflecting the ancient people's production activities, such as hunting, grazing, slash-and-burn cultivation, as well as their world outlook.

3 See *Markam County Annals*, p. 284.

and pray for happiness--"Pema Paden" in Tibetan, commonly known as "marriage festival". It usually falls between the ninth and eleventh of the first month of the Losar Tibetan New Year; the days, not exactly the same each year, are decided by the living Buddha through divination.

During the Marriage Festival, the whole village will be bustling with excitement. The first day is for wedding preparation. In the morning, a respected and eloquent wedding host (errand-man), by holding a village youth meeting, select several most hardworking and beautiful pairs of young people of the year as the groom and the bride, as well as groomsmen and bridesmaids for this fake wedding, and determine the hosting family for the "wedding" by drawing lots. In the afternoon, the "groom" and the "bride" undergo ablution respectively and invitations are sent out; other young people prepare the next day's "wedding" for them.

Early the next morning, men and women, old and young in the village, dressed in festive attire and carrying wedding gifts, go to the hosting family to express congratulations and help organize the "wedding". The groomsmen and bridesmaids also go to the homes of the "groom" and the "bride" respectively to congratulate them, and help dress them up. Each wearing a flat circular hat made of red silk ornaments, an ochre-red Tibetan robe, and a pair of Tibetan boots decorated with bright patterns, the men ride in formation and march into the village with a heroic posture, symbolizing the arrival of an intelligent, capable and upright son-in-law from a rich and distant region. At the village entrance, the bride's escorting procession will be stopped with Khata by some women dancing Guozhuang there. And every 100 meters or so apart, there will be women playing the trick of blocking the way, asking for red packets, and then disappearing. When arriving at the home of the "bride", the procession still can't get in, and will face a series of roadblock questions; a certain number have to be answered before they are allowed to enter the door.

After picking the bride up, and on their way to the public house, some comically dressed "monsters" (known as Arakaji in Tibetan) may rush out. They are either wearing masks, or have painted a lot of strange designs on their faces with paint, ink, and pot ash, etc., or are not properly dressed, with socks, dolls, tsampa bags, tea mugs, and other items of no practical use hanging around their waists. It's said that their various funny words and deeds could not only bring laughter to the villagers, but also please the local gods and get rid of the foul air in the village. Just as these Arakaji were making a great mess of the scene, the host (errand-man) will stand on a table covered with cattlehide by the roadside, reciting aloud a long list of eulogies to mountain gods, earth gods, and village gods. After that, the procession turns and heads towards the public house on the hillside in the village. In the public house, the hilarious and lively atmosphere continues, and when the host announces the beginning of the grand "wedding", the "groom" and "bride" dressed in gorgeous costumes, attended by "groomsmen" and "bridesmaids", walk slowly onto the auspicious pattern painted with the "Tashi daji" symbol (i.e., the Swastika: 卐), listening to the host's "wedding congratulations", which is called "Dangxie" in Tibetan. And he, on behalf of the village, will present Khata, a symbol of happiness and auspiciousness, and cash gift to the "newlywed". Then, representatives of the "groom" and the "bride" will deliver congratulatory speeches to them both. Finally, the "groom" and the "bride" will conduct a brilliant free defense, ending the "wedding" ceremony amidst bursts of laughter from the crowds. After a lively morning, everyone will have "Jiajia Noodles" for lunch, until they are full.

In the evening, men and women, old and young will gather again in public house, and the Guozhuang Dance involving the whole village is about to begin. With the rhythm of beautiful music, the festival reaches its climax. The Guozhuang Dance praying for blessings, with melodious tunes, sweet and agreeable voices, cheerful steps, and fluttering long sleeves, is extremely refreshing and intoxicating. This unique Marriage Festival in Jiada Village will not come to an end until the third day.[1]

The custom of fake marriage in Jiada Village has a long history, and its wedding ceremony may have come from the legend of the mountain god Aninaiqiu. It's said that the mountain god enjoys a abustle and astir atmosphere, so in the village a wedding ceremony has to be organized every year to please him; otherwise, floods will strike the next year, bringing great disaster to the salt pans as well as the villagers. On the other hand, Marriage Festival, as a ritual activity, is aimed to eliminate the participants' mental anxiety to some extent, to achieve their psychological satisfaction and sense of security, and to vent their emotions properly. What's more, it may reflect the cultural identity, religious identity, and public affairs

1 Ibid.

the bride's escorts dressed in festive costumes

the "bride"

the wedding site in the center of a public house, where "Yongliang Baigen" (a master scripture interpreter and chanter for eliciting blessings) is sitting on the high platform above the seats

a moment of the wedding ceremony

concern and consensus of the people living here.[1]

As an important component of Yanjing folk culture, local traditional festivals have been formed, developed, and enriched in its social and historical development. In addition to the above mentioned, there are many other similar festival activities. Carrying Yanjing people's ideological concepts, spiritual outlooks, and value orientations, as well as their forms of expression in the productive, religious, and

1 Ma Qing-Hu. "Ritual and Function Research of Collective Ba Ma Ba Dian Festival, False Marriage--in Jiada, Yanjin, Moreangkam Carounty for Example". Shanxi Archives, 2015 (01).

cultural activities, they are not only a mirror reflecting the local social history, but also a magnificent picture scroll depicting the customs and sentiments of these people.

3) Literary and Artistic Activities

The development of well salt economy in Yanjing has also promoted the rise of local folk art and recreational activities, giving birth to a type of colorful salt art culture. Ethnic songs and dances, as well as traditional dramas, etc. all show that people of various nationalities are on friendly terms with each other and enjoying a good and prosperous life.

a. Xianzi Dance

Xianzi Dance, called "Xie" in Tibetan and translated as "Circle Dance", because in this dance, Xianzi[1] is used as the accompaniment, and the dance is led by a person who plays it, hence the name. As one of the three major Tibetan dances, it is a traditional folk comprehensive art that integrates poetry, *qin*, song, and dance. There are many categories of Xianzi Dance. Among them, Yanjing Xianzi Dance, with the momentum of the Lantsang River, the charm of lofty mountains and flowing water, and the imposing lyrics, may fill people with lofty sentiments and high aspirations, for example, "In March, the peach blossoms are blooming with snow-white salt, and the songs are wafting to the distance along the river; beautiful faces are reflected in the clear brine water, the brine pails are full of colorful clouds, and happiness originates from layers of salt..."[2]

1 A type of erhu, historically known as "huqin", shorter and thicker than those in other regions, is an essential instrument for Xianzi Dance.

2 "Yanjing, Markam: Fragrance of A Special Flavour Wafting Thousands of Miles", see http://mangkang.changdu.gov.cn/mkx/c105860/201610/2f1c8bc7c0ce454da9352e7a6c34c096.shtml.

Xianzi Dance has a long history, which can be traced back to a sacrificial dance known as "*Gezhuo*" during the Warring States period. During the Tang Dynasty, it gradually became a small group dance participated by the local people in their spare time from productive labor, during seasonal festivals, and at weddings and funerals. Mainly performed with a single bowed string instrument and singing, and in the form of a family, this was the very prototype of Xianzi Dance. With the development of "the Ancient Tea-Horse Road" in Batang and Markam areas, new vitality of innovation and development was also brought into Markam Xianzi Dance. After the Song and Yuan dynasties and the Ming and Qing dynasties, in the cultural exchanges and integration with the surrounding areas and nationalities, Markam people, by constantly absorbing other cultural and artistic specialties, nourished the Xianzi Dance, and gradually finalized it through the innovation and refinement of dancers, forming Batang Xianzi and Markam Xianzi with their own fixed basic movements, postures, rhythms, and dance forms, as well as unique ethnic cultural and regional characteristics, and have been passed down to this day.[1]

Now, Xianzi is mostly danced by people in their spare time from labor, on holidays, during wedding gatherings, or at "Dapinghuo" (dinner parties) and "Shuabazi"[2] before the summer and autumn harvests, etc. The melodies of Xianzi are short and concise, soft and smooth, and wonderfully tuneful; lyrics of many different themes can be sung to one music score. And there are no restrictions on Xianzi Dance; regardless of the number of people, and the size of the venue, it can be performed both on and off stage. Dancing in a circle around the bonfire, another circle can be made if there are too many people. Men and women are separated; the former stand at the head of a line playing Xianzi, and lead the crowd to dance with their hands swinging two long sleeves. They sometimes gather, sometimes scatter, sometimes take a roundabout course, singing while dancing. The lyrics are folk songs of a "Xie" style, and may also be improvised. Men and women are divided into groups, singing to each other's tune, one after another, to express their inner

---

1 Shang Jiangchuan. "Reflections on the Inheritance and Development of Tibetan Xianzi Dance in Cross-Cultural Communication: Taking Batang Xianzi and Markam Xianzi as Examples". Sichuan Drama, 2013 (04).

2 Known as "Sarley", when people may set up tents in the meadow, having fun, singing and dancing, and holding a horse-riding race, etc.

emotions. The pace is determined by the music rhythm of Xianzi played by the man, generally slow first and then fast. It starts with a tuneful and gentle melody, continues freely in a smooth and cheerful manner, and ends in a rising and passionate atmosphere.

As a mass cultural and recreational activity, Xianzi, whose lyrics are mostly odes to their hometown, eulogies to the snowy landscape, and reflections of the ethnic customs, or voicing their life philosophy, pouring out the love between men and women, and expressing happiness and auspiciousness through metaphors, is not only a kind of musical enjoyment to people, but also the bold and unrestrained folk free verses with subtle language, providing spiritual nourishment, rich local flavor, and profound educational significance.

b. Guozhuang Dance

Guozhuang Dance, called "Zhuo" in Tibetan, was originally a primitive sacrificial ritual for Markam Tibetans to worship totems, or make a fetish of supernatural beings. With the passage of time, its social function gradually shifted from thanking and entertaining gods to entertaining people, and later developed into an art form of singing and dancing in Markam County. According to its lyrics and folklore in Qamdo County, Guozhuang Dance had come into existence as early as the Tubo period.[1]

In Yanjing, Markam, Guozhuang Dance, without instrumental accompaniment, are classified into Old Guozhuang and New Guozhuang. For the former, with a sacrificial quality, its lyric content and dance steps are relatively old, such as "The

1 "An Elaboration on 'Guozhuang Dance'--One of the Three Major Tibetan Folk Dances", see https://fs.aipingxiang.com/115290.html.

Resplendent and Magnificent Temple", "Good Luck and Wealth Descend Here", and "Harvest, Ah, Harvest", etc. The local religious circles and the elderly mostly prefer this tune. When engaging in this kind of dance, which is generally slow, steady, simple and unsophisticated, and solemn, you are required to sing special lyrics, and cannot make any alteration. In contrast, for the latter, its lyric content and dancing postures are relatively flexible, mostly reflecting productive labor, the development of farming and animal husbandry production, and business and trade activities, such as "The Northern Grassland", "Three Colors Gathering in A White Porcelain Bowl", and "Above the Golden Plain", etc., which is very popular among young people.

When performing Yanjing Guozhuang Dance, people are very particular about the costume. Men should put on fox fur hats, or braid their hair with red silk threads, and string gold, silver, and elephant bone rings embedded with coral and emeralds on the braids around their heads. And their attire may include a woolen or brocade Tibetan robe trimmed with otter fur or leopard fur, a pair of oversized specially-made trousers, and thick soled Tibetan boots, with a pair of silver ornaments hanging on both sides of the hip; thus, they appear to be mighty and handsome. Women comb their hair into small braids and twine them with colorful silk threads, so that they are all connected, draping over the back. As for the apparel and accessories, they, dressed in a brightly colored folding skirt and a top quality buttonless robe, may wear gold and silver earrings in the ears, a silk ribbon and a colored apron around the waist, as well as jewelry, necklaces, and gold and silver Buddha boxes around the neck, with a pair of decorative chain boxes connected by silver bells hanging on both sides of the hip; thus they look as pretty as a flower and jade.

On such joyous occasions as festivals, celebrations, and weddings, men and women, both standing in a horseshoe, form a circle by holding each other's hands in the square or courtyard. When one group starts singing, another joins in, usually led by men and accompanied by women. The sound of singing is loud and clear, hence very penetrating. And the dancers, "swinging arms while stepping forward with shaking knees", walk around the circle. When a section of lyrics is finished, everyone will call out "Ya" in unison, and then speed up. They swing their arms up and down, twist their waists sideways, and jump up with outward crossing steps, now singing, now dancing. And they also change their movements through running and jumping. Among them, the males are characterized by a relatively larger movement range, stretching their arms like an eagle in hovering flight, while the females' movement range is smaller, whirling around with their tiptoes touching the ground, simply like a phoenix flapping wings in the air. Hence this dance has such features as fitness, brightness, and liveliness, etc. of sports. The formation moves clockwise, with circles varying in size and occasionally changing into the pattern of "a dragon swaying its tail". As one of the folk dances, Guozhuang is not only self-entertaining, but also highly technical.

The Tibetan Guozhuang Dance in Yanjing artistically shows us the life or conditions of their primitive ancestors. On May 20, 2006, Guozhuang Dance was approved by the State Council of the People's Republic of China to be included in the first batch of national intangible cultural heritage list.

c. Reba Dance

Reba Dance is a transliteration of Tibetan, which means "tassels" (braided tassels). As a type of drum bell dance, it's a comprehensive performing art integrating singing and dancing, poetic drama, mime, rap, acrobatics, playlet, and *qigong* (a system of deep breathing exercises), etc. Regarding its origin, there is an opinion that: "Reba Dance, originating in the 11th century AD, was created by Milarepa, the great Tibetan Buddhist".[1] To the present day, it has already developed into many styles and schools, and those spread in the Qamdo region of Tibet are characterized by such connotations as "passion, roughness, and boldness" and "highly difficult" skills.

1 Wang Xinghuai, Yang Wenhao, and Sun Jing. "A Study on the Inheritance Mode and Cultural Characteristics of 'Reba Dance'". Journal of Xizang Minzu University (Philosophy and Social Sciences Edition), 2016,37(01).

In Yanjing, there is a strict dress code for dancers when performing Reba Dance. Female dancers should wear a "Bajiao" (silver triangular headdress), and a yellow long sleeveless jacket, plus a red shirt. The two main colors, yellow and red, express a sense of nobility, magnificence, elegance, decency, and religious sacredness, implying the incarnation of Tara. On their feet is a pair of "Rezamarang"[1], which, generally hand-stitched from cowhide or sheepskin, are characteristic of warmth retention, breathability, and comfort, etc. Among the people, there are craftsmen specializing in making such boots. Male dancers should wear a pair of bicolored trousers, i.e., with red and yellow, and black and white on the front and back respectively, specially made for dancing Reba, representing political and religious powers, as well as good and evil. What is more, they'd wear a white shirt, with a black "Chuba"[2] tied around the waists, indicating the suppression of evil spirits. As for the Reba belt, it is a cord twisted from black and white yak hair.

There are lots of props for Reba Dance; for example, drums and tray-shaped bells[3], as well as "Ya'an" (yak tails), are all unique ones with exceptional features. The Reba performance can be divided into four parts. The first part is self-introduction and singing congratulations. The second part is the prelude dance. In this part, male dancers, around whose waists are tied with a multicolored cord twisted from black and white yak hair, each hanging colorful tassels at the lower end, hold a bell in the right hand, and a red yak tail (in token of warding off evil spirits and avoiding disasters) in the left hand; female dancers are dressed in gorgeous Tibetan costume, a drum in the left hand and a bow-shaped drumstick in the right hand. All men and women cooperate well with each other. The third part is the bell and drum dance, which has a total of 5 types of drumbeats, i.e., single beat, triple beats, rim beat, sextuple beats, and nonuple beats, dancing while beating the drums. Every time a drum dance is completed, a wonderful performance of acrobatics and martial arts will be put on; every time two drum dances are completed, there will be a section of witty and humorous performance of rap and spoken parts. The fourth part is the Reba Xianzi dance; while singing and dancing in a warm, passionate, and unrestrained

1 Transliteration, female Tibetan boots for Reba Dance.

2 A Tibetan long robe, whose front buttons on the right.

3 They are generally crafted with great care of precious brass selected by specialized coppersmiths.

manner, the performers receive honoraria and donations from the crowd, and end the performance with their humorous and witty dialogues.

As one of the three major dances in Tibetan areas, Reba Dance in Yanjing has absorbed the essence of rich religious culture, ancient Zhang-zhung culture, and ethnic folk art, etc., coupled with long-term integration and exploration, thus forming the splendid culture and art of the Tibetan people. On November 11, 2014, it was approved by the State Council of the People's Republic of China to be included in the third batch of national intangible cultural heritage list.

d. "Zani" Qin Dance

Zani (the general name of stringed instruments in Tibetan), is a bowed string instrument of chordophones, also called oxhorn *qin*, oxhorn *hu* (a two-stringed bowed instrument), and horsetail *hu*. With a long history and characterized by the shape and structure of primitive simplicity, it's quite different from the common stringed instruments in the inland region. The little-known "Zani" *Qin* Dance, as a folk group dance accompanied by Zani *qin* and integrating dancing, singing, and strumming, has a unique form of expression, and is thus very popular in the villages of Yanjing and the surrounding areas in Markam County.

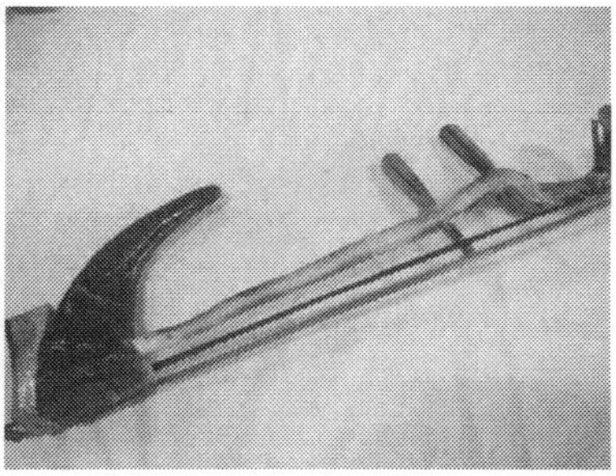

A "Zani" *qin* is about 80 centimeters long; its body and box are made of a whole walnut tree trunk. Generally, three strings are fixed on it (it's also said that some may be fixed with five strings), and there are three pegs for fixing the strings, one on the left and two on the right. What's more, there are three small sound holes in the box.

When strumming, the "Zani" *qin* is hung diagonally in front of the waist, and its neck, placed between the thumb and the index finger of the left hand, is firmly held with the former; the other four fingers either tap, or vibrate, or press it according to different melodies, and the plectrum is sandwiched between the right thumb and index finger to produce sound by plucking. Its range is not very wide, but the tone quality is rich, soft, and crisp. There are many tunes for "Zani" *qin*; some have both melodies and lyrics, such as "Let's Go and Enjoy the Sight", and "Spreading Blessings", etc., while others have only melodies but no lyric, such as "Ah! The Rabbits Are Bouncing", and "Golden Wild Geese", etc.

It's said that "Zani" Qin Dance, with a history of ten generations in the local area, is usually performed during joyous festivals. On such occasions, men and women, assembling into formations separately, dance in circles. Generally, each man holds a "Zani" *qin*, singing and dancing while strumming. Women, just singing and dancing joyously, do not use "Zani" *qin*, but are more particular about their attire; they usually wear a long-sleeved pink shirt inside and a short-sleeved white shirt outside, thus creating a color contrast between red and white. While dancing, the men will sing a section, and the women will repeat it; for those parts without lyrics, there will be only the "Zani" *qin* accompaniment. The movements of this dance are wide in range, and very open, with lots of variations. Generally, "Zani" Qin Dance has three parts: adagio, moderato, and allegro. Through these three parts, the whole dance is pushed to the climax, one wave higher than another. Each dance consists of at least five combined movements. As for such movements as "tiptoes touching the ground left, right, and forward before bouncing", "T-shaped outward crossing steps", "zigzagging kick steps", "tiptoes touching the ground left and right, with a step out", and "bouncing with a step out while turning round", not only don't they exist in the Tibetan Zhuo Dance[1] and Duixie[2], etc., but also they are rarely seen in the

---

1 Zhuo Dance, called "waist drum dance" in Tibetan, is generally performed at the beginning and end of major festival activities. "Zhuo" means auspicious.

2 The literal translation of "Dui" in Tibetan is "upper" or "west", and "Xie" means "song and dance"; hence, Duixie is the song and dance of the upper or the western region of Tibet. As it has gradually evolved into a form of foot tapping, and is most popular in Lhasa, people also call it "Lhasa Tap dance" or "Tap dance".

movements of the popular Xianzi Dance in the local area.[1]

However, due to the difficulty of dancing, singing, and strumming in "Zani" Qin Dance, and the fact that the villages where it's very popular are located on the relatively remote banks of the Lantsang River, not only cannot it be popularized in the vast area, but also it has not attracted the attention of the art circles.

Anyway, the Markam area, of which Yanjing is a part, is known as the "ocean of songs and dances"; in addition to the above-mentioned folk songs and dances, there are also rich folk love songs, mountain songs, pastoral songs, wine songs, labor songs (*haozi*, i.e., a work song sung to synchronize movements, with one person leading.), Three-stringed Dance, Pipa Dance, Zhega Dance[2], Zanian Dance[3], and ballads, etc. Popular in certain regions and populations, they are all miraculous among the folk songs and dances in Eastern Tibet.

4) Specialty Foods and Drinks

Yanjing, located in the ethnic corridor, has been gathering a lot of delicacies here for thousands of years. The various famous dishes evolved from salt are characterized by unique flavors and lingering aftertastes. And the following are only some of the most distinctive and representative ones with local features.

a. Pig-killing Feast

Every August or September of the Tibetan calendar, i.e., around October of the solar calendar, Yanjing people are expecting the bountiful harvest of two crops a year; they busy themselves in the fields, and begin to prepare for the autumn harvest. After a month-long tiring and joyful farming season, they can finally relax a little bit, visiting relatives and dropping around. They, while enjoying a happy life to the fullest, attach great importance to gatherings of family members and friends, and the

---

1 Qamdo Regional Office for Rescuing Ethnic Cultural Heritage. "Introduction of 'Zani' Qin Dance in Markam County, Qamdo District". *Tibetan Art Studies*, 1991(04).

2 "Zhega", which can be translated as "white fruit" or "auspicious fruit", consists of two parts: speaking and singing. Generally, speaking is followed by singing, and interspersed with some simple dancing.

3 "Zannian", which means "pleasant sound" in Tibetan, is a plucked string instrument with a long history. It can be classified into six-stringed ones, eight-stringed ones, sixteen-stringed ones, and so on, of which the six-stringed is the most common, widely spread throughout Tibet.

Pig-killing Feast is exactly a unique and vital activity for entertaining guests as well as gathering of family and friends, which already has a history of several hundred years in Yanjing.

Speaking of the Pig-killing Feast, we must first recount a period of history of Yanjing. More than 500 years ago, the Naxi people in Lijiang, Yunnan Province, owing to its rapidly increasing power, began to emigrate and expand to the surrounding areas (the Naxi nationality in the Yanjing area today possibly immigrated from that period). Since then, the fusion of the Naxi nationality and the local Tibetan ethnic group has formed a rich dietary culture, and the Pig-killing Feast has been popular ever since. Due to its typical dry-hot valley climate, the Yanjing area is a very suitable place for raising pigs. And over hundreds of years, the Pig-killing Feast has already integrated with the original Yanjing dietary culture, forming a unique diet custom. Like in the Napa Festival, pork, as an important source of meat for people in Yanjing, is essential to Pig-killing Feast. To prepare for the main course, a cook must choose the best pork and haslet, and thus every household here will raise a pig every year to make dishes for this feast. On such an occasion, dishes prepared by people are both very rich and delicious, from soup to hot dishes, cold dishes, and blood sausages, etc. On this day, people will invite their relatives and friends to their homes; in addition to eating and drinking, the main purpose is to get together, shoot the breeze, and promote the relationships.

Pig-killing Feast, particularly popular in the Yanjing area, is not only a great cultural tradition, but also a very meaningful way to enhance people's emotions. Nowadays, with the rapid development of society, more and more people have become financially independent, and their feelings for relatives and friends are gradually fading, too. As a traditional cultural form, the Pig-killing Feast constantly reminds Yanjing people to strengthen the cultivation of affection between family members and friends, respect the aged and take good care of children, and live in harmony and amity with each other. It is a virtue that needs to be extolled and eulogized, and also requires generation after generation of Yanjing people to inherit, develop, and carry it forward.

b. Pipa Meat

Pipa meat, also known as "Paguo meat" (transliteration), is a favorite delicacy of Yanjing people, and it is also the main ingredient of Jiajia noodles. In the early autumn of each year, people start making pipa meat in preparation for the new year, which has been a Naxi tradition. The method is as follows: After a Tibetan pig is butchered, they bleed its complete body, and remove the hair, bones, and internal organs. Then, with mixed coarse salt, Sichuan pepper, liquor, and other raw materials produced in Yanjing, they evenly smear the inner wall and outer skin of the pork while kneading it continuously. And then, they sew all cuts tightly with strings; and there must be no gap left during this process. After that, they smear a mixture of plant ash and water on the seam to isolate the inside of the pork from the air. As the main component of plant ash is potassium carbonate, a kind of sylvine, it plays a role in the dehydration of the pork. After sewing is completed, they place it on a flat wooden board in the attic below the roof, cover it with another wooden board, and put a large stone on top of them. When the water has dried, they hang it in a dry and cool place, so that it will be air-dried and dehydrated naturally. After more than 3 months, the delicious "Paguo meat" is done.

According to *A New Account of Southern Yunnan*, "(it is) thin and greasy, analogous to bright amber, and shaped like a pipa." Hence the locals call it "pipa meat". Its salting process has a tradition for thousands of years. Once made, it can be preserved for a long time, and will not go bad within three or four years as long as it is not cut open. Every time they want to eat it, they cut off a piece with a knife, and boil it in water before making a stir-fried dish or soup. Transparent in color, it is sweet-smelling and tasty, with a unique flavor, and has thus become a top-notch delicacy for Yanjing residents to entertain guests and their families.

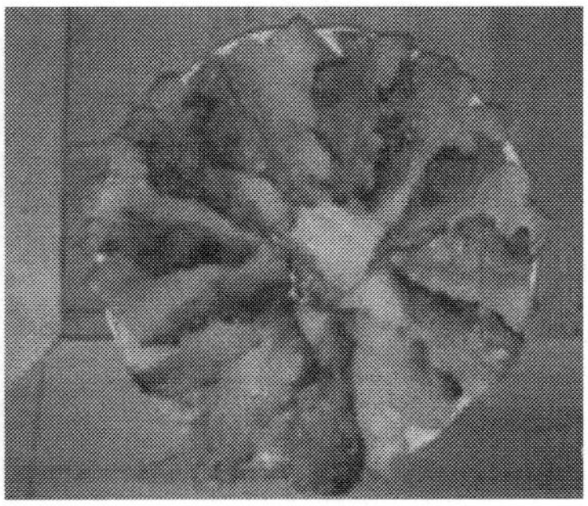

Pipa meat

c. Jiajia Noodles

As a favorite staple food of people in the Yanjing area, made with pure folk craftsmanship, Jiajia noodles is mainly characterized by its chewy and elastic properties, and good taste. It is said that Yanjing Jiajia noodles, made of Sichuan noodles, Yunnan meat, and three local ingredients with great care, has a history of nearly a thousand years, and thus can be considered an antique. In 1265 AD, Hphags-pa, after accepting the conferment from the Yuan Dynasty, returned to Tibet through the Yanjing area, where he was offered "Jiajia noodles". To show his reverence to the guru, and also to suit the latter's appetite, the cook chose to serve the noodles in small bowls. Hphags-pa, having eaten several bowls in succession, praised its delicious taste profusely. Since then, Yanjing Jiajia noodles has been eaten in such a way of "Jiajia" (adding and adding) with small bowls, until the guests are stuffed and satisfied. And this way of eating noodles, which may be called the one and only in the world, has been passed down.

In the process of making noodles, people knead dough with a mixture of plant ash and water in addition to some eggs, which is a unique formula in Yanjing. Due to the alkalinity of the plant ash water, and the uniqueness of the locally produced flour, Yanjing noodles tastes soft, chewy, and elastic. When preparing it, people first cool the boiled noodles off by immersing them in cold water, and then pick them into small bowls, reheat them with thick gravy made from Yanjing cured meat, add some minced cured meat, and sprinkle them with chopped green onion, so that the noodles will become more delicious and flavorful.

Yanjing Jiajia noodles, as a fusion of ethnic cultures and the creation of local people, combined with the innovation of travelling merchants on the ancient road, has met the requirements of travelers along the way; thus, it can be said to be a genuine innovative cultural product. Due to this characteristic of Jiajia noodles, Yanjing people are accustomed to prepare it for weddings, festivals, and receptions, so as to increase the jubilation and joy, the festive atmosphere, and the enthusiasm for hospitality. Whenever someone gets married, a holiday is celebrated, or guests come over, Jiajia noodles will be an indispensable festive food for Yanjing people.

Today, it has become a must-have for travelers coming and going in great number through this area. As a delicacy on the Ancient Tea-Horse Road, it was rated as an intangible cultural heritage of the autonomous region in 2010.

Yanjing noodles (Jiajia noodles)

d. Yanjing Red Wine

The red wine is a "local specialty" in Yanjing Naxi Ethnic Township, brewed by traditional natural fermentation method with grapes as raw materials. The average frost-free period here is more than 340 days, with abundant sunlight, strong ultraviolet radiation, and dry air; the temperature difference between day and night is about 15 °C, and the annual rainfall is less than 300mm, which belongs to the typical plateau dry-hot valley climate. The grapes are grown on gravel soil, all irrigated with the melting snow water of the Dameiyong Snow Mountain. This harsh natural environment eliminates pests and molds, ensuring that the whole process of grape cultivation is almost free of pesticides and fertilizers. Hence, it can be said that the wine produced in Yanjing is not only a healthy organic product, but also a resource-based product.

Yanjing red wine has a relatively long brewing history, which can be traced back to 1855 when western Catholicism was introduced to here. French missionaries, in addition to bringing with them viticulture and winemaking techniques, also taught the local nuns how to grow grapes and make wine. Over more than 200 years, the grape variety called Rose Honey (Madame Hua) introduced from Bordeaux, France has been domesticated by natural selection in the harsh climate of the plateau, and has been fully adapted to the local environment. Today, Yanjing people have generally mastered and inherited the viticulture and winemaking techniques, and have been continuing the habit of planting grapes, as well as brewing and drinking wine. The winemaking process here is traditional, simple, and unsophisticated, with the raw materials and brewing techniques coming down in one continuous line with the French "Bordeaux". Due to its pure taste, palatable flavor, and obvious health efficacy, the wine is known far and wide, and even called the "Bordeaux" of Tibet.

In October 2009, as the first batch of high-quality Yanjing wine that met the national standards was produced, people began to make the highest-altitude wine in the world here, setting a precedent for the commercialization of the wine industry in Tibet. Because of the unique geographical environment here, the wine is clear and transparent, with a deep color, a rich aroma, a full flavor, and a lingering aftertaste, etc. Widely known in Tibetan areas, its supply often falls short of demand. On December 11, 2014, the former General Administration of Quality Supervision approved the implementation of geographical indication product protection for "Yanjing Wine". In March 2017, Yanjing Wine won the gold medal in the "8th Asian Wine Quality Competition". In the same year, it won the great honor of "Famous Brand of Tibet Autonomous Region". With the development of Tibetan society, and driven by market demand and its good reputation, Yanjing red wine, including the "Dameiyong" series, has become more and more popular in the Chinese wine market for the past few years. Just in less than one year from 2021 to 2022, it won a number of international and domestic awards. As the plateau red wine was promoted to a higher platform, it has not only brought better economic benefits to the masses, but also allowed more and more people to appreciate the unique customs of Tibet.

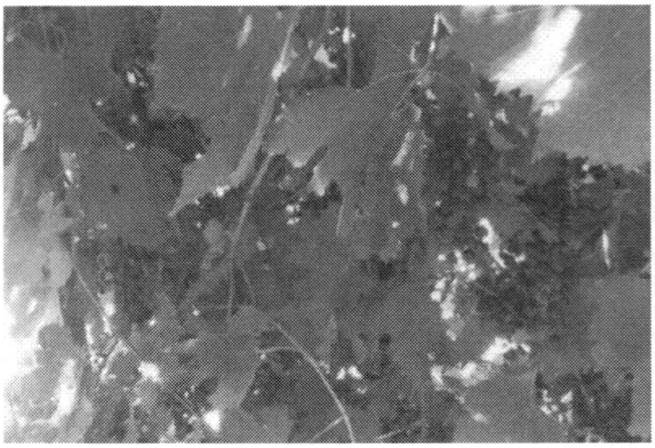

In 2019, a villager of Naxi Ethnic Township is making red wine at home

In addition to the red wine, the highland barley wine is also a major specialty of Yanjing. The highland barley wine, called "Qiang" in Tibetan, which is orange-yellow in color, sweet and refreshing, with a lingering aftertaste and low alcohol content, is quite analogous to beer. After drinking, there's neither headache nor dryness in the mouth, and the drinker sobers up quickly; hence it is always regarded as a unique beverage.

The production process of the highland barley wine is very distinctive. First, it is washed clean and boiled; when it cools off slightly, add some distiller's yeast, and seal it in a clay pot or wooden barrel for fermentation; two or three days later, add some water and cover the lid, and after one or two days it will be done. In Yanjing, all people, regardless of age and sex, like drinking highland barley wine; it is essential for New Year's Day or other festivals, weddings, the birth of children, and greeting and seeing off relatives and friends, etc. When drinking highland barley wine, the local people stress "three sips before draining a whole cup"; that is, first take one sip, and fill the cup up; then take another sip, and fill it up again; and then take the third sip, and fill it up over again before draining. At an ordinary banquet, the host and hostess will make a toast while singing wine songs. At a grand banquet, there are dedicated toast girls, who, dressed in gorgeous costumes and singing songs, take turns to urge the guests to drink until they get drunk. Today, the local highland barley wine is mainly produced by Yanjing Distillery. Most tourists, when passing by here, would drink a few cups before taking one bottle with them.

Of course, the folk culture related to the production activities of salt in Yanjing includes far more than these things mentioned above. But due to limited space, time, and energy, etc., we cannot present any more materials here, such as sacrifices, funerals, taboos, and etiquette, etc. Yanjing culture is a three-dimensional cultural system, which consists of not only local culture but also external culture; not only up-class culture but also mass culture; not only religious culture but also secular culture. Too many legends, historical facts, ethnic groups, beliefs, and modes of life are sealed, stirred, and fermented in Yanjing, undergoing a subtle chemical reaction full of conflicts and fusions. And the fusion of multiple ethnic groups, as well as the harmonious coexistence of various religions, are the most mysterious and impressive characteristics here. The diversity and peculiarity of Yanjing salt culture have become a precious treasure of the local culture.

The unique regional folk culture created by Yanjing people is rooted in all aspects of social life here, its connotations being extremely rich and voluminous, and its accumulation being very profound and dignified. In the long course of historical development, they have been continuously passed down, refreshing the memory of ethnic bloodlines, and reflecting the unique regional culture and regional identity of Yanjing; thus, it has a strong and far-reaching cohesion, affinity and influence in the local area.

CHAPTER FIVE

The Predicament and Survival Crisis of the Salt Industry in Yanjing

As a millennium-old salt capital on the Ancient Tea-Horse Road, Yanjing still retains traditional salt-making techniques. The salterns here have thus become the "living sites" of traditional sun-drying technology in China, as well as cultural relics with strong aesthetic, historical, and artistic characteristics formed in human productive activities. It is a "living fossil" for the study of traditional salt-making technology, where traditional skills developed in the process of the survival and reproduction of such ethnic minorities as Tibetans and Nakhi, etc. This type of traditional salt-making site, with scarcity value, visual value (including authenticity), scientific research value and economic (tourism development) value, vividly demonstrates the scene of ethnic minorities in the Qinghai-Tibet Plateau adapting to the environment and using resources to maintain their livelihoods.[1]

However, with the progress of society and the entry of various modern productive activities and life styles, many problems have arisen in the development of the local salt industry in recent years; the salt economy has begun to shrink, and the ancient salt pans are facing the authenticity challenge and survival crisis. By drawing on the experience of salt industry development in other parts of China and combining relevant investigations, some researchers believe that the reasons for the current salt industry development dilemma of Yanjing are multifaceted, such as: the

1 Li Hechun, Luo Bo. Research on the Multiple Values, Inheritance, and Protection Strategies of Traditional Salt-Drying Techniques on the Qinghai-Tibet Plateau: A Case Study of Salterns in the Upper Reaches of the Lantsang River [J]. Qinghai Social Sciences, 2020 (06).

scattered salt production, the slow logistics economy construction, the contradiction between production and marketing intensified by the external system of salt industry control regulation, the excessive production costs, and the underdeveloped salt culture tourism, etc.[1] But overall, there are no more than three aspects: internal factors, external factors, and human factors.

1. Lack of Self-development Impetus

Mao Zedong said, "The fundamental cause of the development of things lies not in the outside of them but in the inside of them, in the contradictions within them"[2]. Yanjing, Markam County, located in the southeast of the Tibet Autonomous Region, is on the southwestern border of the motherland. Historically, due to long-term closure and conservatism, the infrastructure construction and economic development have been relatively backward, and people today are still obstinately adhering to the ancient traditional method of salt production, i.e., "building salt pans with wood and carrying brine for evaporation". Compared with modern large-scale industrial salt production methods, its technological, technical and economical shortcomings are very obvious. Especially in recent years, with the gradual depletion of brine resources, the shrinking market, the weak competition mechanism, and other problems, the development of Yanjing has gradually lost its vitality and vigor. The author believes that the main reasons could be summarized as follows.

1) Harsh Production Environment

Through field investigation of Yanjing, it can be seen that along the 4,900-meter-long Lantsang River, the section in Yanjing, Markam County, which completely passes through the Hengduan Mountains, is the narrowest and steepest part, and also the most difficult area for humans to survive. With high mountains and deep valleys, the complex topography and landforms have greatly restricted the scale and technology of salt making here. Thus, local people could only carry out their production by constantly adapting to the harsh natural geographical conditions.

1 Chen Tianpei, Yang Jie. Strategies for Predicament of Zigong Salt Industry in Development [J]. *Journal of Chongqing University of Arts and Sciences (Natural Science Edition)*. 2008,27 (06).

2 See "On Contradiction", Chapter 18 of Vol. 1 in *Selected Works of Mao Zedong.*

First of all, the terrain of Yanjing being rather rugged, except for in Jiada Village, it's almost impossible to find any open and flat place for evaporating brine in the sun; so salt makers could mostly put up patches salt pans along the hillsides. As is well known, the method of evaporating brine in the sun requires a large number of salt pans and covers a wide area. The data in Table 5-1 is a preliminary statistical result, but objectively reflects the land area occupied by salt pans. This data only calculates the salt pan area, in addition to the brine wells, brine pools, roads, etc., which should be increased by 1/3 on this basis.[1] Yanjing, located by the river, with steep mountains on both sides, its arable land area is less than half a *mu* per capita. For Jiada Village on the west bank, it's no problem; but for Nakhi Village and Upper Yanjing Village on the east bank, it's obviously impossible to build any sun-drying ground like in the coastal areas, due to the steep terrain and large slopes there. According to Zheng Zhenman, who once analyzed the process of changing the salt production method from boiling to sun-drying in Jinmen salterns in Fujian Province during the Yuan Dynasty: "After switching from boiling to sun-drying method, a large area of land must be requisitioned, which has led to profound changes in the ecological environment and industrial structure."[2] Therefore, in ancient times, Yanjing salt makers, based on the local topography, creatively built salt pans with a terrace-like structure, tier upon tier. Only in this way could the maximum number of salt pans be built in a limited space. As in the case of limited land area in Yanjing, salt pans might only be built on the cliffs by the river, so as to overcome the limitation of terrain and land shortage, while ensuring that each salt pan would crystallize into salt under the action of sunlight and wind. On the other hand, the reason why the densely packed sun-drying wooden frames are mostly within the range of nearly 2 kilometers from south to north along the Lantsang River is actually related to the location of the brine exposure. The brine in Yanjing is natural and exposed on the river side, so the brine wells are mostly drilled along the riversides, and the salt pans are built there, too, naturally in order to reduce the conveying distance between the brine and

---

1 Li Hechun. *Impetus and Shackles: Salt and Salt Taxes in the Lantsang River Canyon* [M]. Guangzhou: Sun Yat-sen University Press, 2016:273.

2 Ibid.

the drying sites.[1] However, due to the influence of such narrow canyon spaces, the transportation "bottleneck" remains prominent, and salt production has long been in a closed and conservative state.

表 10－5　各阶段盐井盐田的面积

时间	盐田数（块）	面积①（平方米）	面积（亩）
清宣统元年（1909 年）	2763②	40063.5	60.09
民国二十年（1931 年）	3966③	57507	86.26
民国三十年（1941 年）	2559④	37105.5	55.66
2000 年	2665⑤	38642.5	57.96
2007/2009 年	3000⑥	43500	65.25
2013 年	3650⑦	52925	79.39

Secondly, the well salt production activities in Yanjing area can only depend on the weather, relying too much on wind and sunlight. Historically, the Tibetans have long had the tradition and technology of salt production, such as the salt making in Tubo in the Tang Dynasty, or the so-called "charcoal washing method". According to Fan Chuo's *Yunnan Chronicles* of the Tang Dynasty, the earliest literature on Tibetan salt making found so far, the method was to soak firewood in brine (salty water), burn the wood into charcoal, and then scrape crystallized salt from the charcoal. However, the well salt production in Yanjing Town is obviously different from that in the above record.[2] And the salt making process here is nothing like the boiling method involving deep drilling of brine wells and drawing brine water in Yunnan, Sichuan, and inland China, either, but similar to the sun-drying method in coastal areas. "...Brine is drawn and dried in the sunslight, relying entirely on the power of wind and sun to crystallize into salt." Since the brine only precipitates crystalline salt after being exposed to the sun, salt pans are just sun-drying platforms in accordance with the local environment, and the drying of salt mainly depends

1 Li Zhongdong, Jiang Hai. Yanjing, Tibet: The Legend of Sunshine and Wind [J]. *Resources and Habitant Environment*, 2015(07).

2 Zhu Xia, Li Xiaocen. Well Salt Production in Yanjing Town, Markam County, Tibet Autonomous Region [J]. China Tibetology, 2007(03).

on natural wind and sunlight; thus, the production process appears to be primitive. Although it has been mentioned earlier that Yanjing belongs to the plateau temperate semi-humid monsoon climate, with sufficient sunshine, high temperature, and strong wind, which are all favorable conditions for evaporating brine, it mainly refers to the peak production season from March to May each year. During this period, the wind being strong, the sunlight abundant, the brine evaporates quickly, and in turn, the salt particles precipitate quickly. At this time, the brine is the thickest, and the salt is the whitest and best. From July to September each year, it's the worrying rainy season; due to the lack of sunlight, it would be difficult to produce salt. Therefore, the salt production activities had to be stopped, and there were no more salt makers working in the salt pans. In this kind of seasonal production, there is a break in the continuity of time, and uncontrollable factors have a great impact on it.[1] To solve this problem, salt makers have attempted to adopt other methods by exerting their own subjective initiative. Besides, in the late Qing Dynasty, the government also planned to apply another high-yield salt-making technique--the method of boiling. And through experiments, they found that the refined salt was excellent in color and taste. However, due to the lack of local resources, i.e., the absence of coal mines, and few trees because of the high altitude, not to mention that it was very inconvenient to use mountain firewood only, the cost increased by more than 20 times. So, they had to give up and adopt the original method again. It can be seen that due to the constraints of local resources, sun-drying in salt pans was also a helpless move.

Thirdly, the salt production activities in Yanjing are greatly affected by the fluctuation of river water. According to "Unusual Natural Disasters" in *Important Records of Yanjing County*: "...The local salt-making is far from ingenious; the brine wells are so close to the riverbanks that once the river water rises, the wells would be flooded, and the salt production activities have to stop. Therefore, the amount of salt produced depends entirely on the rise and fall as well as the time of flood, and there is no fixed quantity." And *Historical Materials of Sichuan-Yunnan Border Affairs Archives in the Late Qing Dynasty* also recorded: "But every summer, the water of the Lantsang River rises vigorously, and the wellheads are completely submerged.

1 Li Hechun. *Impetus and Shackles: Salt and Salt Taxes in the Lantsang River Canyon* [M]. Guangzhou: Sun Yat-sen University Press, 2016:272.

The salt makers have to wait for the water to recede in autumn and winter before resuming their business..."[1] Since the brine wells are all on the riverbed by the banks, every year when the rainy season comes, the roaring waves in the Lantsang River soar by nearly 10 meters, and most of the brine wells on both sides will suffer. Although some salt makers built cylindrical enclosing walls with stones and mud on the mouths of some high-yield wells, or covered them with wooden boards, so as to prevent the rising river from flooding in, and ensure that production can be carried out in summer, this is not absolutely safe. Because even if the brine wells are not flooded or washed away, there will still be turbid river water and sediment entering in, resulting in the possible reddish color of the salt, which is slightly inferior in quality. But if the brine wells are unfortunately washed away, they can only rely on the brine in the storage tank to maintain salt production then. Of course, when encountering excessive floods, salt pans and brine storage tanks built on higher grounds are also vulnerable to damages by the river water, and there are also risks of being washed away or filled with muddy water, making it difficult to continue production. And this period is also busy with farming, so there will be only a small amount of salt production or even no production at all. After the rainy season, when the river recedes, people will rebuild salt wells and salt pans that have been washed away, or spend two or three days clearing the sediment in the wells without any complaint, waiting for a bumper harvest in the coming year...

In addition, such infrastructure items as brine wells, salt pans and brine storage tanks, which are also subject to wind and rain erosion, may often require a considerable amount of manpower and material resources for maintenance. The brine storage tanks, for example, need to be renovated every year; apart from regular inspections, it is also necessary to prevent water leakage and clean up all kinds of waste deposited at the bottom. The salt pans need to be repaired every one or two years, too. The wood used for salt shelves comes from local pine trees over 30 years old, and the stones and sand are also drawn on local resources. Moreover, due to the active geological structure of the Yanjing area, various salt production infrastructure items are at risk of being damaged by earthquakes, landslides, and debris flows, etc.

---

1 See *Historical Materials of Sichuan-Yunnan Border Affairs Archives in the Late Qing Dynasty*, Zhonghua Book Company, 1989 edition.

For example, the above mentioned 6.75-magnitude earthquake in 1921 not only caused a large number of collapsed houses and casualties, but also damaged some brine tanks and salterns, and even caused the saline springs to move forward. In the spring of the 11th year of the Republic of China (1922), "an earthquake lasted for about one month...and more than half of the salt pans on both sides of the Lantsang River collapsed"; in 2001, the salt pans at four places in Nakhi Village were washed away by mudslides. Their numbers were constantly changing, with some even abandoned due to destruction or unuse for a long time.

Through the field investigation in Yanjing, it can be seen that the complex topography and landforms make it very unlikely to carry out large-scale salt making here, for its natural environment has limited local people's salt production activities. Due to the excessive constraints of the natural environment, this salt production method has a fragile foundation and low ability to resist natural disasters, and is thus completely dependent on the weather conditions, which is very passive. But most seriously, the planned Gushui Hydropower Station on the Lantsang River will leave these salt wells and salt pans permanently submerged. In the near future, if they cannot find a way out of this dependence, the prospects of local salt production are worrying, indeed.

2) Relatively Backward Technology, Tools and Equipment

When people acquire salt resources, they will adopt different salt-making technologies in accordance with different salt types. In the long development and evolution process of well salt production in China, the scientific and technological achievements and major inventions, which embodied the painstaking efforts and wisdom of the laboring people, have greatly enriched the content of China and the world's history of science and technology. Through a comparative study of salt making technology across time and space, it's not difficult to find that there are obvious differences between Yanjing salt production technology and other salt production methods, which is related not only to the natural geographical environment, but also to the development of social technology.[1]

---

1 Wu Chengli. Investigation and Study on Salt Making technology in Yanjing area of Tibet [J]. Journal of Guangdong Polytechnic Normal University (Social Sciences), 2010(5).

In terms of the time span of salt making technology, whether it is well salt, lake salt, or sea salt, people have relied on manpower, animal power and natural force for a considerable period of history. However, with the development of social economy and technology, most of their collection and production skills have undergone great changes, transitioned from traditional non-mechanical manual production to modern industrial mechanized operation. In contrast, there have been few innovations in Yanjing salt making skill, which still retains manual brine drawing and simple sun-drying crystallization methods. Without meticulous craftsmanship, automatic machinery, and complex processing separation, it is very primitive, or even protogenetic. As a result, the problem of relatively backward infrastructure and salt production technical equipment has become increasingly prominent. This is mainly reflected in the following aspects.

a. The Well Sinking Skill

In the process of well salt production, well sinking skills are comparatively more complicated and technical. And the level of well sinking skills has a significant impact on the salt production area. After a simple comparison with traditional and modern lake salt, sea salt, and Zigong well salt production technology, it can be found that the current well sinking skill in Yanjing is still relatively backward, even basically the same as the early modes in various well salt production areas in Sichuan. The brine wells here are wide ones and large ones. Adopting the method by which ordinary people sank wells to fetch water at that time, it's usually not too deep, and has a large diameter of about 2~3 meters, hence called "large-mouth shallow well".

Lin Yuanxiong et al. (1987) divided the Chinese history of salt well sinking into three stages: manual sinking stage (250 BC-1049 AD), simple mechanical drilling stage (1050-1901), and power machinery drilling stage (1902-modern times).[1] Drawing brine from large mouth shallow wells was just the first stage in the development of Chinese salt well sinking skills. As large mouth shallow wells, apart from low productivity and rough sinking techniques, might easily collapse and cause injuries, the salt production with them gradually headed toward a decline during the

1 Xu Baozheng, Wang Liandi, Yang Jinhu. Development of the Salt Production Technology in China and Europe During 15th-19th Centuries [J]. *Geology of Chemical Minerals*, 1994(03).

reign of Emperor Renzong of the Song Dynasty.

During the Qingli period of the Northern Song Dynasty (1041-1048), the Zhuo-tube well by cable drilling came out, which marked the development of China's salt well sinking technology from large mouth wells to small mouth wells, and was also the earliest small-caliber drilling technology invented by human beings. It's a salt well with an upright and thick bamboo tube to draw brine, whose caliber is only the size of a bamboo tube, and it can drill up to tens of *zhang*[1] deep into the ground. Su Shi wrote in his *On Sichuan Salt*: "Since the periods of Qingli and Huangyou, 'Zhuo-tube' wells have emerged in Sichuan. A hole dug with a circular knife, which is as big as a bowl, could be tens of *zhang* deep...Then the saline spring will gush out automatically." In *Heavenly Creations*, it's also recorded that this technology, by adopting the "cable drilling method", can make small-caliber deep wells with a circular blade and a "wooden shelf", i.e., a foot pedal lever like the rice pounding tool...The success of Zhuo-tube well not only pioneered human mechanical drilling, but also brought about a breakthrough and innovation for brine drawing technology in salt well production. Since the Ming and Qing dynasties, with the development of drilling technology, many deep wells have emerged one after another, increasing from tens of meters to one or two hundred meters deep in the Song Dynasty (Zhuo-tube wells) to hundreds of and even more than 1,000 meters. Unfortunately, this Zhuo-tube well technique, considered by Joseph Needham, the British expert on the history of science, to "have pioneered mechanical drilling", was never borrowed by the salt makers in Yanjing area, owing to various subjective and objective factors, depite the fact that it has been used to this day and is very common in neighboring Sichuan salterns for such advantages as covering only a small area, low cost, and high efficiency, etc.

Because drilling and production tools, as well as the speed and depth of drilling, are important indicators of the well salt production level, the salt production in Yanjing Township appeara to be quite primitive in terms of salt making techniques and sinking technology. For example, compared with salterns in Sichuan and other regions, where the drilling technology has gradually matured and the procedures have become more complete, such as drill bits and drilling tools of different shapes and

1 A unit of length, equivalent to 3 1/3 meters.

functions, and dozens of different types of drill bits used for driving in and expanding holes, the brine wells in Yanjing are mainly made by manual digging with such tools as hoes and pickaxes, etc. What's more, Yanjing salt wells are not only much larger in caliber and relatively shallower in depth, compared with "Zhuo-tube wells" in the Song Dynasty and the small-caliber deep wells in the Qing Dynasty, but also the sinking technology here is much more backward than that of the Pujiang salt wells in Sichuan in the Tang Dynasty. No wonder someone said that the sinking technology in Yanjing Township not only lags behind that of Zhuo-tube wells of the Song Dynasty, but also remains at the level before the Tang Dynasty at most.[1]

In April 2022, when we came here, we found that this was indeed the case; the salt wells were mostly close to the river, with the nearest being only a few tens of centimeters and the farthest only over 10 meters. The lower wellheads were almost level with the river surface, while the higher ones were about 3 meters above. Because the sinking work is easy, salt makers can draw brine simply through renovation and expansion of the spring mouths by the riverbanks. And the time required for well sinking is not long, either, ranging from a dozen days to just one day or a few hours.

b. Brine Extraction Technology

Brine drawing is the first important step in well salt production, while brine drawing skill mainly depends on the type and depth of salt wells, as well as the development degree of brine drawing tools, which have successively passed through the stages of manual brine drawing, animal power brine drawing, mechanical brine drawing, and gas-lift brine drawing. The brine drawing method often varies from place to place and from well to well. In the early stage of well salt production, as the brine drawing technology was not yet mature, people might use well buckets, leather bags, or windlasses, or build frame structures, or go down the wells to bail brine, or carry it out with pails directly. Through field investigation, combined with historical documents, we can see that the brine drawing and processing technology in Yanjing mainly relies on manpower.

Before Ming and Qing dynasties, the power of brine drawing was all humanly.

---

1 Tao Hong, Huang Jian. Research on the Salt Industry in Yanjing Township, Markam County, Tibet [J]. *Salt Industry History Research*, 2002(04).

In the first year of the Baoyou period of Southern Song Dynasty (1253), although "brine was once conveyed by ox carts instead" in Qingjing, it only appeared temporarily in this one place, which was accidental. Since the Ming Dynasty, animal power, mainly oxen, had been introduced into the brine drawing in salt wells, adapting to the requirement of "deep wells with strong force", and thus became popular rapidly, resulting in significant changes in brine drawing power. But in some remote mountainous areas or small salterns, manpower was still used in brine drawing.

Since Ming and Qing dynasties, with the development of sinking technology and the increase of well depths, well calibers and drivepipe inner diameters have both increased. Therefore, the brine drawing machinery has also changed accordingly, such as the wooden derricks, tubes, and traction force, etc., whose progress has been very obvious.[1] The pipes were lengthened and their calibers enlarged; a new type of derrick appeared and gradually increased in height; the winches were replaced by larger ones; animal power, mainly oxen, was widespread in brine drawing, and steam engines began to be used in the late Qing Dynasty, etc. As a result, the salt making capacity of this period was greatly improved, and its brine drawing machinery formed a complete system.[2] According to relevant records, since the 20th year of the Guangxu period (1894), there have already been steam engines in a jointly designed and developed manner. After many setbacks, China's first steam engine for brine drawing was successfully trial-produced in Hanyang, which opened the prelude to the mechanization of bring drawing in Chinese well salt industy. Around the 28th year of the Guangxu period, trials were conducted in the Shixing Well and the Hongzhu Well of the Ziliujing saltern, and finally succeeded, "ten times stronger than the previous ox power in drawing brine"[3].

In the brine extraction process where machines were easy to be used, following steam engines, electric ones also developed rapidly in small and medium-sized wells,

1 Zhang Xuejun, Ran Guangrong. Historical Manuscripts of Sichuan Well Salt in Ming and Qing Dynasties [M]. Chengdu: Sichuan People's Publishing House, 1984:53, 62.

2 Song Liangxi. China's Technology for Brine Extraction in Ming and Qing Dynasties [J]. *China Historical Materials of Science and Technology*, 1990(02).

3 See "An Official Petition by Ouyang Xianrong", Zigong National Comprehensive Archives, Volume No. 475.

opening up the way for brine extraction to realize machinery production. Leshan Changpengrong, Zhengtai, Tongqing, Tianshou and other stoves to lift 94 wells. For example, by 1946, in addition to the extensive use of electric power for brine extraction in Zigong salterns, "such salt maker families as Dexing, Rongcheng, Xieyu, and Rongsheng in Qianwei Saltern planned to drill 68 wells extracted by electric power, and such salt maker families as Pengrong, Zhengtai, Tongqing, and Tianshou in Leshan Saltern planned to drill 94, too."[1] Thanks to steam engines and electric power, machinery brine extraction was basically realized in Zigong, as well as Qianwei and Leshan areas.[2]

Compared with Sichuan and other regions, where the brine extraction technology had constantly changed from the use of human power to the use of animal power and mechanical pulleys, in Yanjing the brine extraction still relied on manpower. As the brine gushed out naturally from salt well bottoms, people would carry it with pails to their brine storage tanks. Cui Kexin, a geologist who had investigated Yanjing during the Republic of China, pointed out: "The first step in salt making is to draw brine, and in the *cayin* (which means salt pans) of Jiada (the present Jiada Village), the brine is drawn with wooden pails from a well and carried to the salt pans..."[3] After going down a flight of narrow stone or wooden steps into the well for brine, as the slope of the steps is close to 90 degrees, a salt maker can only come up by holding onto the steps with one hand while holding the pail tight with the other. This is by no means an easy task, and only one person can go down at a time...

Nowadays, some people also use water pumps to extract brine. Especially after 2007, whether on the east or west coast, diesel engines might be used to pump brine from the wells first to a large brine storage tank. However, due to high power consumption, few people could afford it.

c. Brine Conveyance Technology

Brine conveyance refers to the delivery of extracted brine. In the production of

---

1 See the annual report of Chuankang District Salt Administration Office for the 35th year of the Republic of China.

2 Zhong Changyong. An Overview of the Development of Well Salt Production in Sichuan [J]. *Sichuan Cultural Relics*, 1984(02).

3 Cui Kexin. Investigation of the Geology and Salt Production in Yanjing County [J]. *Xikang Economic Quarterly*, 1944(8).

well salt, whether it is the collection of natural saline springs, or the extraction of underground natural brine, or the solution mining by drilling of rock salt, there's a matter of conveying brine for boiling or sun-drying into salt. Brine conveyance is an indispensable process linking brine extraction and salt making. With the development of drilling technology and the strengthening of brine extraction, the technology of brine conveyance also developed accordingly. However, in Yanjing, brine is still carried by manpower, usually either on the back or on the shoulder, and the skill of brine conveyance is relatively primitive.

In the early days of well salt production, most of the salt stoves were built next to salt wells; thus, the brine conveyance distance was very short, and the skill was very simple. But with the expansion of the brine mining area and the increase of brine conveying volume, the problem of distance between well locations and salt making area became increasingly prominent, resulting in the need to "transfer brine for boiling".[1] Therefore, brine pipelining technology also emerged, gradually forming a complete set of bamboo brine-conveying techniques. The bamboo brine-conveying, based on the principle of pressure difference and communicating vessels, is to transfer the brine with bamboo pipes from a low area to a higher area, over the mountains, and ultimately reach the destination without any pressurization device. For whatever reason, local salt makers did not borrow this technology, but there's no doubt that this brine conveying technology is particularly suitable for the topography of Yanjing. In the development of well salt brine conveying technology, bamboo pipelines became common in many well saltworks, especially the Furong, Qianwei, and Leshan salterns in Sichuan. As the bamboo brine-conveying technology continued to develop, it was the most complete. In the Qing Dynasty, on the basis of summarizing the brine conveyance experience in previous dynasties, people advanced the bamboo brine-conveying technology to a new stage. According to the increasingly complex practical needs, salt makers invented a variety of brine conveying facilities, bringing the application of the fluid mechanics principle and the manufacture of natural communicating vessels to a new level.

Going through a long historical period, especially after the establishment of the

1 Quoted from Wang Shuang's article "Ancient Technology: The Mining of Well Salt", see *Science and Technology Association Channel*.

People's Republic of China, modern brine conveying technology took shape step by step, and gradually replaced the ancient bamboo brine-conveying method. Nowadays, whether natural brine or rock salt brine, it is still divided into two categories: channel conveying and pipeline conveying. Channel brine conveying is used only in very few places, such as Yipinglang Salt Mine in Yunnan Province, while most salt production areas adopt steel pipes or plastic pipes to convey brine. The original bamboo brine-conveying remains only between a very small number of small salt wells and stoves. In modern pipeline brine conveying, new materials, new equipment, and new techniques are widely being used; the brine conveying technology has also developed to a new stage, achieving pressurized brine conveyance.[1]

However, salt makers in Yanjing are still continuing the way of conveying brine on their backs and shoulders. As to which of the two is to be adopted, it depends on the difference in terrain between the two sides of the Lantsang River. The work of brine conveying is mainly undertaken by married women. But whether carrying it on the back or on the shoulder, it's no light task. As far as the former is concerned, local women, with 30 to 35 kilograms of brine on their backs, have to climb up the precipitous cliffs on the east bank. The slope here is both steep and slippery, with an incline exceeding 60 degrees; the distance between the brine wells by the river and the brine storage tanks may vary from several hundred meters to eight or nine hundred meters, or even one kilometer away. From morning till night, at the end of the day, those whose storage tanks are near the brine wells need to carry fifty to sixty pails on their backs, while those whose storage tanks are far away need to carry twenty to thirty pails. Salt makers of Jiada Village on the west bank may carry 30 to 35 kilograms of brine each time on their shoulders, too, and need to make 40 or 50 trips back and forth between salt pans and the brine wells. Although it's very arduous, in order to obtain more, salt makers pour the brine into the storage tanks at a very fast speed (most of the storage tanks are located halfway up the mountain).

Now, electric poles have gradually been erected between salt pans. With water pumps, the hard work of carrying brine to salt pans on women's shoulders may be unnecessary in the future. Locally, this is also a kind of progress of the times.

---

1 Song Liangxi. On the Development of Well Salt Brine Conveying Technology [J]. *Salt Industry History Research*, 1988(02).

In short, the equipment level and technical skills of well salt mining and processing in Yanjing are relatively backward. They cannot be compared with modern well salt production technology (not only reflected in the above aspects), which is sufficient to show that this problem exists objectively. Indeed, it's difficult to judge the superiority and inferiority between the sun-drying method and the traditional common boiling method as well as the vacuum evaporation method in well salt production (because there are too many factors to consider). But for local salt making activities which rely entirely on wind and sunlight, still retain manual brine extraction, and have almost no mechanical or industrial operations, in order to promote the upgrading of traditional salt products and improve labor productivity, it's very likely to bring vitality to the perennially sluggish salt production by actively adopting new technologies and processes with a more open mind.

3) Low Degree of Industrialization

The so-called salt industrialization refers to a type of salt industry that is market-oriented, centered on improving comparative profit, and driven by leading salt enterprises, organically combining production, processing, and marketing to achieve integrated management. It's characterized by marketization, regionalization, specialization, scalization, and integration, etc. However, in Yanjing, due to the long-standing traditional family-style workshops, which have been small in scale, paid little attention to the formation of salt industrial clusters, deveoped insufficient multi-variety salts and salt chemical products, and failed to implement diversification and integration strategies, the product structure is not very reasonable, resulting in insufficient extension and expansion of salt industry chain, and thus the real industry has not taken shape yet.[1] Specifically, the major problems here include a lack of leading salt enterprises, being "small, scattered, chaotic, and weak", excessive production capacity, and the singular variety of product, etc.

a. Small Production Scale and Low Output

From the development of Yanjing salt, it can be seen that although the local brine resources are abundant, the intensity of labor organization and the degree of

---

1 Chen Tianpei, Yang Jie. Strategies for Predicament of Zigong Salt Industry in Development [J]. *Journal of Chongqing University of Arts and Sciences (Natural Science Edition)*. 2008,27 (06).

production concentration remain quite low, and thus the situation of small-scale and scattered production has not been significantly improved. Due to the lack of integration advantages, the scale and cluster benefits of salt industry development have not been brought into play, and salt production has remained at a relatively low level for a long time.

First of all, restricted by the landforms and sinking technology, the supply of brine wells has limited the expansion of salt production to some extent. Early in the 1940s, researchers showed that compared to the number of local salt pans, the number of brine wells here was clearly not enough. As salt maker households had to be allocated with the time and number of pails of brine from the wells, and it could not be drawn freely, the increase of salt pan output was affected. Meanwhile, brine wells are also unevenly distributed among these villages. Taking the brine wells in Jiada Village as an example, except for two or three, they are all located in the same sandstone layer. Though adjacent to each other, the brine is actually dispersed, resulting in a low salinity and thus affecting the production. As for the other villages, there are fewer brine wells, and they are comparatively far from most salt pans, making it difficult to draw brine and affecting the work efficiency.[1]After the formation of brine, it has been buried in a centralizing manner in the pores, crevices, and cracks of rocks ranging from several hundred meters to three kilometers underground, hence of different depths from the surface.[2] However, compared with brine wells elsewhere, which are often hundreds of meters, or even more than one thousand meters deep, those in Yanjing are sunk with the method of digging water wells. Simple tools such as shovels, chisels, and hoes, etc. in hands, well sinkers directly dig into the soil, break the rocks, and take the earth and stone to the ground surface until brine gushes out; hence the depths are quite limited. With this kind of long-term unplanned and free mining, the shallow brine resources are gradually decreasing, and those deep underground may also be easily damaged, affecting the local salt production.

Secondly, the family inheritance model will seriously hinder the development of local salt industry. From the history of Yanjing salt development, it can be seen

---

1 Cui Kexin. Investigation of the Geology and Salt Production in Yanjing County [J]. *Xikang Economic Quarterly*, 1944(8).

2 Song Liangxi, Lin Jianyu, Huang Jian, Cheng Longgang. Dictionary of Chinese Salt Industry History [M]. Shanghai Lexicographical Publishing House, 2010:253.

that the salt-making skill here has always been passed down from generation to generation in the way of female inheritance, i.e., a female inheriting her mother's business, and won't be taught to non-blood relatives. Although this can provide targeted training for the successors, enabling them to experience the family heirloom skills smoothly as early as their youth, and receive uninterrupted training in the craft through co-production and living with their mothers, yet the technology has not been effectively transformed into an industry, for it has been dominated in the history of development by a family workshop style production charaterized by its small and scattered scale. On the one hand, due to its complex and cumbersome production process, which is very time-consuming and needs large manpower, it is difficult to rely solely on the inheritance power of each generation of a family. Besides, their working ideas, inheritance concepts, and self-realms are all rather narrow-minded, and prone to be self-enclosed, which may objectively lead to slow updating of intangible cultural heritage-related skills, or even directly result in scale reduction. On the other hand, the number of inheritors per generation is extremely limited, and not each of them has the passion and is willing to pass on this technology. Simply because this craftsmanship is handmade and conservative, with comparatively low socio-economic benefits, few people are willing to invest energy in learning it. After mastering the skill, some successors may lose the motivation to continue and slack off. This family inheritance development model leads to a shortage of salt technical talents, the instability of industry practitioner number, and a lack of high-quality talent reserves necessary for the development of the industry, which also became the main reasons for its low output.

Thirdly, due to the small population of local salt-making villages, and the limited number of engaged salt-maker households, this area does not have such advantages as spatial agglomeration and economies of scale, etc. Of course, far from being invariable, the scale of salt production in Yanjing has been greatly affected by local systems in different periods. As the industry of Markam County started in the 1960s, according to *Mangkang County Annals*, in 1960, Yanjing salt makers increased brine wells and salt pans along the east and west banks of the Lantsang River, expanded the scale, and established a state-owned salt factory, with an annual output of more than 15 million kilograms of salt. Unfortunately, this situation did not last long. In 1963, Yanjing Salt Factory was abolished and converted into a salt commune. In 1965, local salt-maker households were organized to set up a producers' cooperative. After the organization of people's communes throughout the county in 1974, the salt production commune was changed into a salt production team. In 1985, the household contract responsibility system for salt production was implemented. By 1986, there were 66 households, or 274 people contracted to produce salt in the county. By the year 2000, there were a total of 2,655 salt pans, over 300 salt brine tanks, and more than 60 brine wells in Yanjing, with a total annual output of 750 tons. In 2001, the total salt production in the township was 816 tons, including 344 tons in Jiada Village, 250 tons in Nakhi Village, and 222 tons in Upper Yanjing Village. Today, there are more than 300 salt makers in Yanjing Township, and the salt production is carried out in a workshop style on a household basis, with the annual output of a household ranging from two or three thousand *jin* to four or five thousand *jin* of salt per year. Compared with salterns elsewhere that have realized the transformation of technology to industry, these figures are not really impressive.

In short, due to the limitations of various historical and objective conditions, the production scale of Yanjing well salt, whichhas been small and coarse, cannot form large-scale management and scale benefits, making it difficult to form a large salt industry, and this also restricts the high-level and rapid development of Yanjing salt industry.

b. High Production and Sales Costs, and the Lack of Quality Improvement

At present, the salt production in Yanjing still remains at the traditional economic stage. The saltern infrastructure is outdated, mainly focusing on raw material production and primary processing. Its cumbersome operation process, as well as few high value-added products involving deep processing, and fine finishing, etc., also result in low production efficiency and cheap salt prices, and the product quality cannot meet the market needs, either. Moreover, its relatively remote geographical location has increased the transaction cost of salt products.

First of all, due to the low technology content of the product, the low level of automation, and the low operating level of resource exploitation and utilization, the salt production cost here is relatively high. Being mainly manual, this backward way consumes a lot of manpower and material resources; if local people make salt by themselves, it takes a lot of time and energy to complete the task. For a long time, villagers draw brine from the wells with wooden pails, and carry it up the riversides by manpower, either on their backs or shoulders, before pouring it into the salt pans. There, it is sun-dried, precipitating coarse salt, and turns into edible salt after further sun-drying and being attended to. But the production efficiency is very low. In addition, there are also equipment and materials required in salt production, such as salt pans for sun-drying, tools, and labor force, etc., which are not small costs for common people. This is evident through horizontal comparisons. The cost of sea salt is about 30 yuan per ton, and the cost of lake salt is even lower, only more than 20 yuan per ton. But the cost of Zigong well salt is as high as 150 yuan per ton, which is 5~6 times the cost of sea salt and lake salt respectively[1]. The sun-drying of Yanjing well salt is very difficult, and thus its production cost will only be higher. According to Zhenma Yongzong, one year she avoided the rainy season and made more than 120 bags of salt; each bag was sold for 70 yuan, earning 8,800 yuan. Today, affected by the market, the profit margin is too small, and there are fewer and fewer people in this business. Currently, only a small proportion of local households are still making salt.

Secondly, Yanjing is located in the mountainous area of southwest China, far from economically developed areas, resulting in higher transportation costs. There's a wide market for local salt. In addition to Gonjo County, Zhag'yab County, Zogang County, Basu County, and Markam County in Qamdo, and Zayu County in Nyingchi, Tibet, it is also sold to Batang County, Litang County, and Kangting County, Sichuan, as well as Deqin County, Shangri-La County, and Weixi County, Yunnan, etc. Although in the 1970s, National Highway 214 (Yunnan-Tibet Highway) ran through Yanjing from Deqin County, Yunnan Province, connecting with Sichuan-Tibet Highway in the north, and was later extended from Yanjing to Gatuo the county town, becoming an important channel for Yanjing to exchange with the outside world, and the original mountain roads between villages became asphalt roads, the overall rural road infrastructure construction still lagged behind, which to a certain extent increased the cost of product transportation, and was not conducive to the development of Yanjing salt industry. This is specifically manifested in the fact that local people did not have quick access to the market information for the production

---

1 Zhang Xin. Zigong: The Millennium Salt Capital Broke through the Siege with Three Tricks [N]. China Business Journal, 2002-07-15.

and sales of salt products, and thus could not conduct production according to market demand. Therefore, product sales were disconnected from each other, and limited by road conditions and transportation time, stagnation of outward sales might result in falling prices, and dampen the enthusiasm of salt makers. In particular, it was difficult to fully realize the high quality and high prices of some salt products, which has hindered the development of production and the seizing of market opportunities.

Thirdly, the backward, insufficient research and development of follow-up technologies such as deep processing, etc. has resulted in poor salt quality. Well salt is made by sinking wells and extracting brine from natural salt layers deep underground. The brine in Yanjing area, after filtrations and chemical precipitations many times underground, usually contain more minerals and microelements, but fewer impurities; thus, it tastes more delicious, and are considered to be more nutritious than salt produced by China National Salt Industry Corporation. However, as salt pans here are covered with sand and mud on the surface, after being soaked in brine, impurities will often be mixed into it, resulting in impure salt quality; for example, the third-layer salt, which might turn red due to too much sediment, can only be consumed by livestock instead of human beings. In addition, since iodine is an indispensable trace element for the human body, optimizing the iodization technology in well salt production and improving the process is a necessary way to achieve high-quality production. However, though Yanjing county government established a salt iodization site in 1978, due to inadequate technology and low production, it was closed down later. Since then, the government had to subsidize farmers and herdsmen to ensure that iodized salt would be consumed, but this has seriously affected the sales volume of Yanjing salt.[1] In recent years, it's said that due to the lack of iodine and selenium in Yanjing well salt, the government is preparing to close the salt production. If this happens, the cooperation between sunshine and wind will come to an end. And the living millennium salt pans and ancient salt making techniques here may hence survive as a relic of the salt industry...

The profit of salt depends on a variety of factors such as production costs, market demand, and sales channels, etc. But Yanjing well salt making process lacks

---

1 Li Hechun. *Impetus and Shackles: Salt and Salt Taxes in the Lantsang River Canyon* [M]. Guangzhou: Sun Yat-sen University Press, 2016:212.

competitiveness in terms of technical level, technology content, product quality, and production cost. Despite the efforts of the government to coordinate various aspects in recent years, these problems have not been solved yet.

c. Singular Product Structure and Low Level of Industrialization

For thousands of years, salt has been an important material resource for the local economic development in Yanjing. Yanjing people are full of a sense of superiority, which has also resulted in their narrow and closed mindedness, and their conservative and rigid concepts, failing to break free from the "Yanjing consciousness". For a long period of time, Yanjing economy had always been dominated by "salt", with the singular salt industry as the main body of Yanjing industry, and the idea of economic development was naturally focused on the production of "salt". After liberation, the idea of Yanjing economic development was broadened, but it still centred on "salt". At present, due to the limited variety of products and simple industrial structure, it is facing a huge challenge of lacking sustainable internal driving force for economic growth.

First of all, Yanjing salt industry, lacking in industrial production experience, fails to make strategic adjustment to the original industrial structure, and has not implemented the strategy of diversification and integration of industrial structure to achieve industrial upgrading. Transforming the salt industry from simply selling crude salt to deep processing, stretching the well salt industry chain, and forming industrial clusters can inject new momentum into industrial innovation and development. However, due to the late start and poor foundation of the salt industry in Tibet, the development of emerging industries is insufficient, resulting in a singular industrial structure and slow market cultivation. The proportion of traditional salt production is relatively large; in addition to excessively "living off the past" to develop traditional resources, there are few afterproducts, especially a lack of products and services with both Tibetan ethnic characteristics and modern style. This has also led to insufficient extension and expansion of the salt industry chain, and the salt industry's role in driving and supporting related industries has not been brought into play. In recent years, Yanjing Township has begun to promote the transformation of salt production by relying on salt cultural resources, creating an integrated cultural tourism model of "eating, drinking, playing, living, learning and shopping" step by step, but there's still a long way to go to gradually connect the local salt production and manufacturing industry, catering and entertainment industry, tourism and cultural industry, and cultural and creative industry, etc. The main problems are as follows: the lack of a stable, mature, and perfect coordination mechanism for integrated development; the achievements of integrated development need to be expanded urgently; the incentive mechanism for integrated development has not yet been fully established.[1]

Secondly, the product research and development capacity is weak, and the development of multi-variety salts and salt chemical products is insufficient. With the continuous improvement of people's living standards, some have transitioned from meeting basic demands to moderate prosperity, and the demand for variety salts, especially variety salt products with technical content and high quality, will continue to increase. Therefore, it has become an inevitable trend to strengthen the research and development of safe and hygienic, nutritious and convenient and efficient multi-variety salt, and to initiate the personalized and diversified consumption of salt products.[2] But compared with Zigong, Sichuan Province, and other well salt producing areas, where a series of products such as edible salt, industrial salt, and health-enhancing salt have been developed, at present, in Markam County the development, production, and sales of variety salts have just started, and there's relatively little research and development in the use and variety of salt. Taking Tibet Mangkang Yanjing Salt Development Co., Ltd., which was established on March 30, 2018, as an example, its production still remains in the imitation stage, lacking its own patented products and techniques, its innovation capabilitiy has not yet been explored, and its product levels are still quite low. The variety of well salt produced here is relatively singular, mainly edible salt. In such fields as salt therapy and health preservation products, bath salt, and beauty salt, etc., new technologies and processes have not been independently developed, and the breadth and depth of the use of salt products are not enough, either. Therefore, it's necessary to increase investment and development in the salt industry, strive to create products with technological content reflecting such high added-values as green environmental protection, etc., and

---

1 Zigong Municipal Advisory Committee for Policy Decision. Research on the High Quality Development Path of Zigong Salt Industry [J]. *Decision-Making & Consultancy*, 2020 (01).

2 Jiang Haibin. Discussion on the Implementation of Technologies and Processes in Salt Industry Chain [J]. *Jiangsu Salt Science & Technology*, 2007(03).

innovate in terms of varieties and specifications, packaging, taste, and granules, etc., so as to meet the needs of different consumer groups in the modern market.

Thirdly, the "cultural memory" of local salt is not publicized enough, and the brand building is quite ineffective. In order to leave a clear and favourable impression in the minds of consumers, the existing salt industry needs to realize a transformation of product management to brand management and channel management through marketing innovation concepts and means, and meanwhile explore and create a new commercial profit model.[1] At the same time, it's necessary to strengthen the planning, publicity and promotion of the salt brand in order to continuously improve the influence of salt products in domestic and foreign markets. Taking Wudi County, a major sea salt producing county in China, as an example, the local government not only carried out the industrialization development project of washing and cosmetic salts, but also registered the trademark "Haishuer" for 20 kinds of washing and cosmetic salt products that have already been developed, such as fruit and vegetable salt, detergent salt, and facial masks, and even registered the trademark "Yuyantang" (which means imperial salt hall) for the key salt health products, thus greatly improving the market competitiveness and market share of salt products there.[2] However, despite the establishment of Tibet Mangkang Yanjing Salt Development Co., Ltd. in the Yanjing area, due to the lack of intensive cultural industry groups, the lack of modern cultural planning and intermediary organs, the poor ability of brand building, and the insufficient experience in brand marketing, there's still a lack of influential product brands (the top ten brands of Qinghai-Tibet salt do not include salt produced in Yanjing). Therefore, Yanjing people should strengthen their brand awareness, establish and improve the cultivation and management system of salt industry brand, actively implement the salt industry brand strategy, and learn from modern scientific and technological means to build a well-known brand, so that they will win with the brand, seeking benefit from its high quality and market from its high level.

---

1 Tan Zheng, Lei Tao. The Status Quo and Countermeasures of Technical Innovation of Salt Enterprises [J]. *China Well and Rock Salt*, 2010,41(03).

2 Shandong Wudi County Salt Industry Company. Giving Full Play to the Advantages of Resources to Stimulate the Vitality of Sea Salt:Realizing the Transformation and Upgrading of Salt Industry [J]. *China Salt Industry*,2017(02).

Yanjing failed to adjust its industrial structure in a timely manner with the changes of the times and the environment, and still "retain what is old and outworn", clinging to its past glory, which has led to such problems as low quality of development, poor sales channels, prominent homogenization and inadequate characteristics, as well as lagging branding and lack of technology, etc. Therefore, in the future, it's necessary to strengthen the thinking of industrial layout, intensify overall planning efforts, and promote the development of industrial standardization, quality and branding, so as to get rid of the original resource constraints of the salt industry, and effectively give full play to and release the huge potential of emerging products, and finally achieve "self-transcendence".

2. The Impact of Various External Factors

One important reason why Yanjing salt culture can be preserved to this day is the poor transportation and remote location, and another is the backward economy and the long-term isolated cultural environment from the world. The inconvenient transportation and backward economy have created a closed environment, thus preserving the local culture up to now. But the situation is different today. After entering a new millennium, the process of global industrialization and economic integration has accelerated, the commodity economy becoming ubiquitous, transportation increasingly convenient, information exchange more and more speedy, and cross-cultural communication more and more frequent. It's unrealistic and impossible for any corner of the world today to maintain its past isolation and backwardness. However, with the weakening viability of the local salt industry, which is typically "numerous, small, weak and scattered", it can hardly withstand the market turbulence, and its survival situation is quite worrying.

1) The Impact of Modern Industrial Technology

In recent years, in order to comprehensively build a moderately prosperous society, the state attached more and more importance to the development of remote areas, formulating lots of supportive policies for them. Such policies have directly brought some modern industrial elements into these remote areas. For example, the implementation of the "three connections" project helped to solve power and water supply, as well as road problems, so that industrial products such as televisions and automobiles have gradually entered these areas. At the same time, the government also increased investment in education in remote areas, enabling local people to enjoy a relatively sound level of education, thereby greatly enlightening the younger generation with modern industrial civilization. It can be said that the inflow of modern industrial products has injected modern industrial elements into these remote areas.[1]

The most prominent feature of modern industrial civilization is that it maximizes labor productivity, allowing people to produce more and better-quality products per unit of time. On the other hand, many industries currently take industrial production as the standard, striving for quality and efficiency, but they have caused significant impacts and challenges to the traditional handicraft industry, which focuses on the production of labor-intensive products, posing an unprecedented crisis to traditional cultural heritage sites. For the traditional salt industry in Markam County, due to its low product quality, high labor intensity, low salt prices in spite of high production and sales costs, and rising labor costs, as well as the low production costs in modern salt industry, plus the lack of special economic subsidies for salt makers, it is in a critical period of survival or extinction.[2] The impact on Yanjing salt industry is mainly shown in the following aspects:

Firstly, there's a conflict between the openness of modern technology and the relatively closed nature of the traditional salt industry in Yanjing. The formation of a regional industrial culture is based on a certain scope of living and production environments. Yanjing salt industry, whose formation has much to do with the relatively closed environment, was even more greatly affected by it. Surrounded by such natural barriers as high mountains and rivers, it has developed a salt making civilization with strong regional characteristics, i.e., the industrial culture here tends to be more closed and conservative. Moreover, economic activities are self circulating within the local area, and economic exchanges with the outside world rarely occur. In a situation where the transportation and technology development level is not yet very

---

1 He Qingguang, Mo Lingna. Analysis on the Impact of Modern Industrial Economy on Economic Development in Remote Areas [J]. *Productivity Research*, 2010(03).

2 Chen Yiyong, Yu Kongjian, et al. Ancient Markham Salt Pans in Tibet: Value of Living Heritage and its Survival Crisis [J]. World Heritages, 2013(2).

high, the exchange and interaction between various local economic activities in the local area are relatively slow, mainly relying on war and migration, etc., to achieve the great integration of ethnic groups. But modern technology, by breaking through the traditional geographical, environmental, and other natural conditions, has brought a more effective, convenient, and faster way of life to the Yanjing area. Local people are willing to accept modern cultural elements that are more convenient and efficient, such as mobile phones, cars, and other modern technological achievements, for they have influenced the communication and travel modes of various ethnic groups at a very fast speed. Moreover, the sales of salt products previously established on the backs of mules and horses mainly relied on horse caravans, but now this has been basically replaced by industrialized automobiles in the society; modern production and transportation modes have changed traditional salt production activities. The foundation and environment that formed the traditional salt industry have undergone profound changes, impacting the independence of the regional culture taking shape on the basis of this industry. What's more, the manufacturing industry established by modern technology and modern industry has gradually narrowed its cultural and regional differences, and the characteristics are not significant enough; the openness of modern technology has greatly impacted the forming foundation of traditional Yanjing industry culture.

Secondly, there's a conflict between the innovation of modern technology and the relative stability of the traditional salt industry form in Yanjing. As the salt production in Yanjing was developed through generations of inheritance and continuous innovation, its industrial form and basic cultural core have not changed much. For a considerable period of time, due to such factors as economic conditions, material levels, and regional environment, etc., Yanjing people's minds were relatively closed. It was difficult for them to receive more, better, and newer technologies and processes from other regions, and their ideological concepts, production methods, and living customs remained relatively stable. But modern technology is revolutionary and disruptive in many fields; although the content of regional culture has a certain degree of stability, with the development of the times, modern technology not only affects the generating environment of regional culture, but also promotes its evolution and transformation. It completely destroys the original traditional social structure, resulting in the rapid communication and integration between people, countries, and regions. Various new and old ideas and customs prone to conflict, it's not easy for Yanjing salt production to maintain the unique characteristics of local industries. The industrial production mode based on machine mass production, which produces or replicates material products through human activities, has completely changed all of this.[1] However, the traditional Yanjing salt culture, in the face of globalization, informatization, and technological trends, is difficult to maintain its stability, and some traditional forms of salt culture are facing a crisis of inheritance and development.

Thirdly, there is a conflict between the scientific spirit of modern technology and the cultural ideology of Yanjing salt makers. The scientific spirit taking shape in the development of modern technology, including modern scientific thinking mode, scientific attitude, and scientific values, etc., has deeply influenced the spiritual and cultural life of modern Yanjing salt makers, allowing them to break through their understanding of heaven, earth, and human beings. Under the influence of the modern social scientific spirit, such ideological content as morality, ethics, values, clan culture, etc. generated in the local salt making activities has been gradually dissolved, only some people in the relatively backward and conservative villages still retaining the traditional feudal thoughts and social concepts. But with the passage of time and social progress, these scientific spirits and rationality are bound to be accepted by people in the whole region. Surely, the traditional Yanjing salt-making culture and ideology have its progressive elements, and its spiritual core is a significant symbol that distinguishes the local salt-making civilization from other civilizations. However, its outdated regional cultural concepts and customs, which do not conform to the trend of the times and are contrary to the scientific spirit and concepts, will be gradually melted away by the times and should no longer be adhered to. The rapid development of modern society benefits from the fact that science and technology are the primary productive forces. In the era of globalization, with the deepening of scientific concepts, the cross-integration of various cultural forms and ideas has impacted some traditional cultures. Through the collision of scientific spirit and ideological culture, only excellent and advanced industrial cultural ideas can

---

1 Sun Dezhong. Development of Scientific and Technological Culture: Social Condition [J]. Wuhan University Journal (Humanity Sciences), 2007(02).

play a dominant role, which also brings opportunities and vitality to the cultural development of Yanjing salt industry in the new era.

Fourthly, there is a conflict between the replicability of modern technology and the originality of the salt industry culture in Yanjing. Cultural products based on modern technology and modern industry are mostly produced in bulk and on a large scale, and thus they are characteristically high-yielding as well as time and labor saving. Through large-scale production, the productive capacity has been greatly improved. With the progress of modern science and technology, people can naturally tell which production and lifestyle are more scientific and convenient; for example, in Yanjing salterns, brine is extracted by modern water pumps instead of being carried manually on the back or shoulders in the traditional way, which has become an inevitable trend of the development of the times. Although the traditional brine conveying method is still applied in some salterns, the large-scale production must still rely on electricity and machines. Perhaps in the future, mannual brine conveyance will be more of a cultural form for people to visit. Compared with modern salt products, traditional Yanjing salt is more distinctive and unique, because like many handicrafts, folk crafts, and other artistic products, it is produced little by little mannually in a traditional manner, which is time-consuming and low in yield, yet relatively rich in characteristics. From another perspective, the mass production and standardization of modern technological products can effectively ensure the product quality, but they have dissolved the originality of traditional salt-making culture. Of course, the unique Yanjing salt products have their important cultural and artistic values, too. Only by packaging, condensing, and promoting these handmade salts that reflect the traditional culture, and improving their cultural connotation and taste, can they further play a unique role in enhancing the influence and attractiveness of the local salt industry.[1]

Modern industrial technology is a double-edged sword, which not only brings opportunities for the development of traditional salt industry in Yanjing, constantly promotes the transformation and upgrading of traditional regional culture, and advances the cultural evolution, but also has a huge impact on local traditional

---

1 Huang Yiwu. Research on the Influence and Inspiration of Modern Science and Technology on the Development of Regional Culture [J]. *Regional Culture Study*, 2019(04).

salt production due to its its openness, variability, and innovativeness. In the past decade, with the sharp increase in industrial production of iodized salt, as well as the continuously improving transportation conditions on the Qinghai-Tibet Plateau, it's quite difficult for the traditional salt industry to improve, the prospects of the handmade salt market are declining, and the protection of salterns here is worrying, indeed.

2) Intensified Competition in the Same Regional Industry

The salt resources in China, which are extremely abundant and have a wide range of uses, can be classified into table salt (including iodized salt), fishery salt, salt for farming and livestock, industrial salt, and export salt, etc. Among them, table salt refers to sea salt, well salt, rock salt, lake salt, and soil salt, etc. of different sources, which are one of the necessities of human life. In recent years, the annual output of well salt and rock salt has reached over 1,000 tons, accounting for more than 1/4 of the total salt production in the country. China's salt industry is a huge market, but it's also a very competitive one. With the improvement of people's living conditions and the increasing requirement for healthy eating standards, the demand for mid-to-high end edible salt has increased significantly, and the competition between enterprises and salterns in different regions is becoming increasingly fierce, too.

Thanks to the geological structure, the salt resources of the Qinghai-Tibet Plateau are extremely abundant. For a long time, salt in Kham areas mainly came from well salt and lake salt (pond salt) on the Qinghai-Tibet Plateau, while Sichuan salt, Yunnan salt, and inland salt were generally circulated in the Kham area where Han and Tibetan people lived together. This vast and sparsely populated area, with complex ethnic and religious situations, backward economy, inconvenient transportation, and traditional exchange relationships formed by its unique nature, culture, and history, as well as the diversified salt management and control, has created the distinct regional characteristics of the salt market and salt administration in Kham areas. What's more, salt industry resources from different regions have developed their own marketing zones there.[1] Located in the Sichuan-Yunnan-Tibet

1 Wang Haibing. The Salt Market, Salt Administration and Ethnic Relationships in Khams Area during the Period of the Qing Dynasty and the Republic of China [J]. *Journal of Yunnan Minzu University (Social Sciences)*. 2019,37(05).

border area in the Hengduan Mountains, the traditional Yanjing salt making industry has a history of hundreds of years. Since the Ming Dynasty, well salt made here has been continuously circulating in the area where three rivers flow abreast near the Sichuan-Yunnan-Tibet border; not only do people and animals in the surrounding areas rely on it for daily life, but also it has become an important commodity on the Ancient Tea-Horse Road. With complex political changes and economic development in the adjacent areas of Sichuan, Yunnan, and Tibet, the routes and scope of salt transportation and sale have also varied in different historical periods.[1]

With the development of Yunnan-Tibet trade and Sichuan-Tibet trade since Ming and Qing dynasties, especially the further development of Yunnan-Tibet trade routes in the late Ming Dynasty, Yanjing salt was mainly marketed in the southern part of Kham areas, "east to Yajiang, north to Jiangka, west to Zogang and Sangang, south to rgyal-thang and Weixi, with such counties as Ba, Li, Derong, Xiang, and Dao, etc., as well as Tsawarong, Menkong, Zayu (Chayu?), Adunzi, Heyu (or Moyu?), and other places in the middle"[2]. Yanjing salt "sells widely, indeed. But in this scarcely populated area, there's little need for it; apart from boiling tea, it's only for feeding cattle and horses. Besides, due to the rugged mountain paths, the transportation is not easy, and despite the low price, it does not sell well, either"[3]. In its transportation and sale, salt was mainly carried by people or horses, and horse caravans played a major role; this situation continued until the liberation of Tibet. In general, the vast majority of salt was consumed locally, with only a small amount sold in neighboring counties.

Meanwhile, kong-po, spo-bo, Luoyu, and Menyu, etc. in the western part of Kham areas formed a salt-grain exchange relationship with the pastoral areas of northern Tibet. There are a large number of lakes in Tibet, and those in Qiangtang

---

1 Li Hechun. Analysis of Yanjing Salt Transportation and Its Route Change in Tibet during the Ming and Qing Dynasties [J]. Journal of Qinghai Nationalities University (Social Sciences). 2014,40(03).

2 Ren Naiqiang. *Illustrative History of Xikang: Folk Customs*. Nanjing: New Asia Society, 1934 edition, p. 67.

3 Jin Fei. *Detailed Investigation of Yanjing County*, published in the first volume of Zhao Xinyu and Qin Heping' *Abstract of Tibetan Documents in Kham areas During the Late Qing Dynasty*. Chengdu: Sichuan Ethnic Publishing House, 2003, p. 161.

in the north part, i.e., to the north of Ngari and the west of Nagqu, are mainly salt-water lakes and salt lakes. As for table salt in the northern counties of Kham areas, it mainly comes from Nangqian, Guoluo, and other palces in Qinghai Province. Salt produced from such saline springs as Darenduo, Dagai, and Rezong, etc. in Nangqian is mainly sold to Tongpu, Gonjo, sde-dge, Jiegu, Twenty-five Tribes, Dengke, Shiqu, and other places in the northern part of Kham areas, with an annual output of about 100,000 kilograms.[1] Part of the table salt consumed in Kham areas comes from lake salt in Qinghai Province, too. The lake salt reserves in Qinghai are second only to Tibe in China, mainly distributed in the Qaidam Basin. The salt lake near Qinghai Lake is "extremely rich in salt deposit; under the grass root and soil, salt accumulates like snow, and gatherers only need to bend down and sweep it into a container. Compared with sea salt, its production method is simpler and does not require manual labor. After the gathering job is done, they may slightly remove the sediment and transport it back to Kham areas for sale. Those from Xikang come to gather salt along Qinghai Lake shore, ...wrap it in sheepskins, and load it on sheep backs on their returning journey, ...It requires no manpower, either, so the price is very low. Such counties as Qamdo, Tongpu, Wucheng, sde-dge, Shiqu, Dengke, Baiyu, Zhanhua, Luhuo, and Daofu, as well as Kangting and Luding in the north of Xikang also rely on this for their food preparation". About 250,000 kilograms of Qinghai Lake salt was transported to Kham areas every year, with a fairly light tax; 1 yuan Sichuan Rupee would be levied for every 50 kilograms, only equivalent to about 40 cents of the Chinese silver dollar.[2]

The sources of salt in the eastern part of Kham areas are relatively complex and diverse. During the Qing Dynasty, in Tachienlu Office "most people eat Tibetan salt, which is made locally in Chawa (now Zayu in Tibet), and the natives transport it here for sale by animal power. The Han people occasionally eat Qingxi salt, but as

---

1 Zha Qian's *Records of the Natural Conditions and Social Customs in the Bordering Area of Tibet*, vol. 1, edited by Lin Chao, and published in the sixth collection of the *Another Collection of Tibetology Document Series*, p.19.

2 Anonymous. "A Survey of Salt in Xikang Counties", see in *The Pioneer of Khams-gter*, Vol.2, No.1, 1934, p.48.

its annual sales are not much, it's difficult to know the exact quantity"[1]. Sichuan salt produced in Leshan, Niuhuaxi, Qianwei and Wutongqiao were sold in such palces as Hanyuan, Luding, Yutong, and Jinchuan, etc. to the west of Qionglai Mountains, with Tachienlu as the farthest end. And salt produced from the Black Salt Well and the White Salt Well in Yanyuan County were sold in Muli and other counties.[2] Meanwhile, Darenduo salt and Oula salt were also sold in Tachienlu.

Thus, it can be seen that in history, even in the most prosperous period, though Yanjing well salt could be sold thousands of kilometers away, its sales scope was actually relatively limited, or far from forming a monopoly situation in the surrounding areas. In recent years, with the changes of modern salt policies, the competition in salt industry has become increasingly fierce, and the survival space of Yanjing salt industry has been further squeezed.

In 2017, the *Reform Plan for the Salt Industry System* issued by The State Council was implemented. On the basis of adhering to the salt monopoly system, existing designated salt production enterprises were allowed to enter the circulation field, and salt wholesale enterprises could carry out cross-regional operations. As soon as the new salt reform policy was introduced, some industry analysts believed that the salt reform would have a greater impact on the existing sales system of the salt industry, and that as the price controls was lifted, salt pricing would be market-oriented, and salt products would show a development trend of subdivision and branding. Market participants also believed that with the improvement of consumers' living standards and the enhancement of their health awareness, the existing basic salt varieties could not meet their demands any more, and there might be a wave of seizing the market through variety salts and high-end salts.[3] Sure enough, with the inflow of salt products from other regions, the state's control and management of the

---

1 [Qing] Cao Lunbin, Cao Lunhan: *Salt Administration*, Vol. 5 of *Yazhou State Annals*, photocopied from the block-printed edition of the fourth year of Qianlong period, and published in No. 28 of *Chinese Local Chronicles Series* (Western Region), Taipei: Cheng Wen Publishing House, 1969, p. 135.

2 Ren Naiqiang. *Illustrative History of Xikang: Folk Customs*. Nanjing: New Asia Society, 1934 edition, p. 67.

3 Salt Reform Promoting the Reshuffle of Salt Enterprises: A Supply Chain Built by Yunnan Energy Investment for the Salt Industry. China Salt Industry. 2017(07).

salt industry, and the entry of other production modes, the production and sales of salt has been greatly impacted, and the traditional salt industry has shrunk sharply in Yanjing--one of the major salt producing points in Kham areas.

One important reason for the decline of Yanjing well salt was the low-price impact of Yunnan salt, Sichuan salt, and Qinghai salt, as well as sea salt and lake salt from the neighboring regions. These salt companies or salterns with strong comprehensive strength made use of the increasingly convenient transportation, participated in transportation, sales, and other links by virtue of products, channels and brand advantages, and increased investment one after another, so as to expand to other provinces, expanding and strengthening their own businesses, and entering the terminal consumption field of the salt market. In contrast, Yanjing salt produced in the traditional way was obviously at a disadvantageous position in this competition because of its relatively singular variety, relatively simple packaging and relatively high degree of product homogeneity. Taking Tibet Mangkang Yanjing Salt Development Co., Ltd., established on March 30, 2018, as an example, its businesses include the production, processing, and sales of various edible salts, industrial salts, salts for farming and livestock, salts for breeding and processing, sun-dried salts, salts for hot compress, beauty salts, salts for special industries, and multi-variety salts, etc., but due to such problems as the lack of investment in equipment and packaging updates, and the aging of its process, there's a large gap compared with other domestic enterprises.

On the other hand, Yanjing salt has also been impacted by salt produced through modern iodization technology from major salt producing areas elsewhere. As early as 1978, in order to meet the market demand, the county government established a salt iodization site in Yanjing, and thus the salt production was developed to a certain extent during this period. According to *Mangkang County Annals*, "The iodized salt produced here is not only sold exclusively in areas with the iodine deficiency disease in the county, but also marketed far away to different parts of Tibet, as well as Sichuan, Yunnan, and Qinghai, etc." However, due to the fact that the technology could not keep up and the output was low, etc., the salt iodization site was closed down afterwards. Obviously, the iodization technology introduced into Yanjing from the inland had been greatly restricted under the local productivity conditions. Moreover, according to Wangchuk, the former township head, Canadians were once asked to design a salt iodization process, but for various reasons, it was ultimately unsuccessful; it was not until 1982 that iodized salt from Sichuan, Yunnan, and other regions began to be consumed. Founded in November, 1997, Tibet Salt Industry Company is a state-owned enterprise authorized to produce, supply, and implement the monopoly of edible iodized salt throughout Tibet Autonomous Region. But, due to the late start and immature development, it's still difficult to exploit its advantages to the full in such aspects as brand, quality, technology, and service, etc.

At present, the sales of Yanjing salt are mainly concentrated in the local area and several villages in the neighboring Quzika Township, reaching Foshan Township in Deqin to the southwest, Xuzhong Township to the east, Markam County Town to the north, and Bitu Township to the west. Salt transported from Yanjing to Foshan Township, Markam County Town, and Batang, etc. is no longer carried by humans or horses, but can be delivered within a few hours by modern means of transportation. However, in the past four or five years, with the free distribution of iodized salt in Tibet and the impact of non-iodized salt on human health, the number of people eating Yangjing salt is getting smaller and smaller, and the sales scope is also getting smaller and small. Mainly with Yanjing as the center, it is sold within a radius of 50~100 kilometers, and the farther away from Yanjing, the less sales volume would become.[1]

Therefore, as salt enterprises try to seize the commanding heights of the domestic salt industry market one after another, and hence the salt industry enters a critical period of reshuffling and profit redistribution, if the Yanjing salt industry cannot successfully complete the construction of existing salt production equipment upgrading projects in the next 3 to 5 years and achieve adjustments to the local salt industry production layout and the transformation and upgrading of traditional salt production process, accelerate the pace of industrial transformation and upgrading, and secure resources, it will miss the window phase of transformation and upgrading development.

---

1 Li Hechun. Analysis of Yanjing Salt Transportation and Its Route Change in Tibet during the Ming and Qing Dynasties [J]. Journal of Qinghai Nationalities University (Social Sciences). 2014,40(03).

3) The Influence of a Modern Lifestyle

As a process of social and cultural change associated with industrial civilization, modernization, including a series of contents such as industrialization, urbanization, informatization, marketization, and globalization, etc., has brought about a modern way of life. However, this way of life, with its convenience and safety, is replacing ethnic and rural lifestyles, and will also have a fragmentation impact on the inheritance and development of traditional production skills in ethnic minority areas.

Since the reform and opening-up, especially the implementation of the strategy for large-scale development of western China, it has brought relatively severe challenges to the Yanjing salt industry while bringing unprecedented opportunities for economic development here. The diffusion of modern instrumental rationality and the application of technological achievements have enabled local ethnic minority groups to actively or passively join the wave of modernization, making their foundations more fragile and thus affecting the sustainable development.

a. Industrialization

Since the 20th century, with the development of industrialization and the introduction of machine manufacturing, the production efficiency has been greatly improved, and the cost is much lower than that of manual production, causing a significant impact on the handicraft industry system. Many crafts that have been passed down from generation to generation seem to have lost their market and production values, and various production and reproduction are increasingly controlled by technology. Due to the development of technology, all kinds of products with local characteristics are no longer unique, but can be copied and disseminated in large quantities, becoming secularized and standardized, losing the atmospheres and colors they should have, and repeating the same forms uniformly. In this progress pursuing large-scale mass production, no dexterous hands can compete with fast-moving machines. This phenomenon has also led to the loss of market for traditional handmade products by ethnic minorities.

In Yanjing, the traditional salt-making process with a long history, as the crystallization of the local ancient civilization and culture, paritularly advocates the spirit of craftsmanship, pursuing excellence and great originality. The craftsmanship spirit is a kind of value pursuit and a behavioral system under its control, corresponding to dedication, people-orientation, continuous improvement, and innovation. Because handicraft production is a kind of personality-oriented, personalized, and artistic production mode, its products are full of humanistic feelings, containing the design and creativity of craftsmen. But the industrial society, by emphasizing cooperation between humans and machines, dissolves human individuality and personality, making humans an appendage to machines. The industrial production mode is large-scale and standardized, while industrial products are cold, lacking individuality, as well as the warmth of humanistic spirit and personality.[1] Nowadays, the mechanization, scale at an accelerating pace, and commercialization of industrial production have turned salt and other kinds of daily necessities cheap and at the same time very good, convenient, and more comfortable. Under such circumstances, local residents, while abandoning their traditional salt products and other handicraft products that are time-consuming, with cumbersome processes and high labor costs, have chosen salt products mass-produced in modern production conditions.

Faced with the driving force of technological innovation and the pressure of market competition, many salt production processes used under primitive production conditions no longer have practical value. With the problem "niche nature of market demand" of "time-consuming" and "refined" traditional salt production processes becoming increasingly prominent, it will only be relegated to the back seat in the face of modernization. In this case, it's difficult to stick to the bottom line of pure handicrafts, and their elimination and disappearance seem to be inevitable. Therefore, the local salt industry has also adopted adaptive strategies. While preserving the space for the handicraft development, operators of the salt industry will gradually move towards the modern economy themselves. Although a salt industry with larger scale and higher production concentration has not yet been established, due to the increasing degree of specialization in industrialization, salt makers are gradually exploring new processes and technologies to reduce costs and improve quality and output, and trying their best to introduce various new machinery and equipment such as water pumps...Moreover, there have also been changes in the varieties, functions, and production methods of salt products, incorporating modern techniques and

1 Wang Yongjian. The Revival of Handicrafts and the Spirit of Craftsmen [N]. *China Culture Daily*, 2018-12-24.

fashions, with greater emphasis on nutrition, hygiene, and diversity.

Thus, it can be said that the current production of Yanjing salt industry is very different from that of the past.

b. Urbanization

Urbanization is a dynamic process of social modernization, or essentially the expansion of urban civilization in social life. From the perspective of population sociology, it refers to the process of transforming the rural population from a closed, poor, and backward rural society to a modern urban one.[1] If it might be said that due to the geographical isolation and the relatively harsh production environment, the development of Yanjing is later and slower than that of the eastern region, and the suitable conditions for traditional salt production are thus maintained, then since the 1990s, with the deepening of the reform and opening-up and the westward shift of the economic center of gravity, the living environment of traditional salt production activities here has been further lost or changed.

In the process of urbanization, with the rise of secondary and tertiary industries, the status of the traditional salt industry is declining, and the foundation of salt production culture begins to shake. Nowadays, as the income from traditional salt production cannot meet their expectations, a large number of rural surplus laborers have left the salterns and begun to engage in modern production activities, rapidly integrating into modern society under market economy conditions. People's target standard of living has shifted from adequate or ample food and clothing to a fairly comfortable life; they pursue material richness and abundance, as well as the maximization of economic benefits, making behavior orientations and choices under the guidance of reason. For example, traditional salt transportation methods such as carrying it on the back or shoulders, delivering it by boats, or horse caravans, etc. have basically disappeared with the change of traffic environment. In the past, they relied on mules and horses, but now on automobiles and motorcycles. When going out, they either drive private cars or take taxis or buses. By October 27, 2013, there were already 173 self-employed entrepreneurs in Yanjing, whose business scope included commerce, service manufacturing, catering, service, and other industries.

1 Li Jiayan. Analysis of the Urbanization Process in Ethnic Minority Areas [J]. *Theory Research*, 2010(17).

Among them, shops and restaurants (providing accommodation) play a dominant role, followed by the entertainment business. The service industry, focused on bars and teahouses, accounts for the least proportion. According to the number of self-employed entrepreneurs registered with the industrial and commercial bureau, local people account for about 50%. It's known that many hotels and restaurants here are basically monopolized by the locals, who have been gradually separated from the Yanjing salt-making group.[1] There are bars and shops with modern and postmodern elements everywhere, featuring the ancient Nakhi residential architecture.

In recent years, with the rapid development of tourism, as well as the excessive commercialization, urbanization, and artificialization of traditional heritage sites, the authenticity of the living heritage of ancient salt pans is also facing challenges. Some precious salt cultural heritage has been demolished or abandoned, and the ethnic characteristics and culture of salt making villages are rapidly disappearing. As is reported, the local government is planning to build such tourist facilities as viewing platforms, expressways, parking lots, and hotels near the ancient salt pans. Without adequate evaluation and verification, the authenticity of the heritage is likely to be damaged. According to Dr. Li Hechun's investigation in April 2013, the brine wells in Lower Yanjing Village had already been deserted and no one was making salt. It was learned that the reason for this phenomenon was that in Nakhi Village, the seat of the township government, with the development of tourism, many households began to develop in the direction of hotels, hostels, shops, snack bars, and restaurants, etc., no longer willing to engage in the hard work of salt making; as for other households, when the younger generation found jobs in the county town or Qamdo after graduating from universities, their parents would be invited to live with them, and thus the salt pans were naturally transferred or directly given away to their relatives and neighbors.[2] Zhenma Yongzong also said that due to the large flow of people after National Highway 214 passing through Nakhi Village, many villagers have opened restaurants and homestays, and some have rented out their houses for repair shops, hotels, and supermarkets, resulting in fewer and fewer people making

---

1 Li Hechun. *Impetus and Shackles: Salt and Salt Taxes in the Lantsang River Canyon* [M]. Guangzhou: Sun Yat-sen University Press, 2016:216.

2 Ibid., p. 264.

salt.

In today's fast-paced life, many people, who lack the true love and humanistic feelings for traditional handicrafts and are obsessed with quickly inheriting a certain skill to earn income, almost have no spare time to understand and learn their own culture and skills. So to speak, as many local residents yearn for a modern urbanized way of life, they have lost interest in the inheritance of traditional salt making culture.

c. Informatization

In 1967, a Japanese scientific, technological, and economic research group first defined "informatization": It refers to a historical process of social transformation in which the information industry has made significant progress and gradually gained a dominant position in the entire socio-economic structure[1]. Global informatization is a basic trend of the present age; the popularization of the Internet and the rise of the big data era of e-commerce have brought huge opportunities to the production and business models of various industries, not only giving birth to many new business forms, but also posing new threats to the development and even the survival of some enterprises. The salt industry in Yanjing, which is of no exception, has been strongly challenged and impacted, too.

First of all, the marketing of new media in the Internet environment has gradually broken the relatively closed situation of Yanjing, and is meanwhile rapidly eroding the local market for traditional salt products. The medium-high speed development of Internet informatization and the new economy has created an invisible threshold for Yanjing salt industry. Under the background of the rising Internet e-commerce, Yanjing salt products will inevitably face three problems that cannot be ignored: (1) High-quality salt products in short supply, the prices will keep rising with people's fever for fresh handmade products; (2) As the number of orders increases, salt product shop owners may not strive for excellence in their quality any more, resulting in a decline in the product quality; (3) When elements of e-commerce have been integrated into the pure traditional handicraft, many traditional handicraft merchants without any foundation will blindly imitate them in pursuit of profit, thus

---

1 Wang Xiangdong. Informatization: China's Choice in the 21st Century [M]. Beijing: Social Sciences Academic Press, 1998:1.

causing the purest handicrafts to lose their original flavor. Undoubtedly, this will more or less damage consumers' confidence and attractiveness towards traditional salt products, seriously affecting people's original impression that salt products here used to be both meticulous and exquisite.

Secondly, as highly efficient information transmission tools, such modern media as television, radio, telecommunications, and the Internet will have an unprecedented degree of coverage, transmission speed, and influence on society. On the one hand, they have broadened the horizons of people in Yanjing, strengthening the exchanges between different places; on the other hand, they will also greatly affect their values, behaviors, and views on the entire traditional salt-making culture. Although Yanjing does not belong to the mainstream region of our country, the Western postmodernism, which is impacting the developed areas, has also invaded ethnic minority areas, changing people's production and life. With the development of traditional media such as radio and television, as well as self-media such as mobile phones, young people, influenced by the strong consumption culture, have even developed a sense of inferiority towards traditional salt-making activities, feeling that they are outdated and useless.

Nowadays, local residents' enthusiasm for traditional salt production activities has decreased, the activity ceremonies have been reduced, the time for activity processes have been shortened, and the cultural significance of activities has been weakened. Over time, the application of traditional salt-making techniques is likely to suffer a "catastrophic disaster".

d. Marketization

Under the wave of marketization, commodities and consumerism have created endless material desires and illusions of enjoyment for people. For Yanjing and other ethnic minority areas, the penetration and strengthening of the market economy after the reform and opening-up have gradually transformed their development from an ethnic type to the market-oriented type and the modern type. People are pursuing the maximization of economic benefits on the basis of rational choices, and their material life is becoming abundant. But at the same time, there have also been some situations that are affecting the sustainable development of ethnic groups, as well as social harmony and stability.

In the past, the number of local salt makers was quite large, and the amount of salt people bought at a time was also relatively large, weighing several hundred *jin* or more--some even bought thousands of *jin* of table salt at a time. However, due to the comparatively traditional and conservative salt production in Yanjing, salt makers, who did not fully consider the current market development and lacked innovation, have been impacted by salt produced through modern iodization technology from major salt producing areas. When these salt products, which can also meet people's needs but are more convenient, practical, fine, and inexpensive, flooded into Yanjing ethnic villages, the market for traditional handmade salt has gradually been occupied. From this time on, the salt trade of Yanjing gradually went into a low gear, and the sales scope kept shrinking.[1] In recent years, the demand for salt in the surrounding areas of Yanjing has dropped sharply, and the amount of salt some people bought might range from as little as a dozen *jin* to one or two bags. The buyers are mostly Tibetans near Yanjing, and as they only use it for making yak butter tea and feeding livestock, the amount needed naturally decreases. Since 2010, iodized salt has been generally distributed for free in Tibet, so that the vast majority of households do not need to buy salt.

Moreover, with the decline in the output and profits of the salt industry, many households who have relied on salt production for dozens of generations to make a living have gradually abandoned salt pans, and this situation is particularly prominent in Nakhi Village. For semi-professional salt-maker households, they can strike a balance between grain income and salt production, which alleviates the survival pressure to a certain extent. However, professional salt-maker households are different; with the erosion of the salt market, salt prices have dropped significantly according to the amount of grain exchangeable now, so they have to choose other livelihoods instead. By analyzing the structure of self-employed entrepreneurs in Yanjing, it can be seen that local people have already participated in the commercial industry.[2] Of course, it's also a last resort for salt makers to completely abandon salt production and engage in other industries. When income is not guaranteed, many salt makers may rent their salt pans to relatives or friends. Sometimes, these salt

1 Li Hechun. *Impetus and Shackles: Salt and Salt Taxes in the Lantsang River Canyon* [M]. Guangzhou: Sun Yat-sen University Press, 2016:261.

2 Ibid., p. 213.

pans may be given out directly, and receivers will then give about 50 kilograms of salt in return for the Spring Festival. As people become indifferent to salt pans, most in Nakhi village on the east bank have been abandoned and no one is willing to repair them; thus, the inheritance of traditional salt-making techniques is facing an embarrassing situation.

At present, centering on tourism development, the particularity of Yanjing traditional salt industry has attracted many domestic and foreign tourists, including scholars from South Korea, Japan, and other countries. And the establishment of Yanjing Museum also marks that Yanjing has entered another stage of development. To some extent, at this stage, Yanjing is faced with the fact that the interests of tourism development outweigh those of salt production, and an important issue is thus ignored: without salt production, tourism would lose its own significance.

e. Globalization

Globalization refers to the exchange, interaction, and integration of such aspects as economy, politics, and culture in the international community, which is also a hot topic of general concern and discussion among people. With the wave of globalization dominated by economic integration sweeping the whole world, on the one hand, it enriches the material and cultural life of ethnic areas and ethnic minorities, thus giving rise to new cultural forms; but on the other hand, it also increases the risk of cultural homogeneity and cultural deprivation, which may lead to the gradual replacement, elimination, and even disappearance of certain ethnic traditional cultures. Salt making in Yanjing is no exception.

Firstly, globalization has disrupted the geographical foundation on which the ethnic character of salt-making activities exists, disturbed their ecosystem, and changed the traditional pattern generated by the region of salt-making culture. In the past, the development of salt-making activities in Yanjing, influenced by fixed areas and regions, was a combination of the thinking, ideas, beliefs, and values of residents living in that area, and their limited living scope resulted in the relatively stable development of salt-making culture. But globalization has changed this situation, with its wide coverage realizing the universal and daily communication of people in the world. With the emergence of transnational immigration, the area where salt makers live is gradually globalized, and the worldwide mobility of labor force has made cross-cultural communication a daily phenomenon. Exchanges between Yanjing and the outside world being frequent, many ethnic minorities absorb alien cultures without innovating and developing their own culture, resulting in the cultural loss and extinction in its development. Thus, it's increasingly difficult for local salt-making culture to maintain the purity, aborinigality, and authenticity of its development as well as its own characteristics.

Secondly, the practical-minded, convenient, and fashionable trend of modern life has disrupted the life order of Yanjing people, resulting in the replacement or transformation of the traditional production and lifestyle in the Yanjing area. With the penetration of alien cultures and pop cultures, the behavior mode, aesthetic taste, and even thinking mode, psychological characteristics, and ideology in the traditional salt-making culture are no longer suitable for modern society, and the new cultural forms can better meet people's preferences. As a result, more and more local people, especially the younger generation, believe that there is no future in learning the traditional salt-making techniques; instead, they turn to the pursuit of Western fashion and consumption, and even go to a broader space to study, look for jobs, work, do business, and travel...Whichever way it is, as long as the cultural trends of modernity and postmodernity appear around them, they will leave a certain imprint on them. In the pursuit of a fast-paced modern lifestyle, the phenomenon "favoring one more than another" has emerged, and traditional skills begin to be downplayed by people.

In this situation of conflict, collision and integration between cultures, it's difficult for Yanjing salt-making technique to achieve the so-called "detaching itself from the interconnected world". If it fails to break the shackles and acquire a new lease of life, it may be in danger of being assimilated, alienated, or even "swallowed up" in the face of a strong culture.

In short, under the strong impact of modern economy, Yanjing salt industry has gradually lost its vitality, only having to survive in an awkward situation. With the forceful advancement of "urbanization construction" and "new rural construction", modern industrial civilization has penetrated into every corner of the countryside, and local resources and culture are gradually disappearing with the development of modern economy: forgotten or made into cheap tourist souvenirs.

3. The Influence of Human Factors

There are many reasons for the crisis of traditional salt production in Yanjing; in addition to the above fundamental internal factors and irresistible external factors, human factors, such as the locals' inadequate understanding and inheritance of their own ethnic minority cultures, insufficient government protection, unbalanced fund structure, lack of professional talents, and slow development of well salt culture with a low level, etc., have also caused devastating damage to local salt industry, leaving it on the brink of extinction.

Moreover, excessive intoxication for a long time has has turned into a burden. In the decades since the founding of the People's Republic of China, although there have been several opportunities, such as the government's work report for 2009, in which the Prime Minister clearly stated the need to accelerate reforms in the salt industry, railways, and electricity, and the "Reform Plan for the Salt Industry System" jointly formulated by the National Development and Reform Commission and the Ministry of Industry and Information Technology, which put forward the reform policy "separation of enterprise from administration, separation of government and business, integration of production and marketing, licensed operation, promoting to the market, improving laws, and strengthening management", determining the direction for the future development of China's salt industry,[1] the local government and salt makers have never been able to break the shackle of "Yanjing consciousness", ultimately falling into the dilemma "a sea of troubles". Without breaking the shackle of traditional consciousness, without the guiding ideology of seeking truth from facts, it's impossible to formulate a new and feasible development strategy, so as to get rid of the greatest challenge, the danger of "death".

1) Insufficient Function of the Local Government

The salt industry is not big, but its impact is very significant. From ancient times to modern times, it's closely related to the national economy and people's livelihood, so the government has always played the role of controller in the salt

1 Xu Zhe. Discussion on Reform and Development of Salt Industry in Xinjiang [J]. *China Salt Industry*, 2016(21).

industry, responsible for formulating its development plans and industrial policies, managing salt monopoly work, and ensuring its steady development. Moreover, for the government, salt tax is always an important source of fiscal revenue, which also explains its tight control over the industry. However, in recent years, the government's function in the development of the salt industry has not been effectively exerted, and there are various problems, too, such as imperfect laws, regulations and policies, inadequate management system and mechanism, and narrow investment channels for its construction and development.[1] Some of these contradictions and problems, rooted in history, have existed for a long time, while others are new ones emerging under new circumstances. There are objective causes, as well as cognitive and work-related reasons. By detailed analysis, they can be roughly summarized into the following aspects.

First of all, the planning and guiding functions of the local government in the construction of the salt industry have not been effectively exerted. For the development of the salt industry, the government should play its important value and role in a series of issues such as policy guidance and market environment cultivation. It is required to be all-round, not only in the planning of the industrial development, i.e., in timely planning and macro-control of the development direction, future, and trend of the industry, so as to provide the services the government should provide for it, but also in fostering a fair and transparent environment for the salt industry market. However, some local officials and units in Yanjing do not fully understand the importance, necessity, and urgency of thriving and developing the salt industry, not attaching enough importance to it in their thinking, or there are even some deviations in their understanding. This has resulted in inadequate organizational leadership, inadequate work measures, and inadequate policy support for the salt industry, which may in turn lead to unreasonable layout of the salt industry development, and even wasteful phenomena that seriously violate the market law. In addition, the protection and development of the local traditional salt-making process related to the vital interests of the people, as a feature of the regional ethnic culture, can not only provide them with the opportunity to find jobs in the neighbourhood,

1 Gu Manyi. The Basic Approaches on the Building of Salt Culture Industry in China [J]. *Joural of Sichuan University of Science & Engineering (Social Sciences Edition)*. 2013,28(02).

but is also an important way to effectively consolidate the achievements of poverty alleviation connected with rural revitalization. However, there's a common problem of insufficient guidance and exploration among the township government departments, such as inadequate understanding of the protection and inheritance of salt-making skills, and a lack of effective guidance on the development of salt-making technology in this region as well as targeted and practical training, etc. Therefore, the government and relevant departments need to view and study this issue as a whole from a national or even global perspective.

Secondly, related laws and regulations for the development of the local salt industry still need to be further improved, and law enforcement needs to be further strengthened, too. Perfect laws, regulations, and policies are the fundamental guarantee for salt industry development, which requires a basis of relevant laws and policies, and their strict implementation. In recent years, Yanjing Township has successively issued a number of laws and regulations on salt industry development, but on the whole, the legal system is imperfect, still lagging behind the requirements of the strategic development of local salt industry. There are also some contents contradicting the spirit of salt reform and the new "Measures for Monopoly of Table Salt", which not only seriously hinder the smooth implementation of salt reform, but are also not conducive to the competitive mechanism of the survival of the fittest in the salt market. With the continuous emergence of new problems, some existing laws and policies are no longer suitable for the development of the times; in particular, some existing laws and policies, with traces of outdated planned economy, can no longer meet the requirements of the development of local market economy. As a result, local law enforcement departments, "without laws to abide by" on many issues, have to follow the old regulations. Moreover, in terms of encouraging policy formulation for the salt industry development, it is particularly inadequate. The salt industry should not only be developed, but also protected; this kind of protection is productive protection, which refers to the support and assistance of laws, regulations, and policies.[1] The development of salt industry must be based on the construction of a scientific, perfect, and systematic regulatory system of its own. At present, the

1 Gu Manyi. A Review of Salt Culture Industry Development and Local Cultural Industry Development Strategy [J]. *Journal of Shanxi Institute of Economic Management*, 2013(9).

problems of Yanjing salt industry lie in the lack of perfect and operational laws and regulations, etc.

Thirdly, the reform has lagged behind, and no efficient and dynamic institutional mechanism has been established. Although the monopoly system of salt industry has lasted for thousands of years, it's the trend for the salt industry system to reform towards marketization. However, the current management system and mechanism of Yanjing salt industry are obviously insufficient in the marketization management of the its own development. Hence some management systems and behaviors may inhibit the vitality of salt enterprises or salt-maker households when the market mechanism develops rapidly, resulting in a low degree of marketization of the entire industry and difficulty in adapting to market competition needs. The reality is that a unified salt industry management system and mechanism has not yet been established in Yanjing, which leads to the phenomenon of fragmentation in the salt industry development. So, reform and improvement of relevant systems and mechanisms have become a necessity for its current development. Moreover, compared with the inland region, Tibet is even more deeply affected by the traditional planned economic system; the reform in various aspects, including the salt industry system, started late, facing more particular difficulties. The problems encountered in the reform of the inland salt industry system are more prominent in Yanjing, Markam County, while the problems that did not emerge in the reform of the inland salt industry system still exist in Yanjing area. It must be clearly recognized that in the face of such "hard bones" as tough issues concerning interests and long-standing unresolved institutional problems, the local salt industry cadres' awareness and willingness to implement reforms are rather weak. One or two cadres, with the mentality of "waiting, relying, and demanding", are even unwilling to change, afraid to change, and unable to change. There are some who handle things in a perfunctory manner, some who procrastinate, evade, or take a wait-and-see attitude, and some who hold an office and enjoy all the privileges without doing a stroke of work, etc. Essentially, it's the mentality of mediocre and lazy administration that is making trouble, which is manifested in their lack of attention, "blurring the line between right and wrong", i.e., trying never to offend anybody.

In addition, although the local government has successively formulated and introduced specific preferential policies to encourage and support young people to participate in the inheritance of traditional salt-making skills, on the whole, the cultivation of inheriting talents is still relatively insufficient. Not only is there a lack of unified training mechanisms, but also the incentive and reward efforts are relatively inadequate, leaving Yanjing salt industry in a rather difficult situation.

2) The Problems of Salt Producers Themselves

Salt producers refer specifically to the group that is engaged in the production of salt. In Yanjing, Markam County, besides the salt production enterprises established in recent years, such as Tibet Mangkang Yanjing Salt Development Co., Ltd., the majority are local salt-maker households who have been making a living by salt production for generations. As for the problems of local salt enterprises, for example, their products remain in the imitation stage, lacking their own patented products and techniques, their product grade is still quite low, and their salt product usages are still insufficient in breadth and depth, we have already mentioned above, so the only focus here will be the principal part--salt-maker households. As is well-known, the vast majority of salt-makers in Yanjing have made significant historical contributions to the increase of local fiscal revenue due to the glory of salt production. However, with the development of society and the continuous progress of science and technology, the traditional production mode and industrial pattern that have been inherited for many years can no longer meet the needs of the leapfrog development of Yanjing economy and the growing spiritual and cultural life of local people. And they themselves have also exposed a lot of problems, which are mainly as follows:

a. The Practitioners' Low Comprehensive Quality

The development of salt industry requires a large number of high-quality talents who are both proficient in the knowledge of salt market management and salt products, and well acquainted with the operating rules of the salt product market. However, as the backbone of the living salt culture, producers of Yanjing salt are all natives, most of whom are less educated, and less formally trained in industry and commerce, resulting in a lack of management knowledge and skills. In addition, living in the local area for a long time, they do not have a deep understanding of the market economy, and has been unable to adapt to the transformation of the local salt industry from labor-intensive, resource-based, and rough processing production to technology-intensive production for high-quality, new, and refined products.

Because of the practitioners' low quality, some advanced production equipment is unusable, which has also restricted the local salt industry from upgrading, improving its quality, and expanding its scale. Also because of the low quality of labor force and low technical level, the local salt industry can only choose those industries with low technical requirements as its development direction. Hence, most salt-maker households would make similar choices, resulting in production activities mainly concentrated in low-grade products and non-technology-intensive products, which has not only affected the economic efficiency of the salt industry, but also wasted precious natural resources.

Specifically speaking, practitioners' low quality can easily lead to a couple of problems. Firstly, it may lead to outdated overall management concept and backward management methods, which basically still follow the traditional empirical management ideas and methods, without much consideration for the production level, the product grade, and the development of new products, and there is a lack of modern management concepts, knowledge, and methods, such as capital management and modern marketing strategies, making it difficult to adapt to the needs of modern salt industry development. Secondly, constrained by traditional concepts and the smallholder consciousness, the organization and management system of the local salt industry is mostly characterized by "familization", with a simple internal organizational structure relying on experience and kinship management. Thus, in decision-making and management, it is often subject to subjectivity and arbitrariness, leading to decision-making errors and non-standard management. Thirdly, the lack of modern management concepts, the weak consciousness of market economy, such as brands, integrity, and marketing, etc., and the weak management awareness of development strategies, talent resources, and informatization, etc., may result in a low level of operating management. Fourthly, operators prioritize production over management, lacking unified planning and management methods in the process of production and operation. There are no clear operating procedures, no technical standards, and no labor quotas and consumption quotas for salt makers' production activities. The production efficiency and the product quality depend entirely on their skills and past experience.

The low quality of practitioners is itself a crisis. Practitioners' quality determines the production and operation capacity of the salt industry. The higher their quality, the stronger the adaptability to the market, and the lower the possibility of industrial crisis; on the contrary, the low quality can not adapt to market changes, and the industry is often in a passive state, making it difficult to avoid crises.

b. Conservative Thinking and Weak Innovative Spirit

The well salt production in Yanjing, Markam County is one of the most brilliant components in the historical development process of China's salt industry. With the advent of the era of global economic integration and knowledge-driven economy, the leapfrog development of the resource-based and labor-intensive local salt industry will face very severe challenges.[1] This requires the majority of practitioners to constantly adapt to the development and changes of practice, timely shift and update their ideas, equip their minds with new ideas, new concepts, and new requirements, and solve problems with new ideas, new methods, and new mechanisms.

However, due to the limitations of their own knowledge and skills, many local practitioners have a bias of following old rules in their thinking; they are unwilling to innovate, let alone take risks. This mentality of satisfaction with the existing state of affairs and reluctant to move forward has seriously constrained their conceptions of thinking and behavioral patterns. Likewise, due to the influence of regional, folkloric, economic, and cultural factors, etc., local salt makers have little communication with the outside world, lacking the awareness of new things, and they often stick to traditional aesthetic concepts, lacking the spirit of exploration and innovation. The solidified thinking and unchanged salt production methods also make it difficult for traditional well salt products to conform to the trend of the times and meet the needs of consumers. For some new situations and new problems in the development of the salt industry, they tend to focus on short-term behaviors and immediate benefits, and are thus unable to conduct timely research and think ahead. They believe that salt production is a traditional industry with ready-made standards and processes, without deep thinking about the development of salt industry assets and resources. More importantly, they have not been able to conduct in-depth investigations and scientific analysis, seek countermeasures, and take the initiative to respond to the strategy of development of western regions from the height of "jumping out of

---

1 Li Aiqin. Research on the Dilemma and Countermeasures of Tibetan Township Enterprises in Their Leapfrog Development [D]. Master's Thesis of Xiamen University, 2009.

the industry, out of the region, so as to develop the salt industry".[1] As a result, the product competitiveness is poor, and many salt products with strong local and ethnic characteristics can only enter the market directly in a primary form, with low added values and too short industrial chains. Moreover, due to insufficient innovation, the product structure tends to be identical. Obscure features, backward product designs, low technical contents, and unstable qualities also result in few specialty products, brand-name products, and high value-added products, making it difficult to form advantageous products with ethnic characteristics, and the products are only limited to the regional market or even the local market, failing to enter the domestic and international markets.

To a certain extent, it can be said that none of the ethnic handicrafts that continue to thrive today are immutable and fettered by old conventions. In order to achieve the sustainable development of manual salt production in Yanjing, it's necessary to base ourselves on the new development era, grasp the new development concept, promote the creative transformation and innovative development of local salt process, and fully stimulate the industrial vitality of salt-making technology. In this way, the well salt production can not only retain its traditional artistry, so that the local salt-making skills can be inherited, but also have the economic effect of industrial production, so as to better meet the needs of local people in the pursuit of a better life. If practitioners stand still and refuse to make progress, and blindly adhere to their own views, it will be difficult for them to adapt to modern society, and eventually their skills will be lost.

c. Singular Inheritance Method, Facing Generational Disconnection

The sun-drying salt in Yanjing has a very long history, going back to ancient times. Even today, it still retains its unique production technique, which has become a spiritual symbol of people in the local and surrounding areas. The traditional salt-making technique is mainly passed down within families, from mothers to daughters. The basic methods, skills, and experiences of salt-making techniques taught to children by female elderly salt-makers are acquired and mastered under the long-term influence and guidance of their female elders. Therefore, teaching by word of mouth

1 Wang Jinfa, Reflections on the Scientific Development of Local Salterns in Xiangshui County out of Difficulties [J]. *Jiangsu Salt Science & Technology*, 2010(02).

has become the main form of inheritance of traditional salt-making techniques. These salt-making techniques have become an important criterion for measuring the ability and virtue of a local woman, but the drawbacks of this inheritance method are also quite obvious.

On the one hand, teaching salt-making skills by word of mouth relies on inheritance from generation to generation, which has a certain degree of subjectivity as well as variability. For this style of inheritance, "skills handed down only in a family", how well one learns mainly depends on his/her comprehension ability. The level and effect of passing on skills from mothers to daughters, greatly influenced by such factors as the former's cultural qualities, ideological realms, and skill levels, as well as the latter's comprehension abilities, knowledge levels, and artistic accomplishments, may also limit the direction of talent development. In order to reach a certain level, practitioners engaged in handmade salt production all rely on years of accumulation, from which they gradually sum up and gain rich experience. They are generally not good at expressing and describing the production process, and students generally watch while learning, i.e., following examples, and was thus self-taught by practicing for a long time.[1] In addition, due to the lack of strict regulations and certain ethnic norms on salt-making technology, and no rigid requirements in the process of inheritance and development, there are only a few famous teachers, and many salt practitioners are not highly skilled. This has also led to the fact that today, when there are few people who have received systematic cultural education and art education, it's difficult to cultivate a large number of high-quality and high-skilled salt production talents by means of mother-daughter inheritance.[2] Many elderly female salt-makers, without comprehensively sorting out the relevant knowledge, theories, and techniques, lay great stress on the practical guidance of handicrafts, which seriously restricts the improvement of the inheritance efficiency of traditional salt-making techniques.

On the other hand, with the development of social economy, people are gradually becoming indifferent to the traditional handicraft culture. As handmade

1 Ma Wenjing. On Inheritance and Development of Traditional Handicraft in Minority Areas [J]. *Guizhou Ethnic Studies*, 2009,29(01).

2 Tang Liqun. Inheritance of Traditional Hunan Embroidery Technique and Its Advantages and Disadvantages [J]. *China Market*, 2015(10).

items and their makers could not receive the respect they deserve, it is difficult for the craftsmen to realize their self-value. Many excellent craftsmen have to give up their traditional manual skills and seek a new way of living. With the gradual aging and passing away of elderly people who are familiar with their own culture and have mastered their own ethnic skills, there are fewer and fewer people who truely understand the traditional ethnic handicrafts. As the inheritance of traditional salt-making techniques becomes unsustainable, it may be disconnected, broken, and eventually lost by degrees.

3) Negative Attitude of the Younger Generation

Yanjing salt-making technique is a treasure and an important part of the traditional culture of the Chinese nation, which nourishes the soul of local ethnic groups and also contains enormous potential for modern industrial development. The inheritance and development of this salt-making technology, related to the continuity of local cultural vein, is the common responsibility of the sons and daughters of Yanjing. However, with the advent of economic globalization, information networking, and the era of knowledge economy, the mentality of the local younger generation is constantly changing. Due to the younger generation's unwillingness to continue the traditional salt-making technique, or "there are teachers, but no students", it's facing the problem of no one inheriting it in the future. Moreover, this situation is currently showing a trend of gradual expansion. To sum up, there are mainly the following reasons.

a. Ideological and Conceptual Problems

In Yanjing, Markam County, there are some problems in the ideology, psychological condition, and volitional quality of the young generation, especially the weakening social recognition of the value of salt-making skills. Their weak identity perception has seriously restricted their subjective willingness to inherit and develop salt-making skills, and thus they cannot form enough emotional motivation to be willing to invest more time and resources in the inheritance and protection of salt culture.

The protection of any kind of craft requires painstaking efforts from generation to generation. But in the local area, for the relatively realistic new generation, their values, worldviews, and outlooks on life have undergone a fundamental qualitative change. Most young people do not truly understand Yanjing, know Yanjing, or identify with Yanjing, and they know very little about the salt-making activities. Due to the lack of awareness of the importance of Yanjing salt-making technology and their own advantages, they are indifferent to the protection and inheritance of their ancestors' salt-making skills, believing that the protection and inheritance of Yanjing salt-making culture is the responsibility of the government and relevant departments, and has nothing to do with the villagers, needless to say they should participate in it. Similarly, young people who have not entered this industry do not have enough aesthetic awareness of local salt-making activities, and thus could not appreciate the connotation and value of well salt production. They even believe that salt making, as an outdated profession, is not an option for their own employment, which is also a major reason why the younger generation are unwilling to learn salt-making skills. They have misunderstood and even rejected salt production activities, so they have not fundamentally developed the desire and passion to inherit and promote these skills.

In terms of practical attitudes, although the vast majority of respondents showed a positive willingness to practice, more than half of them said that they would take the initiative to protect and inherit it only when it's needed or helpful to themselves. It can be seen that from the psychological or practical level, the utilitarian orientation of young people's cultural attitude is very strong, which has seriously affected the enthusiasm and behavioral consciousness of the inheritance and development of local salt production culture. Regarding the cultural embodiment of the Yanjing area, most young ethnic masses expressed that they do not follow traditional customs and etiquette in their daily lives, and a small number of respondents even stated that they never abide by them. At the same time, many young ethnic masses emphasized that they have little or no knowledge of traditional music, dance, and handicrafts related to salt industry activities. Although the worrying feedback results are closely related to the vague cognition of the cultural value of regional salt making among young ethnic members, the fundamental reason for the sluggish behavior of most respondents lies in their belief that the inheritance and development of regional salt-making skills is of little or no practical significance to themselves. This shows that the utilitarian orientation not only leads to the cognitive deviation of young ethnic members on the value of regional culture to a large extent, but also seriously fetters the subjective consciousness of regional cultural inheritance and development.

b. Economic Income Problems

In the process of modern transformation of ethnic society, the value orientation of the market economy often makes the pursuit of economic benefits a prerequisite for regional cultural inheritance and development. It is undeniable that the practical acquisition of economic benefits is an effective prerequisite for the inheritance and development of ethnic regional cultures, because "interests (material and conceptual), rather than ideas, directly control human actions"[1]. In fact, the benefits have always been influencing the attitude of the younger generation in Yanjing area towards salt production activities.

In the past, the salt-making skill, as a subsistence craft, could barely maintain local people's existence. But nowadays, with the rapid development of modern industrial products, it's difficult for Yanjing well salt produced by traditional techniques, which is mostly handmade, with complex processes, high costs, and low economic benefits, to achieve mass production. In addition, with the increasing labor costs in recent years, the income advantage of employees in the salt industry is no longer obvious. Since salt production cannot bring more benefits and value to the younger generation in practical life, it is difficult for them to live decently just by "relying on the craft". In this situation, young people who cannot see any hope, or have no sense of achievement, or cannot achieve their self-values, would rather find jobs elsewhere to earn a living than learn the craft that is both time-consuming and unprofitable. Consequently, the restraint of economic interests has led to a serious decline in the number of people engaged in salt production activities; with salt makers becoming fewer and fewer, the craft is in danger of being lost. On the other hand, despite the overall positive orientation, for example, young ethnic people generally believe that the inheritance and development of local salt culture is quite meaningful, yet when asked about its main role, the vast majority of respondents choose economic benefits, while less than 10% believe that value inheritance is equally important. Most young people have clearly expressed their unwillingness to engage in the inheritance of traditional salt-making skills, and the reason is

1 Max Weber. Confucianism and Taoism [M]. translated by Wang Rongfen. Beijing: The Commercial Press, 2003:19-20.

nothing more than the inability to make money and the lack of definite development prospects.

Moreover, various objective conditions have also restricted young people's participation and weakened their enthusiasm. Local villagers live under great pressure, especially the young, who are anxious about betrothal gifts (from the bridegroom to the bride's family), or the quality of their children's education, or the health of their parents, etc. Thus, they have to pursue economic interests, and are strong in will but weak in power to protect and inherit local salt-making skills. Nowadays, the annual per capita income in Yanjing is around 4,000 yuan, and the main economic pillar is labor service export. Out-migrantion for work has become the main source of income for families, while the proportion of the salt industry, which was once relied on for existence, has dropped to 30%. Of course, excessive emphasis on economic benefits while ignoring the inheritance and development of the ethnic salt-making techniques has also resulted in the weakening ethnic culture.

c. Impetuous Psychological Problems in A Materialized Society

While pursuing further enrichment of material life, we should also leave some room for improvement in our souls. However, through an investigation of the youth group, it is found that most of them in Yanjing, Markam County today are drifting with the tide of material things, losing their accurate self-orientations. Filled with inexplicable excitement and restlessness, they all hope to go to cities for a rich and colorful life. Therefore, they cannot calmly treat life and even their beings, and are naturally unwilling to spend decades studying traditional handicrafts. This is an impetuous mood. And impetuousness has become the biggest mentality problem among young people in Yanjing area.

"Impetuousness" is defined in the dictionary as: "impatience and lack of composure"; that is, unstable in behavior and irritable in psychology. Impetuousness, originally a kind of emotion reflected in an individual, belongs to negative psychological characteristics. However, when manifesting itself in many people in a society, or in group cognition and behavior, it becomes a collective emotional characteristic. In Yanjing, impetuousness, as a bad social spiritual climate, has permeated into such daily activities of young people as making friends, dating, marriage, work, and career, etc. Their psychological and behavioral characteristics can be summarized as follows: Firstly, no planning. They have only short-term goals in life and no long-term plans, so the goals are quite erratic; without much thinking before taking actions, they lack the focus on existing goals. Secondly, no persistence. They always seek quick success and instant benefits, without putting effort into anything in a down-to-earth manner. Unwilling to make persistent investments, they always hope to reach the goal in one step, lacking in patience. Thirdly, no stability. Their goals are constantly changing, so they often appear to be a rolling stone, flighty and impetuous, and they often stop after scratching the surface. Fourthly, impatience. They always envy the achievements of others and yearn to become famous or rich overnight like others. But their own abilities can not achieve these, resulting in a lot of troubles, restlessness, and impatience, and they are often moody, throwing tantrums. Fifthly, fear. Anxious about their career and personal future, they are swayed by consideration of gain and loss. Sixthly, unquietness. In daily life, they give people a very urgent feeling all day long. Actually, there is nothing at all, but they are like ants on a hot pot, unable to settle down.[1]

Admittedly, as part of people's personality, an element of impetuousness may exist in everyone's personality characteristics, and cannot be eliminated artificially. However, in response to this pathological social psychology, i.e., impulsivity, emotionality, and blindness intertwining each other, which is spreading in the local society and growing in the hearts of more and more young people, effective measures need to be taken to alleviate it, so as to promote people's mental health, and in turn spur the harmonious development of the local salt industry.

The ancient style of Yanjing remains as before, and people are still using the primitive sun-drying method to make salt. But it is unknown how long this ancient way of salt production can be preserved under the invasion of modern civilization. The continuous disappearance of the traditional salt-making industry in Yanjing, Markam County will inevitably cause changes in its ethnic constituent elements, i.e., economic life and the internal structure of its ethnic culture (mainly referring to business relationship and product relationship), and then affect the stability of its ethnic composition and ethnic culture. How to deal with the extincting traditional skills of minority nationalities, whether to let them disappear or promote their

1 Zhu Li. Recognizing the Impetuous Characteristics of Society and Cultivating a Healthy Social Mentality [J]. *People's Tribune*, 2014 (20).

extinction, or to save them as much as possible, has become an unavoidable problem facing the local ethnic minorities. However, a good grasp of the salt-making life in Yanjing will help us strengthen our understanding of the interaction between the social, economic, and cultural development and the natural environment of Yanjing, and thus make plans that are conducive to the future development here.

CHAPTER SIX

Excavation and Sorting of Salt Cultural Relics in Yanjing

Salt cultural heritage, as a cultural resource with salt as the dominant tone, is a prominent example reflecting the main evolution stages of human civilization. In the long process of salt industry development and evolution, local residents of Yanjing, Markam County have created a rich and colorful salt cultural heritage full of "salt brine flavor" on this magical harsh land, including not only such material content as the saltern infrastructure and salt makers' residences, etc., but also such non-material content as the salt production process, the life experience of salt makers (including horse caravans who sells salt), the legends about salt making, and the cognition of salt making itself, etc. With inestimable historical value and cultural relics value, it is of great importance for studying the history of China's salt industry as well as the industrial history. It contains rich historical information on the interaction, exchange, and integration of various ethnic groups, and is also an important resource for carrying out the education of the Chinese national community.[1]

These historical relics, ruins, and remains of the salt industry are very rare and fragile. It is not easy for them to survive to this day, indeed. However, with the development of economy and the changes of the times, many carriers of surviving cultural elements that have been accumulated for hundreds of years are facing the severe challenges of economic globalization and the fierce impact of market

---

1 Su Xing, Tian Renbo. Yunnan-Tibet Yanjing Culture: A History of Interaction, Exchange, and Integration of Various Ethnic Groups through Taste Buds. (see https://www.163.com/dy/article/G3ETVRI50550B8ZZ.html).

economy, and are thus in danger of disappearing. For a long time, in the face of the danger of gradual annihilation, damage, and loss of the traditional cultural heritage of ethnic minorities, the state has mobilized various forces to strive for rescuing conservation, excavation, and timely sorting and research, in line with the principles of "rescue first" and "effective protection, rational utilization, and strengthened management".[1] Therefore, it is necessary to adopt a scientific and prudent attitude towards the rich "mineral resources" contained in Yanjing area, and on the basis of a systematic and complete understanding of their significance, excavate, sort, refine, and reject the dross and assimilate the fine essence, so as to maintain the authenticity and integrity of the local salt cultural heritage and enhance the overall value of the salt cultural heritage.

1. Categories of Yanjing Salt Cultural Heritage

Cultural heritage is a precious treasure left by history to mankind, which can be divided into material cultural heritage (tangible cultural heritage) and non-material cultural heritage (intangible cultural heritage) in terms of their forms of existence. The material cultural heritage includes immovable cultural relics such as ancient ruins, ancient tombs, ancient buildings, cave temples, stone carvings, murals, and important historical sites and representative buildings in modern and contemporary times, movable cultural relics such as important material objects, artworks, documents, manuscripts, and books and reference materials of various eras in history, as well as famous historic and cultural cities (streets, villages, and towns) with outstanding universal value in terms of architectural style, uniform distribution, or combination with environmental scenery.[2] The non-material cultural heritage, on the other hand, refers to various traditional cultural patterns of manifestation (such as folklore, folk activities, performing arts, traditional knowledge and skills, as well as related utensils, objects, and handicrafts, etc.) and cultural spaces (i.e., places where traditional cultural activities are held regularly or traditional cultural expressions

1 Wang Xien. On the Present Situation and Trend of the Traditional Culture of China's Ethnic Minorities [J]. *Ethno-National Studies*, 2000(06).

2 Cheng Longgang. Research on the Protection and Utilization of Zigong Salt Cultural Heritage [J]. *China Ancient City*, 2011(8).

are displayed in a concentrated manner, such as songs, temple fairs, and traditional festivals, etc.) that have been passed down from generation to generation by people of all ethnic groups and are closely related to people's lives. Roughly speaking, the salt cultural heritage of Yanjing can also be classified in this way.

However, it's not difficult to see that the richness of Yanjing salt cultural heritage, its variety and quantity, and the huge number of preserved objects, ranking among the top in China, are very rare in the history of the world salt industry. Some of them are tangible cultural heritages such as production sites, tools, life areas, roads for transportation, and architectural relics related to edible salt, while others are intangible cultural heritages such as oral legends, social customs, ceremonial festivals, experience and skills, and traditional crafts, etc. formed in the process of salt production, transportation, exchange, and administration.[1] Here are not only ruins and sites reflecting salt industry production, but also ancient towns and streets reflecting salt transportation and trade, as well as processes and skills reflecting traditional well salt production; here are both folk customs reflecting the salt industry economy, and folk arts related to the salt industry and derived from salt. In addition, historical records, archives, contracts, and account books of the salt industry are also an extremely important part of Yanjing salt cultural heritage. These remains, production methods, poems, legends, and beliefs are condensed into the rich well salt cultural deposits of Yanjing.

Therefore, besides the division of material cultural heritage and intangible cultural heritage in a broad sense, different classifications can also be made according to different criteria such as source, function, use, content, attribute, category, time, space, and shape, etc. Through literature sorting, field research, and examining, the author believes that apart from those buried underground which are currently almost unknown, salt cultural heritage here can be classified into four categories: architectural landscapes, historic sites, museum collections and those scattered among the people, and the intangible type, according to the function, role, and content, etc.

---

1 Su Xing, Tian Renbo. Yunnan-Tibet Yanjing Culture: A History of Interaction, Exchange, and Integration of Various Ethnic Groups through Taste Buds. (see https://www.163.com/dy/article/G3ETVRI50550B8ZZ.html).

1) Architectural Landscapes

Salt cultural heritage of the architectural landscape category mainly includes representative historic and cultural cities (or towns, or villages), historic and cultural streets, and historic buildings, etc. of well salt culture related to economic activities of the salt industry. According to the survey, well-known architectural remains, such as salt guild halls, trade clubs, salt merchants' mansions, salt merchants' forts, salt merchants' public memorial halls as well as manors, ancestral halls, towers, and memorial archways, etc., are rarely found in Yanjing, but some historic buildings are still worth mentioning.

a. Yanjing Catholic Church

Yanjing Catholic Church, located next to the Yunnan-Tibet Line (National Highway 214) in Markam County and founded by French missionaries in 1855, has a fairly long history and profound cultural heritage.

The size of the church is not very large, covering an area of about 6,000 square meters, and it was originally a tripartite circular arch-shaped vaulted building made of earth and wood. However, because the church was severely damaged during the "Cultural Revolution", it was restored in 1986 with the approval of the People's Government of Markam County, and finally rebuilt into a Tibetan-style flat-roofed building. Today, the entire church covers an area of 12,225 square meters, with a building area of 2,618 square meters, fully reflecting the characteristics of cultural exchange and integration between China and the West. There's no longer "the top of the main gate protruding into an arch", but a vertical gate with eaves, beam columns, and colored paintings, etc. with typical characteristics of local Tibetan residences, which is completely different from the Gothic church style in other regions outside the area.[1] The exterior is pure white, with a clock tower standing tall. And the courtyard exudes a unique charm; in addition to the images of Jesus and the Virgin Mary, and some decorations, there are also traditional red lanterns and white Khatas. The largest part of the building is the church hall. When you push the door open and step in, there's a lot of room inside, which is hundreds of square meters. It is mainly colored in gold, without complex paintings or exquisite sculptures, and is relatively

---

1 Wang Yanbo, Yang Jianing. Ethnographic Observation on the Spread of Religions in Yanjing Area [J]. Today's Massmedia, 2019(10).

simple. But the ceiling is painted with murals based on the themes of the New Testament and the Old Testament. Because of its unique architectural style, which is a combination of Chinese and Western elements, i.e., the Western architectural style and the architectural style of the local ethnic minorities are fused together, it is quite different from the rest, and was even rated as one of the "Top 10 Most Beautiful Churches in China".

Moreover, it is said to be the only Catholic church in the entire Tibet Autonomous Region.

b. Gangda Temple

From the branch of National Highway 214 up the mountain, surrounded by springs along the way, Gangda Temple is located at the spring source at the foot of the steep Zhaguxi (which means "open the cliff gate" in Tibetan) Canyon. Built in the mid Ming Dynasty, it belongs to the Gelug school[1], and has a history of over 500 years. The original site was on the mountaintop in Juelong Village, but due to its age, the temple almost fell into disrepair. Later, with the help of the government, a new site was built at the foot of the mountain. The temple, also damaged during the "Cultural Revolution", was repaired in 1988. The main building consists of two parts: the assembly hall and the Dharma protector temple, covering an area of 2,300 square meters. As the gilded top of the temple is particularly dazzling, it's not an exaggeration to describe it as resplendent and magnificent. This is also the entrance of the Ancient Tea-Horse Road. Along the trail forward, the six syllable Mantra (Om mani padme hum) left by the horse caravans in those days on both sides of the cliff can still be seen vaguely. As a holy place for Tibetan Buddhism in Yanjing area, this temple is well-known in Markam County, and even has a great reputation throughout the entire Tibetan region.

1 One of the four major sects of Tibetan Buddhism, founded in the 14th century and known for its rigorous academic tradition and emphasis on moral discipline.

c. Princess Wencheng Temple

Many people know that in Yushu Tibetan Autonomous Prefecture, Qinghai Province, there's a "Princess Wencheng Temple", which is said to be built by Tibetans in the Tang Dynasty in memory of Princess Wencheng, and has become a famous tourist attraction. In fact, there is also a "Wencheng Princess Temple" in Qamdo, Tibet Autonomous Region. It's located at the entrance of the Zhaguxi Canyon in Juelong Village, about 110 kilometers aways from the county seat of Markam County. This is an eye-catching building with strong Tibetan colors.

The entire internal space of the temple is not very large, and in the middle, the main object enshrined is a huge rock. On the rock, there are naturally formed statues of Tibetan King Songtsen Gampo and Princess Wencheng of the Tang Dynasty. With exquisite shapes and smooth curves, they give a deceptively lifelike impression. The wonderfulness of the nature is thus highly praised. As local Tibetans believe that this large rock has a divine aura, they'd come to worship it one after another, and the surrounding area is often covered with offerings. According to the staff, this large rock was originally on a cliff next to the canyon, far above the ground, but later it tumbled down the slope along with many other rocks. While clearing the rocks, people discovered this magical big one, and hence built the current "Princess Wencheng Temple" on the site. There is also a stone carving of Shakyamuni Buddha on the right side of the temple; according to the research of the famous Tibetan scholar Chen Qingying, it was carved during the Trisong Detsen period of the Tubo Dynasty.

Due to its profound historical and cultural heritage, this temple has become a key cultural relics protection unit at the district level in the Tibet Autonomous Region.

2) Historic Sites

Salt cultural heritage of the historic site category mainly includes ancient ruins, ancient tombs, stone carvings, tablet inscriptions, murals, and tombs of modern and contemporary celebrities, etc. According to the *Convention Concerning the Protection of the World Cultural and Natural Heritage* promulgated by UNESCO on 17 October, 1972, a historic site refers to a place for human engineering, or natural and artificial joint engineering, as well as archaeological engineering, with outstanding universal value from the historical, aesthetic, ethnographic, or anthropological point of view.[1] The resources of Yanjing in this respect mainly include the following parts.

a. Millennium Ancient Salt Pans

With a history of 1,300 years, Markam ancient salt pans are distributed in a narrow strip of nearly 500 meters on the east and west banks of the Lantsang River, starting from Lajiuxi Village in the north and ending at Muxu Village in the south within the Naxi Ethnic Township. There are 2,655 salt pans, with a total area of 7,965 square meters; there are more than 300 brine ponds and more than 60 brine wells of different sizes.

Most of these salt pans, built leaning the mountains, seem to be hung in the air in spite of multiple wooden pillars; arranged neatly on top of each other they look like checkerboards. On top of each shelf, a layer of soil is spread and compacted, and a low mud bank is built around it for the purpose of drying brine; in addition, a small number of salt pans are made of soil.[2] Under the background of the blue sky, together with sunshine and wind, the colorful dazzling graphics are presented to realize the dreamlike "eyeshadow palettes" stretching across the vast ground. These salt pans represent an ancient culture type that is still well preserved in the world. And the unique primitive way of drying brine and the exceptional natural environment in the local area have also formed one of the world's wonders, i.e., nature and humanity are perfectly integrated. The most primitive artificial salt-making site is still intact here, and the oldest salt-making technology is still being used, known as the "living fossil"

---

1 *Convention Concerning the Protection of the World Cultural and Natural Heritage*.See Zhang Juan's *Environmental Science Knowledge*, published by Popular Culture & Arts Publishing House, 2008 edition, p. 172.

2 See *Mangkang County Annals*, p.202.

on the Ancient Tea-Horse Road.

In June 2008, Yanjing ancient salt pans were included in the second batch of national intangible cultural heritage list. In 2013, they were listed as a national key cultural relics protection unit.

b. Zhaguxi Stone Carvings and Statues

Zhaguxi stone carvings and statues are situated in the Zhaguxi Canyon, 3 kilometers west of the Juelong Village; with majestic peaks facing each other on both sides, the drop height of the cliffs is about 600 meters, and widest section of the canyon is only about 100 meters. Tibetan scriptures, six syllable Mantra, and Buddha statues, etc. of all sizes can be found everywhere on the cliffs; some of them, overlapping each other and vaguely indistinguishable, seem to have been through the long ages of the past. Moreover, the cliffs are densely covered with karst caves, which are said to be the place where eminent monk and living Buddhas used to devote themselves to cultivation.

This stone carving group, facing south, covers an area of about 150 square meters. There are a total of 4 statues, respectively carved in two rocks. One of them is 2.4 meters high and 4.1 meters wide, with the statue of Songtsen Gampo carved in the middle of the bottom and the statues of Princess Wencheng and Princess Bhrukuti on either side. The statue of Songtsen Gampo, with an overall height of 1.34 meters and a sitting height of 0.28 meters, sits cross-legged on a lotus throne, wearing a long turban hat and a long robe with a large lapel collar. Its right hand is in a gesture of expounding Buddhist doctrine, while its left hand is placed on the knee pointing to the ground, and the back is shining with Buddhist light. The statues of Princess Wencheng and Princess Bhrukuti, dressed in long skirts with large lapels, stand respectively on the left and right of Songtsen Gampo, each with a height of 1.1 meters, and each decorated with a peach-shaped headlight and an oval backlight. The former's left hand holds a lotus branch in front of her chest, the right hand hanging down, while the latter's right holds a lotus branch in front of her chest, the left hand hanging down. The other is 1.40 meters high and 0.90 meters wide, in which is carved the statue of the Tibetan minister blon-stong-btsan sitting cross-legged on another lotus throne, with an overall height of 0.96 meters and a sitting height of 0.21

meters...

Zhaguxi stone carvings belong to the high relief statues of the Tang Dynasty. But in addition to years of exposure to the weather, pilgrims often apply yak butter on them, so the heads and faces are already seriously damaged. However, judging from these rock carvings, they can still reflect the style of statues in the Tubo period.[1]

"a thin strip of sky" in the Zhaguxi Canyon

a part of the cliff carvings in the Zhaguxi Canyon

c. Yanjing Martyrs Cemetery

Yanjing Martyrs Cemetery, located in Upper Yanjing Village of Nakhi Ethnic Township, is right behind Yanjing Middle School, facing west. A total of 54 martyrs were buried in the cemetery, all of whom died a glorious death when quelling the Tibetan rebellion in 1959. Built in 1960 and destroyed during the "Cultural Revolution", the cemetery was rebuilt in 1988, and now it has become an important historical revolutionary relic, a county-level cultural relics protection unit, as well as a patriotic education base for Yanjing people.

The cemetery once covered an area of 2,505 square meters, with a length of 51 meters from east to west and a width of 50 meters from north to south. On both sides of the gate, such an inscription was written in running script: "To lay the foundation of everything, those died in the revolution are still living on in spirit; To destroy the thousand-year shackles, how can people's soldiers be afraid of death?" Near the gate inside, there's a monument, which is 8 meters high and 1.3 meters wide. On it, such seven characters were inscribed in running script: "Monument to Revolutionary

1 See *Mangkang County Annals*, p.203.

Martyrs", and there's a five-pointed star at the top. Behind it stands a small stele, 2 meters high and 1.2 meters wide, the front of which is written with "Monument to Revolutionary Martyrs", and the back of which bears the designations of troops. The tombs in the cemetery are all made of earth and stone piles, and there are a total of 9 rows of such tombs. There are 9 martyrs buried in the first two rows, 5 martyrs in the third row, 11 martyrs in the fourth row, and 11 martyrs in the fifth and sixth rows respectively. In 2017, Yanjing Martyrs Cemetery was renovated again, and currently it covers a total area of 3,768 square meters.

Yanjing Martyrs Cemetery in Markam County

In addition, there's an iron-chained wooden bridge in Yanjing that only allows pedestrians and horse caravans to pass through. Since the opening of the "Ancient Tea-Horse Road", it has always been the most direct passage between the east and west banks, as well as the main channel of political, economic, and cultural exchanges between Tibetans and the Han people. Today, there are already cement bridges passable for automobiles, but as a heritage type with deep humanistic connotations, it's still very appealing to many people.

3) Museum Collections and Those Scattered among the People

Salt cultural heritage of this category mainly includes the salt industry historical relics collected by various museums, libraries, and archives, and cultural relic scattered among the people. In Yanjing, it is currently unknown what important cultural relics are scattered among the people, but the collections are relatively rich, mainly as follows:

a. Yanjing Historical and Cultural Exhibition Hall

The full name of Yanjing Historical and Cultural Exhibition Hall is "China Tea-

Horse Road Tibet Yanjing Historical and Cultural Exhibition Hall", which is the first exhibition hall on the history of the Ancient Tea-Horse Road in China. Here, a large number of historical records, relevant documents, and salt-making objects related to the development of well salt are preserved. They can help clearly understand the mining and production processes of well salt, the salt-making daily life of ancient salt makers, small-scale production simulation devices, and modern well salt production techniques.

As a symbolic building for Yanjing tourism development, its layout reflects a sense of hierarchy: the first part introduces the Ancient Tea-Horse Road, the second part introduces the life of the horse caravans, the third part is about the history and culture of Yanjing, and the fourth part is about the folk tradition folk customs in Yanjing (including ethnic costumes, foods, and religions, etc.). The exhibition hall also displays a large number of books related to Yanjing, such as *Local Geographical Records of Yanjing, Batang*, *Yanjing County Annals*, *Annals of Prefectures and Counties in Tibet*, *Archives of Sichuan-Yunnan Border Affairs in the Late Qing Dynasty*, *The War between the Jiang and Ling Kingdoms*, *King Gesar*, *Relationship between the Naxi and Tibetan Nationalities*, *Markam County Annals*, and *Tibet Travel Records*, etc., as many as more than 50. (see figure 6-5)[1]

Among them, the compilation of *Local Geographical Records of Yanjing, Batang* began at the end of the Qing Dynasty. As the only local chronicle in Tibet at that time, the recorded area in it had not been detailed in any other local chronicles before, and there were abundant historical materials preserved. Therefore, it has a position worthy of attention in the history of Tibetan local chronicles in the Qing Dynasty. The Tibetan epic *The War between the Jiang and Ling Kingdoms*, known as *Defending the Salt Sea*, is not only a legend handed down based on the battle for salt, but also one of the "Four Great Histories of Demon Conquering" in the Tibetan heroic epic *King Gesar*.

1 Li Hechun. *Impetus and Shackles: Salt and Salt Taxes in the Lantsang River Canyon* [M]. Guangzhou: Sun Yat-sen University Press, 2016: p. 224.

the facade of Yanjing Historical and Cultural Exhibition Hall

some local documents of Yanjing displayed in the exhibition hall

b. Horse Caravan Folk Museum

Established on June 8, 2020, Markam County Yanjing Horse Caravan Folk Museum is a horse caravan museum with unique characteristics and viewing value in the entire eastern Tibet region. Compared with the horse caravan museums in Yunnan and other regions, it is not very big, and even a little old and shabby, looking more like a former caravanserai. However, this horse caravan museum is a small courtyard worth visiting, because there are not only relatively well-preserved ruins of the Ancient Tea-Horse Road, but also a systematic introduction to its starting time, route, and important historical events on it.

Changing from the boring showcases plus text arrangements in a traditional museum exhibition, this museum has adopted the internationally popular open scene layout, and is divided into several spaces, displaying a series of horse caravan items, i.e., some historical relics and relics such as horse lamps, saddles, horse bells, horseshoes, leather strips, headstalls, shelves, stone troughs for horses to drink water, and other historical remains and relics, and showing the history of horse caravan culture, horse feeding, and the making of horse gear, etc.(see figure 6-?) Here, visitors can closely observe and experience the way of life in the caravan transportation era. Through the dust-laden years, various household items, which have withdrawn from the historical arena, tell people about the adventurous spirit demonstrated by the horse caravan ancestors in coping with harsh natural environments, the spirit of solidarity and mutual assistance in long-distance journeys, the spirit of honesty and integrity in business exchanges, and the spirit of assiduousness demonstrated when leaving home, often exposed to cold and wet on journeys.

Today, this museum has become an important window for studying and excavating the horse caravan culture and inheriting the historical culture of the Ancient Tea-Horse Road.

Figure 6-? the outside and inside of the horse caravan museum

c. Literature and Archives Remains

The literature and archives mainly include official documents, letters, inscriptions, manuscripts, drawings, leaflets, slogans, cartoons, newspapers, books, and magazines, etc., especially material evidence related to important events. In Yanjing, there are indeed many such relics, and it is worth mentioning those classic documentaries about Yanjing.

Among them, there are not only "Journey to Yanjing" (part of the documentary series *Tibetan Temptation*) brought out by the local Tibet Satellite TV in 2010, and "Tibetan Code: the Millennium Ancient Salt Pans" (also part of the documentary series *Tibetan Temptation*) brought out by it in 2012, but also several documentaries filmed by CCTV, including: "The Ancient Tea-Horse Road", as the fifth episode of *Heavenly Tibet*, the earliest documentary about Yanjing shot in 2007, is very vivid and worth watching; *tshwa-khron*, shot in 2011, which has more in-depth content and involves more interviews, is more realistic, with pretty good dubbing and visuals; and *The Faraway Home*, the 2013 CCTV documentary series exploring the millennium old salt pans...

Moreover, there are also some documentaries shot abroad, such as "Yanjing Salt, Daughter Xin", the fifth episode of the Yanjing documentary *The Ancient Tea-Horse Road* jointly produced by South Korean KBS TV and Japanese NHK TV, which is first-class in terms of content, plot and visuals; and *The Last Salt Caravan*, shot and aired in the United States in 2009... In addition, documentaries such as *Yanjing Nakhi People,* and films related to the theme of Yanjing, such as *One Mile Above* and *Ticket*, are also worth paying attention to. And *Yanjing Nakhi People* even won the mid-length documentary Grand Jury Award of the Al Jazeera International Television Festival for its perfect cinematography and the theme of focusing on the social survival of female ethnic minorities.

"Yanjing Salt, Daughter Xin", the fifth episode of *The Ancient Tea-Horse Road*

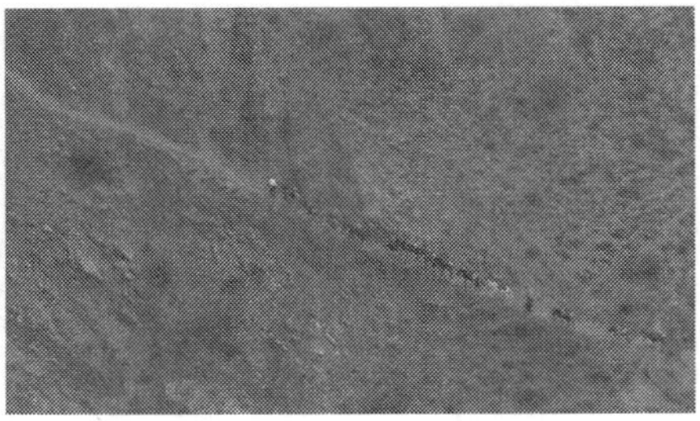

a still of *The Last Salt Caravan*

4) Intangible Salt Cultural Heritage

Intangible salt cultural heritage mainly includes various types of salt-related folk literature, performing arts, traditional handicraft skills, special foods, folk festivals, ritual activities, and production and life experiences. In the hundreds of years of salt production history of Yanjing Township, Markam County, a rich and enormous cultural heritage of well salt has been left behind. They were born and flourished because of salt, all imbued with unique regional customs. In addition to various religious cultures, stories and legends, local festivals, literary and artistic activities, and special foods, such as the Legend of Dameiyong, the "Napa" Festival, the Guozhuang Dance, and the Jiajia Noodles, etc. mentioned in Chapter 4 above, there are mainly the following distinctive types:

a. Murals and Sculptures

Murals are a very popular form of painting in Yanjing Township, so named because they are often painted on the walls. The content of the murals is very rich; in addition to the religious subject matters in monasteries and temples, there are also folk architectural decorations and other subject types. The walls of the main temple halls of various sects of Tibetan Buddhism in Yanjing are painted with figures of

Buddhas and gods, etc., and hundreds of painted images constitute a large mural; each painting is separated by auspicious clouds, mountains and rivers, and forests, and each represents a complete Buddhist legend. These murals have intricate and varied lines, and the figures are very diverse and lifelike, with profound meanings.

With strict artistic norms, the scales, compositions, lines, and colors, etc. of these temple murals are required to be consistent with the relevant content of the Buddhist scriptures. The types include Exotoric Buddhism[1], Esoteric Buddhism[2], bla-ma[3] Inheritance, Dharma Protectors, and so on. The Exotoric Buddhism type is benevolent and compassionate. The subject sits majestically and brightly on the lotus throne, surrounded by lotus decorative flowers and halos; for example, the figures of Sakyamuni Buddha and Maitreya Buddha give people a sense of kindness and stateliness. The Esoteric Buddhism type is quiet, mighty, and romantic; for example, the figure of White Tara[4], with its eyebrows like a crescent moon, eyes like double stars, a fair face born delightful, lips like a little vermilion, shoulders draped with necklace of precious stonesand ribbons, is stepping on an auspicious cloud; her upper body is naked and plump, eyes looking down, just like a shy young girl. The bla-ma Inheritance type, as the portraits of the Buddhist masters and pupils in all ages, is mostly realistic and rarely exaggerated. And the images of Dharma Protectors are downright weird, hideous, and terrifying.

In addition, it is also customary to draw decorative patterns, lions, tigers, cranes, and eight treasures, etc. on the partition boards, kitchen walls, and window frames in Yanjing residences (Tibetan-style houses); in their sutra recitation halls and in front of the Buddha niches, the enshrined Buddhas should also be painted, which are both gorgeous and dignified. These murals are highly skilled; the most exquisite ones need to be "outlined in gold", and for some, even such techniques as "dry pasting" and

---

1 A school of Buddhism that emphasizes the external dissemination of the Dharma and the outward manifestation of rituals and beliefs.

2 This school is a secret Dharma mystery, without being empowered, without being taught, it cannot be arbitrarily transmitted and revealed to others, hence the name.

3 In Tibetan Buddhism, it is a respected name for a person who has high moral integrity and deeds and is a model for the world.

4 Also known as the Mother Buddha of Longevity and Salvation, it is the incarnation of Avalokitesvara Bodhisattva.

"gold grinding" are used to keep the brilliance of the picture unchanged for hundreds of years, or thousands of years.

Sculptures in Yanjing Township are also commonly seen in temples. The raw materials are mainly mud, wood, and metal (gold, silver, copper, or iron). The sculptures of Sakyamuni, Amitabha, Manjusri Bodhisattva, Avalokiteshvara Bodhisattva, and Tara, Arhats, and king kongs of all colors, as well as the Dalai Lamas and Panchen Lamas in history are exquisite in craftsmanship and vivid in image. The Buddhist patriarchs and Buddha immortals, civil officials and military generals, mountains, rivers, trees, flowers, plants, cranes, dragons, elephants, lions, rocs, and towers and pavilions, etc. carved with yak butter are both meticulous and lifelike. On the fifteenth day of the first month of the Tibetan calendar every year, yak butter sculptures are arranged into groups with vivid storylines for display in Gelug temples, commonly known as the "Yak Butter Sculpture Exhibition", which may attract monks and laities in an endless stream.

a mural in a Yanjing temple

a golden Buddha in Yanjing

b. Engravings and tsha-tshas

The engravings in Yanjing are mainly woodcuts used for printing scriptures, Thangkas, Mandalas, the Wheel of the Law, wind horses[1], religious ritual patterns, and other mascot patterns. Similarly, there's a special kind of carved wooden board, called "Seba" in Tibetan, with such patterns as the twelve chinese zodiac signs, animals, insects, and plants, etc. on its four sides. By pressing down the four

1 A small square, multicolored piece of paper with a picture of a deity riding a horse.

sides of a "seba" on the kneaded tsampa dough, patterns of different shapes would be imprinted, and these tsampa patterns can be used as offerings or in religious ceremonies.

"tsha-tsha", the transliteration of Tibetan Sanskrit, meaning "truth" or "replication", is commonly known as the small clay Buddha. As a clay iamge shaped using a pre-carved "concave" mold, it's a sacred object for worshipping gods and Buddhas. This kind of clay Buddha can be divided into haut-relief, bas-relief, and circular engravure. A big one is higher than one foot, while a smaller one is only about 1 inch. It might be round, oval, square, triangle, and even of irregular shapes, etc. Among these tsha-tshas, there might be as few as only one figure, or as many as dozens. These mud Buddha, Bodhisattva, and deity figures of different shapes are all made in bold outlines, with different expressions, and absolutely lifelike. Despite their very small sizes, they can be distinguished at one single glance.

Ordinary clay "tsha-tshas", popular among the people, can be seen everywhere in the countryside. Some "tsha-tshas" are pasted with highland barley or other grains on the back as a sign of hope; some are made by mixing gemstone powder into soil to be more valuable; some are placed in pagodas or inside Buddha statues to add more divine aura; and some are made by mixing the ashes of a cremated living Buddha with soil, and then put in a pagoda as sacred objects to show that the living Buddha's soul is still immortal, and those who worship this pagoda will receive blessings. As a kind of religious sacred object, it is usually enshrined in famous or ancient temples, pagodas, or holy mountain caves, or on sacred lakeshores, or on the way for the turning scripture. However, some people may put it in his/her "Gawu"[1] as an amulet, while some Buddhists may carry it with them as an object for meditation and cultivation.

---

1 Gawu, a transliteration of Tibetan, usually refers to a small shrine or box of a protective Buddha hung on the neck or carried on the back. Its texture could be gold, silver, copper, and other metals, and wood.

a "tsha-tsha" in Yanjing area

the local Lamaist devil dance in Yanjing

c. Lamaist Devil Dance

According to legend, the Lamaist devil dance of Tibetan Buddhism originated during the reign of Trisong Detsen in the 8th century A.D. And in 779, after the completion of Samye Monastery--Trisong Detsen had presided over the laying of its foundation stone, Padmasambhava, the eminent monk, was invited from Tintu to consecrate it. According to the content of the Vajra Dance section of the Yoga Tantra in the four teachings of Esoteric Buddhism by Buddha Shakyamuni, the great guru Padmasambhava, by organizing a dharma assembly of Lamaist devil dance to expel the evil spirits and thanks gods, held a consecration ceremony for Samye Monastery. As is recorded in *Padmasambhava*: "When the Sutra translator finished translating the scriptures in the 'Cishi Islet' of Samye Monastery, the elders circled around the 'Wuzi' Hall three times holding the translated scriptures, lined up, put on masks, and drummed and danced, so as to consecrate the translated scriptures, which has been followed to this day." And this is the origin of the Lamaist devil dance in Tibetan Buddhist monasteries and temples.

Each Tibetan Buddhist monastery or temple in Yanjing Township may hold different forms of dharma assembly of Lamaist devil dance every year, but the most ceremonious is probably the "Guqing" Lamaist devil dance held by the Gelug temple. Every year from December 22 to 29 in the Tibetan calendar, all the lamas of the temple would gather in the main hall when the rooster crows for the first time, chanting in unison and respectfully inviting the Dharma Protectors and the earth immortals to gather here, so as to sort out the fortune, misfortune, auspiciousness, and inauspiciousness of the past year and pray for blessings from the gods for happiness in the coming year. On the 27th, preparations for the Lamaist devil dance begins, and on the 28th and 29th, people dance all day long. At the beginning, monk performers, wearing various masks and colorful costumes, and holding demon-subduing dharma-vessels, come in and dance merrily to the vigorous sound of mang horns, the loud sound of *suonas*[1], and the lively sound of drums and cymbals. Their movements are steady and powerful, and their singing is beautifully loud and sonorous, with tempting charm. On the afternoon of the 29th, an exorcism ceremony is also held after the Lamaist devil dance. When the "Ningga" (a devil body made of tsampa) has been carried by dozens of monks and laities to the center of the palce for dance, the 24 Dharma protectors would aim their dharma-vessels at the "Ningga" to perform the exorcism dance. Afterwards, the monks and laities jointly carry the "Ningga" outside the temple and burn it, to show that the devil has already been eliminated, and the coming year will be favourable climatic weathers, the domestic animals will be all thriving, and everyone will be healthy and blessed. Not only that, on the morning of the thirtieth day (i.e. Tibetan New Year's Eve), some temples may perform the children dance and the 24 Dharma protectors dance in front of the figure of Shakyamuni to wish the new year "Tashi Delek" (good luck).

During the "Guqing" Festival, people from nearby and distant villages, dressed up in costumes, go to the temple to watch the Lamaist devil dance, and participate in the ritual of exorcism and sending away ghosts, so as to prevent and eliminate disasters, and pray for peace and happiness.[2]

Moreover, some poems, folk rhymes, and novels, such as *The Land of Water and Milk*, about the suffering of Yanjing people are also popular in the area. It is a book about Yanjing in the "Tibetan Trilogy" by writer Fan Wen. In this book, we can see the splendor and richness of multiculture, and the tempering and collision, as well as the persistence and integration, of multiculture, multiple nationalities, and multi-faith. In addition, the saltern music, painting, handicrafts, calligraphy, and photography, etc., all embodying the salt cultural gene, are telling the changes in social formations and people's production and lifestyles in various times, reflecting the unique and profound connotations.

---

1 *suona*, a woodwind instrument, which was introduced into China from Eastern Europe and West Asia in the 3rd century.

2 See *Markam County Annals*, pp. 235-236.

2. The Unique Value of Yanjing Salt Cultural Heritage

Yanjing historical and cultural heritage vividly records the historical facts of extensive communications, comprehensive exchanges, and in-depth integrations among various ethnic groups with distinctive characteristics, showing the harmonious relationships of brotherly love, mutual help, and good rapport among them, and carrying their common historical and cultural memories. Thus, it is of great significance to explore, protect, and inherit Yanjing historical and cultural heritage for guiding people of various ethnic groups to better recognize and identify with their common historical and cultural memories, and to forge a strong sense of community of the Chinese nation.[1] Some scholars believe that its value is reflected in four aspects: scarcity value, authenticity value, popular science education and archaeological research value, and social value.[2] However, considering that the value of these aspects is self-evident for the above mentioned types of salt cultural heritage, there's no need to elaborate on them one by one. The author will only discuss the salt production method which is their common inner core.

a. Scarcity Value

Historically, the Tibetans have had a tradition and craft of salt production. The salt production in Tubo during the Tang Dynasty was the so-called "charcoal washing method". As is recorded in Fan Chuo's *Annals of Yunnan*: "There was a large salt pond in Kunming City (now known as Yanyuan), later occupied by the Tubo people. Not knowing the boiling method, they poured the salty pond water on firewoods, and burned them into charcoal, from whose surface salt could be scraped off." This is the earliest literature on Tibetan salt production to date. The method involves soaking firewoods in brine (salt water), burning them into charcoals on fire, and then scraping the crystallized edible salt from them. According to the inlanders in the Tang Dynasty, it was simply because of Tubo people's "not knowing the boiling

1 Su Xing, Tian Renbo. Yunnan-Tibet Yanjing Culture: A History of Interaction, Exchange, and Integration of Various Ethnic Groups through Taste Buds. (see https://www.163.com/dy/article/G3ETVRI50550B8ZZ.html).

2 Habib, Zhang Jianlin, Yao Jun, et al. Investigation Report on the Salt Pans in Yanjing, Markam County, Qamdo Prefecture, Tibet Autonomous Region [J]. *Cultural Relics in Southern China*, 2010(1).

method". However, the well salt production in Yanjing Town, Markam County, Tibet Autonomous Region is obviously different from the above record.[1]

From historical records, it can be seen that for a long time, well salt production has mainly relied on the boiling method, and there are few records of salt making by the sun-drying method. The well salt production in Yanjing Town, Markam County, for its unique historical tradition, production process, and production order, has distinct ethnic characteristics, quite different from that in Sichuan and Yunnan. Moreover, different from such salt areas as Nangqian, Qinghai Province, etc., where "'salt pans' were made of stones", i.e., people directly build salt pans with stone and red clay on hillsides or terraces in river valleys so as to make salt, in Yanjing salt area, local people have adopted the method of "putting up wooden frames to build salt pans". This is because local salterns are located in the Lantsang River valley with dangerous terrain, where the land is extremely limited, and there is no flat land for people to build salt pans. Therefore, local Tibetan and Nakhi salt makers, by making full use of the terrain, climatic conditions, and local materials, set up house-like salt shelves with wood on the terraces or hillsides on both sides of the Lantsang River River. At the top of each completed salt shelf, they lay a layer of wood, make it level, and then spread a layer of sand and gravel, with a thickness of 15 to 20 centimeters. After that, they lay a layer of clay and compact it with a salt hammer before smoothing it flat. In this way, the brine can evaporate upwards and penetrate downwards, which is simple yet very practical. Local villagers first draw brine from brine wells on the Lantsang Riversides, then carry it in wooden pails onto the high salt shelves, and then pour it into the salt pans to be sun-dried. After two or three days' exposure to sunshine and wind, it will be crystallized into salt.

The salt-making technique in the ancient salt pans in Yanjing, Markam County has been passed down for hundreds of years. As the only place in the world that still completely maintains the original manual sun-drying salt production, it was selected into the National Intangible Cultural Heritage List in 2008, and was listed as the seventh batch of national key cultural relics protection units in May 2013. It can be seen that this authentic and unreplicable living heritage still exudes fascinating charm

---

1 "National Intangible Cultural Heritage List: Salt Technique in Yanjing, Markam County", see "Tibetan Cultural Memory Column" at https: //www.tibet3.com, 2011-03-30.

to this day.

b. Authenticity Value

Known as the "work of sunshine and wind", Yanjing ancient salt pans have formed a world wonder where nature and humanity are well integrated because of its unique geographical environment. The living heritage of the ancient salt pans, truly demonstrating the traditional salt-making technique, including traditional salt-making technology and experience, salt-making process and systems, production tools, and traditional beliefs, etc., is more in line with the requirements of authenticity than the general static heritage display.

The terrain of these salt pans is both precipitous and majestic, truly a particular scenery in the world. As a unique landscape of sun-drying salt production on the Qinghai-Tibet Plateau, the biggest characteristic of Yanjing saltern is that it has not been deliberately refined by human beings, but presents real scenes of human beings using natural resources, giving full play to regional advantages, and engaging in salt production so as to maintain collective and individual lives. In a narrow strip of nearly 500 meters along both sides of the river, thousands of salt pans are neatly arranged like chessboards in accordance with the large curve of the river valley and mountains--one on top of the other in a picturesque and orderly manner. In different seasons, different scenery can be seen here: in autumn and winter, it's difficult to tell whether it is snow or salt pans, while in spring and summer, it presents a dazzling array of colors. Overlooked from a high altitude, the brine wells are steaming hot, and the brine in salt pans shimmers like mirrors, reflecting the blue sky, white clouds, and mountain ranges. The two complement the azure Lantsang River water and the flowers and trees all over the mountains, creating a breathtaking view. Not only can you enjoy the entire unique and primitive process of salt making here, but also you can see the industrious and honest salt makers, as well as the unsophisticated folk customs. In these densely packed salt pans, the shadowy figures of the busy female salt makers are scattered sporadically in these squares. The figures and reflections moving in pairs are dotted in the incredibly beautiful picture scrolls, which have also become a humanistic landscape with unique charm. As for the timber that supports the salt pans, since it has been used over the years, the saline water has already seeped through and condensed into a thick layer of crystals on its surface, shining brightly under the sunlight. And the stalactite crystal salt below is in a variety of forms. When being there, you feel as if having entered the Crystal Palace; shuttling between the dense columns, you may feel it complicated and confusing, bringing people an unimaginable sense of surprise.

The brine water, brine wells, salt pans, salt makers, river valleys, villages, farmlands, ancient roads, horse caravans, humanity, history, culture, customs, emotions, and feelings, etc. here, all are impacting the concept of modern people, telling about the plateau, telling about the land of snow, telling about nature, telling about history, telling about culture, telling about the people, and telling about the feelings, which is reminiscent of return, nature, simplicity, purity as well as innocence.[1]

c. Popular Science Education and Archaeological Research Value

The ancient salt pans in Yanjing, Markam County are both an important cultural relic with world heritage value and the essence of folk handicraft of well salt production in China. About when the sun-drying technology began, where it came from, the technological changes in history, the geological and stratigraphic characteristics, the relationship between the ancient salt pans and the nearby hot springs, why white salt and red salt are produced respectively on two sides of river, and the role of the ancient salt pans in trade, cultural exchanges, and ethnic relations on the Ancient Tea-Horse Road, etc., it has important scientific popularization education and archaeological research value in traditional salt-making technology, salt industry history, the Ancient Tea-Horse Road research, and so on.[2]

Having made a significant contribution to the history of Chinese salt industry, Yanjing salt production provides valuable first-hand materials for scholars to further understand Chinese salt culture. It can be seen that the historical and cultural heritage in Yanjing, Markam County, with distinctive characteristics, vividly records the historical facts of extensive contacts, comprehensive exchanges, and in-depth integrations between various ethnic groups, showing their harmonious and friendly relationships as well as mutual assistance, and carrying their common historical and cultural memories. Through archaeological analysis of the traditional salt-making

1 "National Intangible Cultural Heritage: Yanjing Ancient Salt Pans", quoted from Markam County People's Government Network.

2 Chen Yiyong et al. Ancient Markham Salt Pans in Tibet: Value of Living Heritage and its Survival Crisis [J]. *World Heritage*, 2013(1).

sites, it may help to grasp the situation of early human production, organization and social systems. It is also of great significance to explore, protect, and inherit Yanjing historical and cultural heritage for guiding people of various ethnic groups to better recognize and identify with their common historical and cultural memories, as well as forging the sense of community of the Chinese nation. At present, archaeologists are very interested in the infrastructures, production technologies, production tools, and exchanges, etc. of these salterns on the Qinghai-Tibet Plateau. For example, Su Bai, a professor of archaeology at Peking University, believes that the importance of Yanjing salt pans in history, culture, cultural relics, landscape, nature, nationality, and religion, etc., as well as its huge potential heritage and cultural relic values--especially as an existing living history, is extremely rare today.[1] Some scholars have also pointed out that regarding the salt pans in Markam County, "the relevant cultural heritage and cultural relics departments, as well as governments of different levels, should do their utmost to ensure that this precious heritage can be preserved in place, so that the traditional salt-making process can be continued."[2] It can be seen that the Markam Yanjing Saltern is of extremely high value in academic discussion.

Standing at a new historical starting point, we should deeply study and fully explore the connotation and value of Yanjing historical and cultural heritage, strengthen its rational development and utilization, and deepen its publicity, education, and guidance, so that it will bloom with the glory of the times.[3]

d. Social Value

For Yanjing, salt is not only a special product, but also plays a major role in the historical and cultural development process of the local society. And it is of important practical significance in shaping the sense of identity, cohesion, and spiritual home, as well as promoting the regional modernization development of local society. It

---

1 "Leaders of the National Cultural Heritage Administration Replying to Mr. Su Bai on the Protection of Markam County Brine Wells" [N]. *China Cultural Relics News*, 2009-09-09(2).

2 Habib, Zhang Jianlin, Yao Jun, et al. Investigation Report on the Salt Pans in Yanjing, Markam County, Qamdo Prefecture, Tibet Autonomous Region [J]. *Cultural Relics in Southern China*, 2010(1).

3 Su Xing, Tian Renbo. Yunnan-Tibet Yanjing Culture: A History of Interaction, Exchange, and Integration of Various Ethnic Groups through Taste Buds. (see https://www.163.com/dy/article/G3ETVRI50550B8ZZ.html).

not only has important historical value and cultural value, but also has great modern social and economic value.

In the past, salt making, as the way of survival for Yanjing people, was one of their main sources of economic income. Yanjing, Markam County was already very famous in the tea-horse trade era; salt sellers drove mules and horses through the canyons which seemed to have been whittled with a knife and hacked with an axe, following the market and exchanging goods, thus making Markam County a famous prosperous place in Tibet. Today, limited by the current prices of salt, the villagers do not earn much. After a year of hard work, the income of a household may be 20,000 to 30,000 yuan. Despite the insignificant economic value of the production of ancient salt pans, the potential economic benefits brought by them are immeasurable. Because of their scarcity and authenticity, these salt pans have extremely high tourism and visual value, "rich saline spring resources, with great economic value, will play a positive role in promoting local economic development as long as they are well developed and utilized."[1] In fact, in an era of increasingly prominent consumer culture, it is completely necessary and feasible to transform the sun-drying salt-making site into a tourist development scenic area. At the beginning of the 21st century, motivated by governments at all levels in the Tibet Autonomous Region, the county began to attract investment and enterprises to develop salt pan scenic area. First of all, as an important means of income for the scenic area, entrance tickets can not only be used to pay wages to workers and ensure its operation, but also be used to pay a certain amount of taxes, which has the effect of increasing the financial revenue of the local government. Secondly, after the development of the scenic area, the improvement of infrastructure can attract tourists to visit and promote the development of consumption sectors such as catering and accommodation in Yanjing Townlet. Thirdly, in Yanjing townlet and Jiada Village, by continuously creating an experiential tourism mode of "Happy Tibetan-style House", tourists can not only visit the salt pans for sightseeing, but also drink yak butter tea, eat Tibetan meals, and even stay at the homes of salt farmers, which has helped to increase their income.

As one of the most representative tourist attractions in Qamdo, Tibet, the

---

1 Wang Yue, Chen Liang. Investigation and Research on the Salterns and Salt Industry Culture in Nangqian County, Yushu Prefecture [J]. *Cultural Relics in Southern China*, 2019(1).

"Millennium Ancient Salt Pans" was rated as a national 4A-level tourism area in 2012, and then listed in the national 5A scenic spot list in April 2017. At present, they are actively applying to be a protection unit of "world-class intangible cultural heritage" and a national 5A-level tourist attraction. Moreover, it is said that eating local white salt can enhance immunity and expel toxins from the body; the more red salt is eaten, the slimmer one will become, and it can cure diseases, too. Combined with tourism, such products as bath salts, body massage salts, and foot soak salts, etc. have already been produced in the local area, and now there are also shower gels and skin care products made from Yanjing salt. These are also a form in which the economic value of the salt-making site can be reflected.

In short, as a traditional industrial heritage that has lasted for hundreds of years, the value of these ancient salt pans is multifaceted. From the perspective of space, Yanjing salt culture can promote people's understanding of the well salt culture, supplement Chinese salt culture, and improve the possibility of integrating national salt cultural resources.

3. Measures for the Collecting and Exploring Salt Cultural Heritage

The salt culture of Yanjing, Markam County, both remote and ancient, has long been in a "wild" state of "self-weakening", and there are still many treasures that have not been excavated and have not been fully utilized yet. In addition, under the erosion of natural disasters and time, lots of salt cultural heritages, including the techniques and customs of salt production in the region, have died out, or begun to die out or to be lost gradually, and the history and culture contained in them will also disappear. Therefore, according to the specific realities and requirements of the new era, it is extremely important to excavate and rescue the unique relics of Yanjing salt industry and folk customs, and and to sort out and refine them. The following effective measures can be taken:

1) Field Investigation

Field investigation method, derived from the discipline of ethnology, also known as anthropological fieldwork, is the most practical working method in the field of ethnological research.[1] Mr. Wu Zelin, a well-known ethnologist in China, pointed out that the field investigation method is "a method for researchers to conduct in-depth systematic investigation and research in practice or on site, so as to obtain first-hand information, also known as field survey or spot investigation".[2] Here, it refers to exploring various forms of salt cultural heritage, including ancient buildings of salt industry and activities related to salt making, salt transportation, and salt sales, etc., by delving into the field under the guidance of the theoretical methods of cultural anthropology, ethnic history and other disciplines, and through on-site investigations, collation of classics, and the investigation of oral non-physical culture. To conduct field investigations well, the following tasks must be done:

a. Plan Formulation and Preliminary Preparation

In field investigation work, the formulation of a plan is often targeted or of special significance, including the outline of field investigation, the outline of cultural relics collection, personnel composition, location selection, investigation methods, time arrangement, and so on. Since all ethnic events contain complex cultural significance, which cannot be easily discovered in the field, it is necessary not only to have certain experience in field investigation, but more importantly, to have a thorough field investigation plan; otherwise, the lack of clear investigation objectives in a field investigation may result in uneven or missing content. Therefore, it is very important to formulate a comprehensive investigation plan according to the needs of the topic selection before the field investigation. Among them, the formulation of the field investigation outline is the most basic and critical, and priority should be given to cultural projects that are on the brink of disappearance or severe loss.

On the other hand, since the effectiveness of the investigation is closely related to the researcher's choice of field location, the feasibility of conducting an on-site investigation in the local area should also be considered, including the understanding of the local language and translation issues, the researcher's familiarity with the local culture, and the acceptance of the researcher by the project stakeholders

---

1 Yang Yuan. Reports from the Field: Ethnic Fieldwork and Intangible Cultural Heritage Protection [J]. *Chinese Museum*, 2006(4).

2 Wu Zelin. Methods of Ethnological Field Investigation [J]. *China's Ethnic Groups*, 1982 (6).

(such as inheritors, local villagers or officials), etc. There are many ethnic groups in Yanjing, Markam County, and the local dialects and ethnic languages are very different from those in the inland region. If the researcher does not have a certain degree of understanding of the local language or does not have suitable translators, the investigation result will be directly affected. After all, as "the other" of the local culture, it is difficult for the researcher to conduct an in-depth investigation on cultural items when they are completely unfamiliar with the local culture. The differences in language and culture will lead to the rejection of the local people. Therefore, the knowledge of the local language and culture is an essential condition for conducting practical regional research. Only by learning the local language of the field, or at least having no listening barriers, can we truly delve into the customs, culture, and other aspects of the field, and truly conduct in-depth observation and interview to obtain more valuable information.

As for the preparatory work of field investigation, firstly, it is to have as much understanding as possible of the ethnic groups and cultures involved in the topic selection by collecting and studying literature materials; at the same time, it is also necessary to pay attention to the research results of predecessors, so as to supplement their shortcomings, and find the best entry point. Secondly, it's the necessary material preparation. This preliminary preparation work is also very important, as it is related to the effectiveness of the field investigation, as well as the completeness and accuracy of field notes. Since Yanjing salt making activities and techniques have always been closely related to the overall local culture, the collection of field data must be comprehensive, and the search for data will involve many disciplines, including ecological environments, historical origins, social organizations, religious beliefs, life customs, cultural arts, and demography, etc. related to the investigation site, which must be inquired and studied before conducting field research. Only in this way can many phenomena in Yanjing salt culture be correctly interpreted.

b. Participant Observation and Oral History Recording

Participant observation is to integrate into a culture, which requires a researcher to delve into the community where the investigation objects are living, work and live together with them for a period of time, and repeatedly and comprehensively observe their production and life practices for data collection. The "participant observation" proposed by Malinowski requires a time period of more than one year, while the time for the field investigation of Yanjing salt culture can be more flexible, adopting a combination of concentrated time period on-site investigation and intermittent fixed-point tracking investigation. At the beginning of the investigation, it is necessary to concentrate on entering the community for a period of time, get close to the investigation objects, comprehensively observe their “daily” living conditions, and personally experience the intangible cultural heritage practice they engage in. After that, special time points (such as traditional festivals, folk festival activities, and large-scale exhibitions or events organized by the government or non-government institution, etc.) will be selected for fixed-point tracking investigation. In this way, it is possible to observe both the normal state of the salt cultural heritage and its state in special situations.

In terms of observation techniques, “unfocused observation” and “focused observation” should be adopted. Unfocused observation refers to the comprehensive and indiscriminate observation and recording of all information related to the investigation objects, while focused observation is to focus on the key or special events in which the investigation objects participate. Based on comprehensive observation of the investigation objects, it is particularly important to focus on the events in which the they participate. For example, when investigating such cultural heritage items as traditional music and dance, it is important to focus on observing the ceremonial activities they participate in, such as marriage customs, sacrifices, traditional festival customs, and folk competitions, etc. These key or special events should be observed and recorded in a focused manner.

Oral history, as an essential part in cultural heritage investigation, is also a key task in the practice of participant observation. In the field investigation of Yanjing salt culture, oral history records should be made focusing on the main inheritors or groups of intangible cultural heritage projects, and presenting their inheritance lineage and inheritance mode. Therefore, when making oral history records, it is necessary not only to list the interviewees’ working experience and awards in a resume style, but also to deeply explore their life experiences and opportunities to enter the intangible cultural heritage industry, as well as their concepts and attitudes towards the intangible cultural heritage projects they hold. And it is necessary to combine the life history and career history of the investigation objects, and delve deeper into the spiritual level to explore their relationship with the intangible cultural heritage projects they hold, so as to outline the relationship between individuals and the intangible cultural heritage as well as local social and cultural development.[1]

The working method of field investigation should be comprehensive, including written recording and image shooting. Written recording is an indispensable method in the investigation, including diaries, work logs, conversation records, document transcriptions, cultural relics descriptions, and so on, in all of which one should pay attention to the authenticity. Besides, these written records must be marked with the date, the place, and the name and age, etc., of the respondents. As for image shooting, which includes picture data, and movie and video materials, a number of themes should be determined in advance so that there will be some emphases in shooting and the investigation process can be recorded completely and truthfully. In addition, in terms of itinerary scheduling, it is necessary to link work at selected spots with that in an entire area. "Working by travelling about in an entire area" means that the researcher doesn't stay in each place for a long time, but he/she can visit more places in order to have a broader understanding of the content of the investigation. "Working at a selected spot" means staying at the selected location for a longer period of time to complete the thematic investigation. The latter is very important, and the former is also indispensable. The researcher can work by travelling about in an entire area first and then carry out thematic investigations. Or he/she can work at selected spots first and then in an entire area. By comparing with each other and complementing each other, he/she can collect comprehensive and systematic field investigation data and cultural relics data.

c. Collation and Analysis of Field Investigation Data

The collation of field investigation data refers to the course of work of processing and sorting out various cultural relics and documents obtained from the investigation, including manuscripts, printed copies, notebooks, ritual books, inscriptions, and other related textual materials, as well as physical objects such as clothing, utensils, furniture, and statues, etc. by applying scientific methods, so as to reflect the overall characteristics of the investigation objects in a systematic, organized, and standardized manner. As an important basis for analysis, it can not

1 Wang Xin. Ethnographic Research on Intangible Cultural Heritage [J]. *Journal of Guangxi University for Nationalities (Philosophy and Social Science Edition)*, 2021,43(06).

only improve the quality of salt cultural heritage data in Yanjing area, but also facilitate the long-term preservation and research of these data. Specifically, it may include: (1) the confirmation of salt cultural heritage data, i.e., to review relevant original data or second-hand materials, find out problems, take remedial measures, and ensure data quality; (2) the processing of salt cultural heritage data, i.e., to classify and summarize the original data provided by questionnaires or lists for investigations, or to reclassify and adjust second-hand data; (3) the data presentation of salt cultural heritage, i.e., to present the processed and collated data in the form of statistical tables, statistical charts, databases, and data reports, etc. The steps for collating the field investigation data of salt cultural heritage are as follows: firstly, to work out a plan for data collation; secondly, to review the investigation data; thirdly, to group the investigation data; fourthly, to summarize the investigation data; fifthly, to compile an investigation collation form, and store it in the database. The reviewed data will be classified into appropriate categories, and made into relevant statistical tables or charts for easy observation, analysis, and application. It should be noted that all kinds of salt cultural heritage materials should be investigated and sorted out simultaneously, and it is best to sort them out in a timely manner on that very day. In this way, it is possible to find out what aspects of the investigation are insufficient, so as to supplement it at any time, lest that important information be missed when the work is over and the situation has changed.

Field investigation data analysis refers to the process of conducting a comparative study of various collected and sorted salt cultural heritage data by various analysis methods according to the purpose of field investigation, as well as drawing investigation conclusions, carrying out countermeasure research, and writing field investigation reports through synthesis, refinement, induction, and generalization. Its essence is leading to conclusions from data and strategies from conclusions through deep processing of data. The analysis of salt cultural heritage data can be divided into: (1) background analysis, i.e., to get to know the causes and backgrounds of the problems, and to grasp the purposes and directions of analysis and research; (2) state analysis, i.e., to describe and evaluate the quantitative manifestations of various aspects of the salt culture phenomenon, and to summarize the various characteristics of the phenomenon; (3) cause and effect analysis, i.e., to identify the internal and external factors that affect the changes of salt culture, revealing the problems and reasons; (4) countermeasure research, i.e., to put forward countermeasures to solve the problems according to the conclusions as well as the enlightenments, problems, and causes of salt culture investigation. The analysis methods of salt culture investigation data mainly include qualitative analysis and quantitative analysis. The former, starting with the qualitative aspect of salt culture, is to judge and reason about the qualitative prescriptiveness of salt cultural heritage by such thinking methods as experience-based judgment, dialectical thinking, logical thinking, and creative thinking, etc., mainly for defining the size, the direction of change, and the speed of development of salt culture, as well as the strengths and weaknesses of salt cultural heritage, people's attitudes, and the nature of the problems; that is, thinking about "what is left here". The latter, starting with the quantitative aspect of salt culture, is to conduct comparative research by using certain statistical analysis methods, so as to explore the essential characteristics and regularity of salt cultural heritage; that is, to think about "how much is left here". Generally speaking, when investigating salt folk culture, we should first conduct qualitative analysis to make clear the nature of the investigated cultural heritage; then, through quantitative analysis, we will further understand their quantity and combination modes.

The data obtained from the field investigation of Yanjing salt culture need to be collated and analyzed, in order to extract deep insights about human society and culture. Only if it is based on this, can a field research report help anthropologists discover laws, explain phenomena, and construct theoretical frameworks. And we have realized that a vast world for the protection of such intangible cultural heritage as well salt is in the field. To fundamentally protect the precious Yanjing salt cultural heritage, we should study it in the same way that ethnologists study ethnic groups and their histories and cultures, i.e., stepping forward and conducting on-the-spot investigations and records in the vast fields of the local area.[1]

2) Comprehensive Social Investigation

In view of the fact that both the tangible and intangible cultural heritages of

---

1 Yang Yuan. Reports from the Field: Ethnic Fieldwork and Intangible Cultural Heritage Protection [J]. *Chinese Museum*, 2006(4).

Yanjing, Markam County have left the imprint of salt culture on the regional culture, we should also pay attention to the overall perspective of regional research so as to find out the current situation of salt cultural heritage in the local area. In addition to the existing singlular old salt well site, former production site, life area, old storage site, salt industry regulatory agencies, and stores and former residences of salt merchants, etc., it is also necessary to conduct a comprehensive investigation and extensive collection of local cultural heritage resources by starting from the aspects of time, space, heritage functions, and so on, and comprehensively grasp such basic materials as the distribution and current situation of salt cultural relics and historic sites, so as to obtain an interpretation of the overall characteristics and connotation of Yanjing salt culture. Moreover, it is necessary to combine the salt culture study with the interaction and development of the social culture in the region, deeply understand the relationship between the two, and be aware of the relationship between the salt industry and the salt production area, transfer area, and sales area, as well as the social and cultural changes along the routes.[1]

This kind of comprehensive investigation work is not to excavate antiques, but to understand reality; not nostalgia, but learning. Our observation must go deep into its inner spirit, and there must be a structural plan or structural observation; otherwise, it is easy to see the minute details but miss the major issue. At the same time, in order to avoid being vague and general, our investigation must go deep into some key "points", through which we may understand the specific situation and vivid reality of ethnic culture.

Yanjing people have lived long and prospered in this mysterious land for generations; by laboring, living, and creating, they have left behind a thick and diverse cultural accumulation and cultural landscape, preserving a very rich cultural heritage. However, prior to this, there was a lack of systematic sorting and research on the history and current situation of the local well salt culture, as well as the cultural relics along the routes. Specifically, the stories and legends of salt merchants and old salt workers are relatively less widely circulated among the people, the collection and refining of salt-related folk culture and art, such as poetry, opera, and

---

1 Deng Jun. Preliminary Study on Ancient Road of Sichuan Salt--Based on the Field Survey of Ancient Road of Sichuan Salt [J]. *Salt Industry History Research*, 2015(2).

fine arts, etc., are not enough, and the way in which people's living customs reflecting the salt flavor are widely recorded and passed down has not yet been systematically presented...Therefore, it is necessary to first conduct in-depth exploration of the local salt industry history and the salt culture. In addition to searching for "cultural blank spots" in the ocean of historical documents, we should also form a cultural relics general survey team (or archaeological team) to conduct a planned and step-by-step general survey of the aboveground and underground cultural relics in the township. Only in this way can we make sense of the local cultural relic situation, and have a specific understanding of the distribution and characteristics of the entire Yanjing salt cultural heritage in various areas.

A cultural heritage general survey focuses on the overall situation of a local culture; that is, through on-site visits, informal discussions, video recordings, and audio recordings, etc., and adopting advance screening, on-site general surveys, and scientific collection, classification, registration, and sorting, it adheres to the principle of seeing the original, checking the original appearance, and recording the original state, so as to gain a detailed understanding of the project's historical origin, inheritance lineage, technical characteristics, and apprenticeship situation. The purpose of a general survey is to "determine the main types and forms, reserves, distribution areas, inheritance scope, inheritance vein, and evolution of the intangible cultural heritages passed down in a certain region, as well as the history of their previous collections; discover storytellers, inheritors, and performers with unique talents who carry a large number of intangible cultural heritages, and to record and collect representative works of the intangible cultural heritages from their narrations or performances; record or transcribe various folk works, skills, and folk activities".[1]

The general survey of Yanjing salt cultural heritage is a meticulous and arduous task, which should not only focus on the historical evolution, overall process, inheritance lineage, and some related customs of each project itself, but also grasp the entire "cultural ecology" environment. Therefore, the following three principles need to be adhered to: (1) Comprehensiveness. The investigation area should include each

---

1 See *Handbook for the General Survey of China's Intangible Cultural Heritage* compiled by the Chinese Academy of Arts under the China Intangible Cultural Heritage Protection Center. Beijing: Culture and Art Publishing House, 2007, pp. 5, 7.

village, and the investigation objects should include different groups of people. What is more, attention should be paid to the investigation of various traditional folk culture carriers hidden and inherited in the investigation area. It is necessary to investigate and record traditional folk culture according to its original living conditions, instead of setting a framework before the investigation and subjectively and a priori discarding certain aspects. (2) Representativeness. On the basis of a comprehensive general survey, we should strive to grasp the mainstream or main forms, works, types, and folk customs of the cultural phenomena, lest efforts be equally divided in the survey work. While discarding the dross and keeping the finer part, we should also pick out representative projects that have a greater impact in the local area, and conduct in-depth investigation. (3) Authenticity. We should respect the history and original appearance of Yanjing salt cultural heritage, and record it truthfully, without any modification or distortion, so as to ensure that the content and results of the general survey are authentic and reliable, preventing the provision of false materials.

The general survey can be divided into three phases: The first phase is a preparatory one, which mainly involves formulating a general survey plan, drafting a investigation outline, making a investigation form, drawing a general survey map, and conducting training for the general survey personnel. The formulation of a general survey plan should be combined with the actual situation, carefully clarifying the objectives, specific tasks, general points, personnel allocation, work steps, investigation methods, and phased achievements, etc. of the survey. On the basis of the general survey plan, a general survey outline should be prepared according to the general survey tasks for the reference of general survey takers in specific investigations. In terms of personnel training, it is necessary to select a group of people, who are enthusiastic about public welfare cultural undertakings and have certain professional expertise in relevant departments and society, in line with the wide coverage of Yanjing salt cultural heritage general survey, to form a survey working group and conduct professional training. In addition, a reasonable division of labor should be made among the personnel participating in the general survey according to the tasks of the survey and their personal expertise. The second stage is to conduct on-site investigations based on the general survey plan and outline. Besides the field investigation mentioned earlier, which involves in-depth understanding of the form and connotation of salt cultural heritage on site, it also includes adopting different investigation methods according to different situations. While doing a good job in the general survey, key visits and sample surveys can be conducted, and small fact-finding meetings can be held, too, so as to give full play to the role of warm-hearted local people. It is necessary to focus on the investigation of towns and villages with relatively rich resources; focus on the selection of more talented and creative individuals who carry a significant amount of salt cultural heritage, for conducting face-to-face interviews, questions and answers, performances, and demonstrations; and it is also necessary to seize the opportunities of local folk festivals (such as solar terms, temple fairs, etc.), lose no time in conducting on-site interviews, and collect relevant pictures, videos, audios, and other materials by various means. The third phase is to summarize the findings of the general survey and write a survey report. The results of a general survey include a survey report, local intangible cultural heritage distribution atlases, lists of local intangible cultural heritage projects, recorded texts of works and investigation charts, audio and video recordings, pictures, as well as collected folk objects and materials.[1] Based on sorting, classifying, and coding the collected general survey data of salt cultural heritage, statistical and analytical methods are used to analyze and summarize them, forming a general survey report that summarizes the types, quantities, distribution, and characteristics of the salt cultural heritage.

Due to the integration and symbiosis of salt-making culture, ethnic culture, and regional characteristic culture, in terms of research methods, a combination of "bottom-up" and "top-down" observation perspectives should be used to reveal the historical and cultural connotations of typical communities in the county through ethnographic "deep description", examine the rise and change of settlements, analyze the deep correlation between salt-making activities and the historical course of settlements as well as the characteristics of social and cultural development, and grasp the overall appearance of the salt cultural heritage on the basis of numerous detailed case studies. Among them, special attention should be paid to the study of "people", particularly strengthening the study of salt transportation groups such as carriers, porters, boatmen, horse caravans, and salt merchants, etc., as well as salt

1 Wang Xin. Ethnographic Research on Intangible Cultural Heritage [J]. *Journal of Guangxi University for Nationalities (Philosophy and Social Science Edition)*, 2021,43(06).

businessmen families, salt shops, stores, and other salt transportation organizations in the area. Firstly, the descendants of the feudal salt businessmen families. Through such true records as their family trees (genealogical trees), etc., as well as memoirs, novels, and other literary forms, they have not only shown the vibrant life and rise and fall history of feudal salt businessmen families to the public, but also vividly reproduced to us the folk culture, language culture, architectural culture, and dietary culture related to salt. Secondly, the living old salt workers. They have created the technology and art of well salt production in their labor practice, and invented many tools for well salt production. Spreading the ancient and long-standing salt culture through oral history or making models of salt production tools, they have become witnesses and observers of the historical development of local salt industry. In addition, it is necessary to collect as many folktales, legends, and proverbs as possible circulating on the salt roads, and make them fresh and full, so as to reconstruct and present the history of local social life.[1]

By carrying out a detailed and comprehensive general survey of the salt cultural heritage, collecting back those salt history and folk cultural relics with strong local and ethnic characteristics, taking photos and video materials that truly record the salt production and the life of salt makers in the area, and grasping the existing condition of local salt cultural heritage, we may provide basic data and information for the protection and inheritance of such heritage. This is also of great significance for formulating local salt cultural heritage protection plans, systems, and policies.

3) Construction of Various Salt Cultural Heritage Archives

Archivization of folk culture refers to the process of transforming folk culture into archives for preservation through texts, audio recordings, photography, video recordings, and digitization, etc., according to the principles of archival science, and using them as a basis for reproduction, restoration, and creation.[2] On October 17, 2003, UNESCO adopted the Convention for the Safeguarding of the Intangible Cultural Heritage, which put forward the requirement for the "archiving" protection

1 Deng Jun. Preliminary Study on Ancient Road of Sichuan Salt--Based on the Field Survey of Ancient Road of Sichuan Salt [J]. *Salt Industry History Research*, 2015(2).

2 Kang Li, Zhou Ming, Wang Xianfa. Motivation and Principle Analysis of Archival Protection of Folk Culture [J]. *Heilongjiang Chronicles*, 2013(03).

of intangible cultural heritage. China formally acceded to the Convention in 2004, starting the work of identifying, recording, and archiving the intangible cultural heritage, and in 2011 promulgated the Intangible Cultural Heritage Law of the People's Republic of China, which legally clarified the requirements for the "archiving" of intangible cultural heritage. At this point, the recording and sorting activities of folk culture in China have officially entered the stage of archiving management, and the archival theory and practical experience have become one of the important pillars of the folk culture protection system.[1]

In Yanjing Township, Markam County, despite the profound accumulation and rich connotation of the salt cultural heritage, a large part of it has not yet appeared in formal documentary records, and there are few records of it in Chinese historical books; thus, its long-standing historical culture might only be slightly reflected from some folklore and epics, etc. Moreover, due to the delayed identification, many cultural relics with significant value have not yet been included in the scope of cultural relics protection, and consequently have not become cultural relics protection sites. Therefore, no matter for the requirement of cultural originality, or the need of social memory construction, or the inevitable requirements of cultural preservation and utilization, it is necessary for us to integrate cultural resources and build more stereoscopic and complete archives of salt cultural heritage. First of all, on the premise of fully understanding the current preservation situation of Yanjing salt cultural heritage, we should be able to identify priorities, grasp the key points, and strengthen its archival transformation in a targeted manner. Among them, the salt cultural heritage with great historical value, which is facing the survival predicament and is mainly inherited orally and behaviorally, should be given priority to the rescue and the establishment of archives. Secondly, drawing on the management experience and relatively perfect information technology of archival institutions, the collected salt cultural heritage archives should be sorted out into categories, and a salt cultural heritage archives database should be established in accordance with the principles of normalization, standardization, and scalability.[2]

1 Tian Liyuan. Research on the Construction of Folk Culture Archives Resources in China [J]. *Zhejiang Archives*, 2015(10).

2 Zhou Yanqing, Zhang Shuniang. A Brief Analysis of the Archival Protection of Hani Culture [J]. *Lantai World*, 2015(29).

Diverse and dispersed, the salt cultural heritage archive includes not only traditional textual archives, but also photographs, audio and video recordings, and multimedia, etc. It manifests as a situation of concentration in a small area and dispersion over a large area; although centralized and sorted out to a certain extent, it is still widely dispersed among various departments and the folk society. Archiving mainly includes two aspects: one for the local well salt culture, and the other for the inheritors (ethnic groups or groups) of the salt culture. The archival materials of the well salt culture include its formation, regional distribution, cultural characteristics, manifestations, endangered degree, and problems existing in the protection, etc. The archiving for the salt culture inheritors is also of great importance. The inheritors, retaining a lot of tacit knowledge in their brains, are the carriers as well as props of Yanjing salt culture. As living carriers of the local cultural inheritance, they directly participate in the inheritance of Yanjing salt culture. However, under the influence of subjective and objective factors, this inheritance mode has its obvious uncertainty and vulnerability. To this end, we must establish inheritor archives, whose collection scope may include the inheritors' basic information, customs, etiquettes, and so on. At the same time, it is necessary to pay dynamic attention to the inheritors and constantly enrich the data, so as to deeply understand their development and evolution.[1]

At present, digital humanities, with its distinctive interdisciplinarity, provides new tools, new perspectives, and new ideas for the research on the construction of archival resources of Yanjing salt cultural heritage. Through the construction of a digitization and informatization system, all collected archival data can be digitized. The text materials can be tranformed into electronic versions by means of high-speed scanners or manual input, and the electronic versions of old photos or other pictures can be obtained by digital camera reproduction or scanner scanning. In addition, all images could be further processed, and according to the requirements of relevant business standards and specifications, they might be adjusted in terms of size, resolution ratio, clarity, color saturation, and brightness contrast, etc., so as to achieve the best effect of image preservation and display. At the same time, a preliminary screening of all types of archives and materials related to the salt cultural heritage, both the existing and the newly collected, should be conducted. All

1 Ibid.

relevant archival materials that are open and unclassified can be picked out, and then sorted, organized, and cataloged for classified preservation, so that the huge amount of data will be quickly retrieved and extracted through a clear directory. Of course, the construction of a digital resource database of Yanjing salt cultural heritage is a project of continuous innovation and long-term accumulation, which needs to be built and improved constantly by relying on the Internet in the future work, following the principle of "practicality, accuracy, timeliness, and standardization", and drawing on the experience from other cases of archival digital resource database construction at home and abroad.[1]

It is worth noting that since the construction framework of the salt cultural heritage archival resources involves many elements, the full play of each element's role requires the active participation of the main body; the archives center alone cannot undertake all the contents, so seeking social cooperation has become an effective way at present. In the process of an archives center's seeking social cooperation, there are usually three multi-subjects cooperation forms. (1) The archives center playing the leading role, supplemented by social forces. In a project, while making independent decisions, the archives center may introduce social forces in a timely manner, and its major objective is to obtain professional guidance. (2) Social forces playing the leading role, supplemented by the archives centrer. Not directly involved in the design of a project, the archives center usually provides support in terms of venues, cultural creativity, etc. (3) Adhering to the principle of consultation and joint construction, the archives center and social forces, each responsible for a certain stage of the work, jointly complete an entire project.[2]

In conclusion, archives are established to preserve the important elements of cultural memories. As salt cultural heritage archives can provide a suitable carrier and "an effective means" for inheriting the salt culture, its archival construction will ensure the originality of Yanjing salt culture, preserve the social memories of local ethnic groups, and serve the local cultural inheritance.

---

1 Li Muwang, Jiang Shibin. On the Construction of Archival Digital Resource Database of "Shun Culture" [J]. *Archives Space*, 2017(08).

2 Huang Xinrong, Ma Yun. Construction of Regional Cultural Archive Resources Based on Holistic Governance [J]. *Beijing Archives*, 2021(02).

CHAPTER SEVEN

Protection and Inheritance of Yanjing Salt Cultural Heritage

In December 2005, the State Council issued "Notice on Strengthening the Protection of Cultural Heritage", which specifically stated that: "Protecting cultural heritage and preserving the inheritance of ethnic culture is an important cultural foundation that connects ethnic emotional bonds, enhances ethnic solidarity, and maintains national unity and social stability, as well as a prerequisite for safeguarding world cultural diversity and creativity, and promoting the common development of mankind." In January 2010, the Fifth Tibet Work Symposium held by the central government put forward the idea of "making Tibet an important cultural protection area with Chinese characteristics", which reflects that the state has attached great importance to the protection of Tibetan characteristic culture. As for Yanjing salt cultural heritage, it's not only the heritage of Tibetan culture, but also an important part of Chinese cultural heritage. Therefore, the protection of Yanjing salt culture and cultural relics there is to protect the cultural heritage of the Chinese nation, and it is also an important part of inheriting and carrying forward the excellent culture of the Chinese nation.

However, despite the fact that the ancient salt pans of Yanjing, Markam County have been included in the national intangible cultural heritage list, and even announced as a cultural relics protection unit of Tibet Autonomous Region, for a long time, due to the lack of correct understanding of the value of Yanjing cultural heritage, its protection work still faces many problems, such as insufficient attention, shortage of protection funds, inadequate protection measures, and a lack of relevant management talent, etc., thus the future is worrying. How to protect Yanjing salt culture and carry it forward for sustainable development, and how to ensure the legend of Yanjing well salt to continue shining in the river of history without being buried, these are the problems we urgently need to face, and the solutions must be found without delay. Based on the previous field investigations and research results, the author intends to discuss from three aspects.

1. Protection Principles

On how to protect the existing ethnic cultural heritages in China, experts and scholars have put forward many principles, mainly including the "people-oriented" principle, the authenticity principle, the integrity principle, the "tangibility" principle, the "living protection" principle, the principle of equal emphasis on protection and development, and the high-quality protection principle, etc.[1] As Yanjing salt culture is a cultural heritage still living in people's real life, the applicability of these principles may need to be further discussed. The author believes that we should adhere to the "live transmission" principle, that is, to protect the "living" salt production mode and the life style and customs of the salt makers in the environment of its survival and development, so as to achieve its inheritance and development. Lest the traditional salt-making craft disappear when the older generation of salt makers pass away, the protection and inheritance of Yanjing salt cultural heritage should stick to the following couple of points:

1) The Principles of Authenticity and Wholeness

Authenticity, sometimes referred to as "facticity", was introduced into the field of cultural heritage conservation in the 1960s. The International Charter for the Conservation and Restoration of Monuments and Sites (also known as the Venice Charter), adopted at the second International Congress of Architects and Technicians of Historic Monuments, which met in Venice in May 1964, stated that it's our duty to pass on ancient monuments truthfully and wholly[2], thus establishing the

---

1 Wang Jushan. Analysis of the Protection Principles of Intangible Cultural Heritage: A Re-understanding of the Authenticity Principle and the Integrity Principle[J]. *Social Science Journal*, 2008(3).

2 Ibid.

significance of authenticity for the cultural heritage protection. There is also a general international consensus on its role as an essential element in defining, assessing, and monitoring cultural heritage.

In the process of protecting, rescuing, and inheriting cultural heritage, no matter what theory and decision-making measures, the only criterion for testing its correctness is to see whether it has well kept its originality and wholeness in practice. The material basis of Yanjing salt culture consists of numerous salt cultural resources that have been preserved through many years of salt industry development, especially the salt industry buildings, production facilities, and relics closely related to the development process of the well salt industry. As a synthesis of various natural and human factors such as local geography, climate, history, customs, etc., they are the product of long-term historical accumulation and changes. Once destroyed, assimilated or alienated, they will lose their uniqueness and even vitality. Therefore, to protect the traditional well salt culture in the original ecological environment, we must first preserve these physical relics, including the original architectural form (including the plane layout, modeling, artistic style, etc.), architectural structure, building materials and technology, etc. In the maintenance and protection of salt cultural heritage, we should unswervingly and completely adhere to the principle of "safeguarding the original state of cultural relics" and the guideline of "prior protection, rescue in the first place". The so-called principle of "safeguarding the original state of cultural relics" means that in the process of cultural relics protection as well as technical treatment and restoration of cultural relics, in addition to comprehensively preserving their appearance characteristics, more importantly, all the historical, artistic, and scientific values inherent in cultural relics cannot be changed or damaged at will. Even if they are still not fully recognized today, they can be left for future generations to study.

Meanwhile, the authenticity principle requires that the preservation of cultural heritage landscapes be premised on their wholeness, including: (1) to preserve the wholeness of the distribution range and internal spatial pattern of cultural sites; (2) to preserve the wholeness of the surrounding environment of cultural sites; (3) to preserve the wholeness of the main architectural ruins and relics of cultural sites; (4) to preserve the wholeness of the historical information and historical environment of cultural sites. In order to understand authenticity scientifically, it is necessary not only to respect the truth of seeking the intact "original state" in local salt cultural heritage protection, but also to ensure the true "original state" of salt cultural heritage reflecting the continuation and change of history.[1]

Moreover, the authenticity protection of Yanjing salt cultural heritage includes not only the authenticity of the heritage itself, but also the authenticity of its surroundings. With regard to the protection of the surroundings, it's necessary to persist in restoring the true historical appearance to the greatest extent. As required by the World Heritage Convention, we should give equal importance to protecting the authenticity of Yanjing salt cultural heritage and its surrounding environment.

2) The Humanism Principle

Worldwide, the idea of "object-oriented" once dominated the cultural heritage protection, but with the progress of cultural heritage theory, the relationship between cultural heritage and people has undergone a transformation from that of "object-oriented" to "people-oriented". In 1975, the European Conference on Architectural Heritage adopted the Amsterdam Declaration, which included stakeholders in the decision-making process of cultural heritage, marking the beginning of the international community's attention to the value and significance of cultural heritage for people. Subsequently, the international cultural heritage community paid more attention to the relationship between heritage and people. In 2019, the World Heritage Committee decided to include the concept of "human rights-based protection of cultural heritage" in the Operational Guidelines for the Implementation of the World Heritage Convention. It can be seen that in the relationship between cultural heritage protection and human beings, the latter is becoming more and more important, and receiving more and more attention. China has also undergone this conceptual change, entering a new stage of development, and in the protection of cultural heritage, the significance of the human element has become even more prominent. The Guidelines for the Opening of Cultural Relics and Buildings and the Guidelines for the Utilization of Large Sites (Trial) issued by the National Cultural Heritage Administration highlight the people-centered, inclusive and participatory sharing

1 Ruan Yisan, Lin Lin. Authenticity in Relation to the Conservation of Cultural Heritage[J]. *Tongji University Journal (Social Science Section)*, 2003(2).

concept of cultural heritage protection in the new era.[1]

As far as Yanjing salt cultural heritage is concerned, to adhere to the humanism principle, on the one hand, we must protect the "owners" who have created and possessed it. As the materialized form and visualized representative of "intangible cultural heritage", "material" heritage such as local salt cultural landscapes and architectural remains is of course very important, but those who have created and inherited such unique salt cultural heritage, as well as their expertise and exquisite craftsmanship in creating these things, are closer to the true "intangible" nature of the salt cultural heritage. And with deeper connotation of "intangible cultural heritage", they are the real spirit of it. If we only stop at the collection and preservation of the material achievements of the salt cultural heritage, but fail to grasp those who are of more fundamental value and can continuously create these classic works, and their artistic wisdom and outstanding craftsmanship, from a certain point of view, it's just like putting the cart before the horse.[2] Hence it is the principal line of Yanjing salt cultural heritage protection to focus on "people" first and highlight their core functions and important roles in the whole mission.

On the other hand, it is also necessary to pay attention to and respect the practical needs of Yanjing people and consider their right to development. This means that the protection of Yanjing salt cultural heritage must not be conditioned on impeding local economic development and lowering salt makers' life quality. The protection should meet the needs of the contemporary people without jeopardizing the ability of future generations for long-term development. What's more, all sentient beings in the world have the right to survival and development, and the right to enjoy human material civilization and social progress. No one with a conscience should set up obstacles or create trouble for the development and progress of other nationalities, or cause them to stagnate in a backward state for a long time with beautiful empty words and become "specimens" only for some people to "watch" and explore. After all, it is rather abstract and empty to talk about the inheritance, protection, and development of ethnic cultural heritage without the foundation of

---

1 See People's Daily (2021-08-10, 13th edition).

2 Ouyang Shuli. A Preliminary Study on the Protection and Inheritance of Intangible Cultural Heritage under the "People-oriented" Principle[J]. *Case File*, 2012(8).

economic development, and it is impossible to last long, either. Therefore, in the face of historical heritage seeking new vigor and vitality, we should give Yanjing people enough space to keep up with the trend of social development, and create conditions for them to enjoy the comfort and convenience of modern life.

As the irreplaceable active subjects in the protection of Yanjing salt cultural heritage, local people are its creators, users, inheritors as well as guardians. So, we must rely on them, fully mobilizing the power of the public, and giving full play to their wisdom and talents.

3) The Integrity Principle

As another important principle of cultural heritage protection, the integrity principle is to protect the connotations and formation elements of cultural heritage, its surrounding environment, etc., and the "content concept" and "environment concept" advocated by it play an important guiding role in the preservation and restoration of tangible cultural heritage such as monuments, historic sites, etc. In the practice of intangible cultural heritage protection, the integrity principle is also used as a "yardstick" by protection workers to measure and require the preservation of intangible cultural heritage in various categories and forms.[1]

Since the 20th century, the integrity principle, as an important concept in the field of cultural heritage protection, has often appeared in documents or proposals formulated by UNESCO and other cultural heritage protection organizations, while the concept of coordinated protection of heritage and its environment has emerged as early as the 19th century.[2] The Venice Charter, adopted in 1964, divides the elaboration of the integrity principle into four aspects: protection, restoration, excavation, and publication, among which protection and restoration are the most detailed parts and the core content of the principle. "The protection of historic sites includes the protection of the environment of a certain scale"; and in the restoration part, it is emphasized that the restoration of monuments and sites should be based on respect for original materials and conclusive literature, the missing parts must be

---

1 Ye Xuan. An Analysis of the Relationship between the Protection of Intangible Cultural Heritage and the Integrity Principle[J]. *Ability and Wisdom*, 2015(15).

2 Liu Kuili. On the Integrated Principles of Non-Physically Cultural Legacy and Protection[J]. *Journal of Guangxi Teachers Education University (Social Science Edition)*, 2004(4).

filled in, and the overall harmony must be maintained.[1]

According to the meaning of the integrity principle, Yanjing salt cultural heritage contains rich and diverse contents and forms, and is interdependent with the specific ecological environment around. Therefore, when protecting a specific object, we cannot take into account the object itself only, but must protect it together with the ecological environment that is closely related to the eco-culture on which it depends, including the social environment and the natural environment. "We need to protect not only the heritage itself, but also the environment around it... This is the way to reflect heritage's style and features." Integrity also includes the historical, scientific, emotional and other connotations of cultural heritage and the elements of cultural heritage formation; for example, the neighborhood should include Yanjing residents' life activities and all related environmental objects. As mentioned in the Venice Charter, the preservation of monuments includes protecting the environment of a certain scale; unless there are exceptional conditions, monuments cannot be separated from the history they bear witness to and the environment in which they are produced, and generally cannot be moved.[2] Three notable features of the protection of Yanjing salt cultural heritage are as follows: the first is wholeness, involving lots of aspects and levels; the second is systematicness, involving lots of elements and links; and the third is its protracted nature, involving the historical destiny of the present and future generations. All these directly determines the strategy of the protection of Yanjing salt cultural heritage.

Behind the integrity, it implies our desire for the return of the place spirit and the continuation of context in Yanjing. Thus, to protect the wholeness, perhaps more emphasis should be placed on the local salt cultural inheritance behind it, rather than on the solidification of the physical environment.

---

1 Legal Affairs Division under the National Cultural Heritage Administration: Suggestions on Protecting the Style and Features of Landscapes and Historic Sites. Selected International Legal Documents for the Protection of Cultural Heritage [G]. Beijing: Forbidden City Press, 1992.

2 Wang Jushan. Analysis of the Protection Principles of Intangible Cultural Heritage: A Re-understanding of the Authenticity Principle and the Integrity Principle[J]. *Social Science Journal*, 2008(3).

4) The Principles of Inheritability and Innovativeness

As a bridge between the past and the present, traditional cultural heritage records the historical and cultural memory of a region, which is of great significance for recognizing and understanding its own origin. Therefore, inheritability is one of the characteristics of culture, and it's necessary for survival to protect Yanjing historical and cultural heritage, ethnic cultural style, and traditional features according to local conditions, and to preserve local cultural memory continuously.

As far as the Yanjing area is concerned, the inheritability principle is to preserve the treasures of salt cultural heritage left by older generations for hundreds of years from the aspects of cultural form and spiritual connotation, so as to maintain its endless life. And it can be divided into four main aspects: (1) The inheritance of traditional customs, which refer to the fashions, etiquettes, and habits established by usage and passed down from generation to generation among local Yanjing people, and which, having a lasting impact on their material life and spiritual life, is one of the basic forms of traditional culture. (2) The inheritance of traditional architecture (known as the solidified art), which is an important symbol of the traditional salt-making culture in Yanjing. (3) The inheritance of traditional literature and art (known as the torch of ethnic spirit), which, represented by ancient literature, traditional operas, traditional painting, etc., is an essential part of the splendid culture of Yanjing ethnic minorities. (4) The inheritance of traditional ideology, which, including theoretical viewpoints, academic thoughts, moral concepts, etc. formed in the long-term historical accumulation, has a profound impact on the values, lifestyles, and development directions of Yanjing people today. Although the traditional Yanjing salt culture, which is relatively stable, retains its basic characteristics in the intergenerational transmission, at the same time it's essentially a living culture, whose specific connotation may change with time. Hence, the protection of Yanjing salt cultural heritage is not an invariable maintenance, or a simple preservation and reproduction, but needs to be given new connotations while integrating classics into modernity through a second-time creation, so as to bring tradition into a new era.

We should not focus on the historical forms of Yanjing salt cultural heritage only, believing that its "past tense" is the most reasonable existence, ignoring or even discriminating against its current state and future development, dividing its development and change, or even artificially "fossilizing" the cultural heritage that is still living. After all, it is fluid in development, and as a living content existing in the life of a particular group, it is a traditional way of development, which is difficult to be forcibly solidified and preserved.[1] If the forms are inherited only, Yanjing salt cultural heritage will lose its historical and cultural value, becoming an empty shell. Take a closer look at the life history of Yanjing salt culture, it is permeated by the laboring people' creativity in the past, present, and future. Thus, to ensure and continue the vitality of Yanjing salt cultural heritage, the most critical is to adhere to the principle of innovativeness. The so-called heritage innovation means that the Yanjing salt cultural heritage, with the vitality of creative transformation, can adapt to the ever-changing objective environment, so as to have an important impact on the reality. Heritage innovation is the source of its sustainable development. In other words, the inheritance means accumulation and quantitative change, while the innovation means leap and qualitative change; the former is connected with the past, and the latter communicates with the future, constituting the inherent contradiction between the formation and continuation of the heritage. An advanced culture rooted in cultural heritage, like water with a source or a tree with roots, has both the continuity of heritage inheritance and the vitality of heritage innovation.[2]

"Creative transformation and innovative development" has been the Chinese expression of Yanjing salt culture in its contemporary inheritance and development. Without innovation, there will be no impetus for development; without innovation, there will be no way forward. Only by taking an innovative path can the millennium salt cultural heritage adapt to the new environment, rejuvenate, and prosper again.

5) The Principles of Scientificity and Sustainability

For the protection of cultural heritage, scientificity is another principle that cannot do without even for a moment. Not only different objects of protection require implementing different means, but also different cultural forms need corresponding professionals to provide scientific protection guidance. From the rescue of endangered forms, to the collection and research of extinct forms, and to

1 Liu Kuili. On the Integrated Principles of Non-Physically Cultural Legacy and Protection[J]. *Journal of Guangxi Teachers Education University (Social Science Edition),* 2004(4).

2 Bao Zhanbin. Protecting Cultural Heritage and Developing Advanced Culture [J]. *Social Science of Beijing*, 2002(4).

the support, revitalization, development, and utilization of living forms, almost every link of conservation work is inseparable from the participation of corresponding professionals and and the necessary guidance of theoretical knowledge. Otherwise, the effect will only be the opposite of what was desired, and it may even lead to unnecessary regrets and disasters under the good intentions of protection.

The protection of Yanjing salt cultural heritage should take scientific research as the basis for legislation, protection, utilization, management, planning, decision-making, etc. It is necessary to make full use of the research results of relevant disciplines and adopt all effective technology and methods to better protect and manage local salt cultural heritage. Of course, the protection and utilization of Yanjing salt cultural heritage is a complex system, and its management requires a combination of qualitative and quantitative approaches, theory and practice, multidisciplinary knowledges, macro and micro researches, all of which cannot do without the scientificity principle. Moreover, scientific protection and rational utilization of salt cultural heritage should be carried out according to market demand, so as to realize the coordinated development of the two. Practice has proved that on the basis of protection, scientific and rational utilization of Yanjing salt cultural heritage is an important way to achieve its sustainable development in the new era.

As a concept about the coordinated development of nature, technology, economy, and society, sustainable development originated from the "World Conservation Strategy" jointly issued by the International Union for Conservation of Nature, the United Nations Environment Program, and the World Wildlife Fund in 1980. The report "Our Common Future" issued by the World Commission on Environment and Development in 1987 made a relatively systematic definition of the concept of sustainable development, which became the most influential among many definitions: development that meets the needs of present generations without jeopardizing the ability of future generations to meet their own needs. In other words, sustainable development is a kind of development that embodies intergenerational fairness, harmony, and sharing in human society. It takes fairness, continuity, and commonality as the three basic principles.[1] The "sustainability" of Yanjing salt

---

1 Song Junhua. The Concept of Sustainable Development and the Systematic Protection of Intangible Cultural Heritage [J]. *Cultural Heritage*, 2023(3).

cultural heritage protection refers to the long-term consideration of its protection and development, and the path of sustainable development; not for the stagnation of history, but for its continuation. This would satisfy its contemporary development without jeopardizing its ability to develop in the future. The local salt cultural heritage is the common wealth of Yanjing people, and we have the responsibility to preserve it intact and leave it to our future generations. It is a far-reaching and wide-ranging systematic project, and guiding the public to correctly understand the significance of its sustainable protection and development is the premise of carrying out related work.[1]

In the long run, only by scientifically and rationally protecting the excellent salt cultural heritage of the Yanjing area, so that local residents can not only benefit from the protection and utilization, but also realize the significance of the sustainable development of local well salt culture, can the sustainable development of salt cultural resources be achieved on the basis of protection.

In short, the ideas, principles, and theoretical understanding of local cultural heritage protection should be in line with international standards. Adhering to the concepts of holistic protection, authenticity protection, in-situ protection, and dynamic protection is the correct attitude for the protection and inheritance of cultural heritage at present. And the above principles provide important reference for the protection and utilization of Yanjing salt cultural heritage. From the perspective of the overall attributes of cultural heritage, the above principles can also be widely applied to the practice of cultural heritage protection.

2. Protection Methods and Strategies

The protection methods of traditional natural heritage and cultural heritage are relatively mature; as long as the protection area of the heritage project is designated, the construction projects that affect the heritage are relocated, the human activities that damage the heritage are eliminated, and technical restoration is supplemented, everything would be all right. However, it is obviously impossible to adopt such a

1 Bao Rong and Zhang Li. Research on the Sustainable Protection and Development of Intangible Cultural Heritage of Ethnic Minorities in Yunnan [J]. *Art Evaluation*, 2017(13).

simple closed protection mode for the salt-making industrial heritage in Yanjing, nor can we simply copy the protection methods of traditional industrial heritage. Since the protection of this traditional salt-making "living fossil" is a project involving a wide range of aspects, strong professionalism, complex content, and long duration, it has become an important guarantee for promoting the sustainable development of Yanjing salt cultural heritage to clarify the responsibilities of different participants and establish a co-construction and sharing mechanism of "the Party committee leads, the government is responsible, departments cooperate, and society participates".

At the same time, in view of the urgency and severity of salt cultural heritage protection, we put forward the following operational protection strategies to build a cultural heritage protection and inheritance system with Yanjing characteristics while drawing lessons from the protection of ancient and modern Chinese and foreign ethnic cultural heritages.

1) Strengthening the Formulation of Policies and Regulations, and Improving the Establishment of Relevant Institutions

To promote the protection and inheritance of local ethnic cultural heritage through policies and laws is the basic guarantee for achieving its sustainable development. The promulgation of the Intangible Cultural Heritage Law of the People's Republic of China in June 2011 shows that the national level has attached great importance to inheriting and promoting the excellent traditional culture of the Chinese nation. However, due to the wide range of ethnic cultural heritage involved, there will be great differences in each item. And with the passage of time and changes in the social environment, the protection and inheritance work is also faced with some new problems. This requires the Yanjing local government to constantly revise and improve the policies and regulations for the protection and inheritance of salt cultural heritage according to the development of the situation and local conditions, and give full play to the functions of relevant departments.

Firstly, laws and regulations should be promulgated to institutionalize protection. The Yanjing People's Government included salt cultural heritage protection in the strategic goals of economic and social development at an early stage, but the lack of perfect supporting policies has led to people's insufficient understanding of the importance and urgency of protecting and rescuing ethnic and folk culture, thus making inadequate protection efforts. The formulation and improvement of relevant laws and regulations can not only pave the way for the protection of salt cultural heritage, so that the protection work of cultural heritage can move towards the path of legalization and standardization with laws to abide by and rules to follow, but also greatly improve local residents' enthusiasm and effectiveness in practical salt cultural heritage protection and utilization. For example, the Regulations on the Protection of Ancient Salt Pans in Yanjing, Markam, Qamdo City, implemented on June 1, 2023, as the first local regulations formulated by the city for the protection of cultural heritage, involves the protection and management, rational utilization, legal liability, etc. of the local ancient salt pans, and meanwhile stipulates the well salt history and culture, as well as the inheritance, promotion, and sustainable utilization of salt-making skills. Its promulgation and implementation will provide an important legal basis for the management and protection, rational utilization, and live transmission of the ancient salt pans.[1] It is worth noting that, while strengthening responsibilities, Yanjing local government should follow the "principle of minimum intervention" advocated by the Venice Charter in order to maintain the authenticity of the physical state and information landscape, so its role is mainly to provide macro guidance to ensure that the benefits of laws and regulations are effectively played, and that good laws are enacted, development is promoted, and good governance is ensured.

Secondly, while formulating the overall protection plan, the legislation should be further refined to enhance operability. Due to the lack of detailed and specific laws and rules to regulate the various specific problems encountered in the protection and management of Yanjing salt cultural heritage, it is still difficult to operate in accordance with them and ensure the effective and permanent protection work. Therefore, it is also necessary to determine the protection objects, principles, and scope, and formulate more detailed laws and regulations. In terms of protection and management, the Yanjing Cultural Heritage Protection Plan can be promulgated, the supervision and guarantee systems for cultural heritage projects can be established

---

1 Yi Wenwen. Tibet Observation: With Legislation for Salt Pans, Qamdo protects "living fossils" of Local cultural resources. China Tibet Network, 2023-05-29.

as soon as possible, such as a publicity system, a protection responsibility system, a periodic protection notification system, and an expert consultation system, etc., and experts and scholars can be regularly organized to evaluate and inspect the protection situation of the project.[1] In order to ensure that the protection measures of the relevant departments are in place and the enforcement is strong, it is necessary to further standardize the protection and management system, especially to define the rights and responsibilities of the management body, clearly stipulate the responsibilities for violating the laws, etc., and "punish action subjects engaged in activities that affect the safety and environment of the ancient salt pans in Yanjing, Markam County." In addition, the current legislation, which is basically an administrative law in nature, mainly regulating the protection and management of salt cultural heritage, should be supplemented with provisions on the rights of inheritors, such as maintaining the resource property rights of salt pans and salt wells, and giving certain financial compensation to salt makers according to the property rights and outputs, etc., so that they can continue to obtain economic benefits from salt production. Only in this way can their enthusiasm for protecting the local cultural heritage be fully aroused.

Thirdly, special protection and management agencies responsible for the protection of salt cultural heritage should be established. As early as 2003, a county-level cultural relics protection unit for ancient salt pans was established in Markam, so as to reinforce the wooden columns of salt shelves and other infrastructure in local salterns. But for now, that's far from enough. In order to meet the needs of the protection and development of Yanjing salt cultural heritage in the new era, the local government should take the lead in establishing multi-level and three-dimensional protection agencies to coordinate and plan its protection and utilization. First of all, a county cultural relics management committee or cultural heritage protection committee can be established with the leaders of Markam County as the supervisor, which is responsible for formulating policies, planning and supervising the protection of salt cultural heritage, and at the same time is responsible for studying and dealing with major issues in the work of cultural relics in the region. Secondly, a special management agency of the cultural relics section or protection unit can be established

1 Yang Siqi, Li Minzhi, Dong Zhengyang. The Protection Mode of Intangible Cultural Heritage in Shaanxi Province and Its Perfection [J]. *Research on Heritages and Preservation*, 2017(7).

in the local Bureau of Culture, which is headed by the county Department of Culture to supervise the protection and management of cultural relics in Yanjing Township, and is responsible for specific cultural heritage protection work, including protection zoning, heritage registration and management, restoration, maintenance, etc. At the same time, the Cultural Relics Section may also set up a "Yanjing Salt Cultural Relics Rescue Office", forming a team of researchers and managerial staff familiar with the operation of Yanjing salt culture, and reserving special funds for subsidizing the protection and restoration of salt cultural heritage.

Only by clarifying the scope and standards of protection work through the above measures, and strengthening the supervision of relevant agencies and individuals, can we ensure that Yanjing salt cultural heritage be properly protected.

2) Raising the Public Awareness of Salt Cultural Heritage Protection and Inheritance

In accordance with the basic requirements of the Party and the state to "put cultural heritage protection in the first place", while strengthening the protection of various types of cultural heritages, we should continue to raise the awareness of their preventive protection in addition to their rescue protection. The inheritance and protection of Yanjing salt cultural heritage is not only the business of the government and relevant departments; as the real creators, users, inheritors, and guardians of local salt cultural heritage, Yanjing residents' joint participation and the joint efforts of the public are even more important. Therefore, the audiences can be cultivated and expanded through education, guidance, training, etc., so as to enhance people's tolerance and cognition of salt cultural heritage in a subjective sense, and root the awareness of salt cultural protection and inheritance deeply in their hearts.[1]

a. Strengthening the Publicity of Salt Cultural Heritage

In 2005, the Notice of The State Council on Strengthening the Protection of Cultural Heritage proposed that "all kinds of news media should introduce cultural heritage and protection knowledge through special topics and columns", so as to give full play to their due responsibility in the protection of cultural heritage. The

1 Luo Jing, zhang Mu. Protect or damage: discussion on the authenticity of intangible cultural heritage [J]. *Heilongjiang National Series*, 2010(4).

existing more than 3,600 salt pans along the cliffs and the most primitive manual salt-making method in the millennium salterns are unique cultural landscapes in the world, embodying the hard work and wisdom of the local people. In 2008, the millennium salt-making technique (well salt sun-drying technique) was rated as a national intangible cultural heritage; in 2012, the millennium salterns were included in the "China World Cultural Heritage Reserve List", and rated as a national 4A level tourist attraction in the same year; in 2013, the millennium salterns were announced as a key national cultural relics protection unit. Moreover, the Yanjing Ancient Saltern has four major advantages in applying for World Cultural Heritage status: firstly, the integration of nature and culture; secondly, the integration of cultural heritage and intangible cultural heritage; thirdly, its combination with the Ancient Tea-Horse Road; fourthly, witnessing the original ecological civilization.[1] By actively sorting out and deeply publicizing local salt cultural heritage items, such as establishing special websites and public accounts, and regularly disseminating relevant information, a unified image of the salt cultural heritage in Yanjing is likely to be built. On the other hand, new media and we-media, with such characteristics as network interaction, illustration with both text and images, a strong sense of involvement, etc., are easy to resonate with the public in communication. Thus, Yanjing people will not only better remember their own history and understand their own local culture, but also feel the excellence and superiority of their own salt culture, gradually forming a unified cultural recognition and cultural pride in the old and new periods. At the same time, by strengthening the publicity and education of heritage protection law, popularizing the knowledge of salt cultural heritage protection, and stimulating the enthusiasm of local communities and the public, more people will be joined in salt cultural heritage protection, nurturing the strength for its inheritance.

b. Strengthening the Popularization of Salt Industry Knowledge

For hundreds of years, Nakhi and Tibetan salt makers in Yanjing, Markam have been making salt, transporting salt, managing salt, selling salt, using salt, etc. The

---

1 Chongqing Aid to Tibet, People's Livelihood Feelings|Markam: Promoting the Application of Millennium Salterns for World Cultural Heritage Status, http://www.changdu.gov.cn/cdrmzf/c100359/202311/1cdb92aaccdb431ba38ee3bf506f055f.shtml.

salt production here, crucial in the history of the salt industry development in Tibet, will certainly play a positive role in promoting the development of local economy as long as it is well protected and utilized.[1] However, as far as Yanjing local residents and ordinary people are concerned, their understanding of this ancient production technology, production scale and distribution, ethnic salt industry customs, etc. is fragmented and one-sided. It's precisely for this reason that under the influence of the market economy, more and more young people, instead of engaging in this ancient salt industry, want to go out for business or other more profitable things. As a result, only some families are still making salt, and the traditional technology is in danger of being lost. The local government and publicity department should organize and implement various public welfare publicity and education activities on salt industry knowledge, such as setting up advertising columns throughout the township to introduce the development process of Yanjing salt industry and its characteristics with concise words and slogans, so that the local people will subtly accept the impact of salt manufacturing on their lives, and the salt culture industry can be truly deeply rooted in their hearts. What's more, experts and scholars should also be invited to the rural area for popularizing the salt industry knowledge and giving special lectures, so as to deepen local people's correct understanding and rational evaluation of the value orientation of salt cultural industry. The popularization of salt industry knowledge is helpful to cultivate the villagers' homeland feelings, and is of great significance for promoting the protection and inheritance of salt cultural heritage.

c. Widely Mobilizing the Masses to Visit Salt Industry Sites and Museums

The foundation of an ethnic culture and the inheritance of a spiritual civilization can't do without carriers. A long-standing culture is carried on ancient ruins, ancient buildings, and various relics of hundreds, or even thousands of years. As sleeping witnesses, the relics of Yanjing salt industry not only carry rich historical information and cultural connotation, but also reflect the vicissitudes and changes of the region. Therefore, it is necessary to mobilize local people to visit these historical sites. By visiting the ancient saltern sites and ancient dwellings in Yanjing, local people can better understand and recognize the wisdom and art contained in their own cultural

---

1 Tao Hong, Huang Jian. Research on the Salt Industry in Yanjing Township, Markam County, Tibet [J]. *Salt Industry History Research*, 2002(4).

background, and at the same time learn lessons from them to avoid repeating mistakes of the past. Similarly, various salt culture museums, including the Markam Yanjing Horse Caravan Folk Museum and the Yanjing Nakhi Township Ethnography Museum, also showcase the plain and long-standing Yanjing complex of the local people through rich cultural relics and artworks. While paying a visit, local people can travel back in time to understand the historical origin, cultural traditions, and the evolution of values of Yanjing area. By personal contact, they will also more truly feel the charm of history, enhance their sense of identity and pride in the salt cultural tradition, and have a deeper understanding of the inheritance and protection of the local salt cultural heritage. When local people are able to understand and appreciate their own unique and outstanding art, architecture, or other forms of expression through salt industry sites and museums, they will cherish and protect these precious resources more, and actively inherit and promote them in their daily life.

In short, the protection of salt cultural heritage must be based on the conscious internal will of local people, rather than solely on issuing executive orders and preaching great truths. Only by attaching importance to multiple means of propaganda, and deeply rooting its value implication in people's hearts, can their active initiative to protect, inherit, and consciously practice it be mobilized, and in turn transformed into a sense of responsibility and mission for protection.

3) Innovating Technologies for Cultural Heritage Protection and Inheritance

With the development of society and the changes of cultural and ecological environment, the number of cultural heritage protection projects is also increasing year by year. This poses a challenge to existing protection methods, including traditional measures such as text descriptions and collection of physical materials, etc., as well as modern technological means such as audio recording, photography, and camera shooting, etc. In order to record and preserve Yanjing salt cultural heritage as much as possible, and make a three-dimensional and vivid confirmation of salt cultural history, we should fully combine the characteristics of the development of the Internet era, and use modern technologies such as digitization, Internet+, artificial intelligence, big data, virtual reality, and cloud intelligence, etc., to promote the technological innovation in the protection and inheritance of local cultural heritage.

a. The Application of Virtual Reality Technology

Virtual Reality(VR) is a new type of technology based on digital technology, which can create a virtual environment similar to the real world through computer graphics and sound effects, etc., and use various specialized devices to immerse users in the environment, achieving direct natural interaction between users and the environment.[1] In terms of the protection and inheritance of cultural heritage, virtual reality technology, with significant theoretical value and practical significance, has great potential.

As for Yanjing salt cultural heritage protection, the application value of virtual reality technology is mainly reflected in several aspects. First of all, virtual reality technology can effectively improve the technical level of salt cultural relics protection. By pre-displaying images of salt production sites and salt cultural relics after restoration, it can be used to test the feasibility of restoration techniques and means, and examine various links in the restoration process as well as durability after restoration, etc., so as to effectively improve the accuracy of cultural relics restoration and pre-judgment and selection of protection methods, while also shortening the restoration period. Secondly, virtual reality technology can achieve the scientific, high-precision, and permanent preservation of endangered salt cultural relics. Through image data collection, a three-dimensional or model database of ancient brine wells, ancient brine storage tanks, etc., might be established to preserve such important resources as the original data and spatial relationships of various types of cultural relics. For those destroyed ruins and relics that cannot be restored to their original appearance, virtual reality technology can digitize them and permanently store them on storage media, leaving a rich wealth for future generations. Thirdly, virtual reality technology can stimulate visitors' interest in participating in the protection and inheritance of salt cultural heritage. Through high-precision surveying and mapping, digital reconstruction, cultural relics image processing, and other techniques, well sinking, well repairing, brine extraction, salt making, and performing arts, etc. in living states can be presented to people in a more intuitive way. While having more fun and interactivity, visitors can break through

1 Sun Fangfang. Application of Virtual Reality Technology in Tourism [J]. *Jiangxi Science*, 2007(3).

the space, time, and other objective limitations, and gain feelings that cannot be experienced personally in the real world, thereby attracting more people's attention to the protection and inheritance of these cultural relics.[1] Furthermore, as it allows viewers to quickly understand the full picture of local salt production and life, virtual reality technology can greatly reduce the passenger flow in these ancient salterns and shorten the duration of tourists stay, thus achieving the purpose of protection.

Although the application of virtual reality technology in cultural heritage protection and utilization in China is still in its infancy, with many problems at both macro and micro levels[2], yet its advantages are naturally in line with the integrity, authenticity, living state, etc., emphasized in the protection and utilization of Yanjing salt cultural heritage. It may inject fashion elements and maintain the youth of local salt cultural heritage forever.

b. The Application of Artificial Intelligence Technology

Artificial intelligence technology (abbreviated as AI), is an intelligent agent that comprehensively uses computer technology, image technology, simulation technology, language technology, multimedia technology, network technology, data technology, virtual reality technology, etc. to help people complete tasks.[3] As a core force in the new round of technological revolution and industrial transformation, it plays an important role in the protection and inheritance of cultural heritage both domestically and internationally. For example, the inheritance and innovation research of Nantong blue calico patterns, the protection of Dunhuang murals, and the inheritance and development of Xiaogan carving paper-cut are all successful cases using AI technology.

Similarly, in response to various threats such as natural disasters, human destruction, and natural aging, etc., faced by some salt cultural heritage sites in Yanjing area, the intervention of artificial intelligence can also provide effective

---

1 Li Deren. The Application of Virtual Reality Technology to Cultural Heritage Protection [J]. *Journal of Yunnan Normal University (Humanities and Social Sciences Edition)*, 2008(4).

2 Zhang Dongning. Research on the Application of Virtual Reality Technology in the Protection and Utilization of Cultural Heritage [J]. *Journal of Zhengzhou University of Light Industry (Social Science Edition)*, 2020(3).

3 Yi Shanbing. Application of Artificial Intelligence in the Protection of Intangible Cultural Heritage [J]. *Science· Economy· Society*, 2020(1).

solutions to achieve goals that traditional methods cannot. For example, by combining drone technology with artificial intelligence algorithms, it is possible to achieve three-dimensional scanning and modeling of heritage projects such as ancient salt pans, salt shelves, and ancient residential buildings, etc., in order to better understand their structures and current states. Meanwhile, artificial intelligence can monitor the changes in these salt industry cultural relics in real time through image recognition technology, identify their defects and damages, and take corresponding protection measures in advance. In addition, artificial intelligence technology can also be applied to the restoration and protection of salt cultural relics. Through deep learning algorithms, cultural relics can be intelligently analyzed, assisting cultural relics protection institutions in formulating more scientific protection plans, and providing reference and guidance for restoration work, so as to improve restoration efficiency and accuracy.

And through machine deep learning, its new role of inheritor of Yanjing salt production cultural heritage is one of its important means of protection, too. Nowadays, the inheritance of most intangible cultural heritage mainly relies on inheritors to continue. However, as some precious intangible cultural heritage, including Yanjing salt culture, lack attractiveness in the inheritance process, many young people are unwilling to learn, which may cause difficulties in inheritance, and in the long run, it's likely to gradually disappear due to the lack of inheritors. In order to solve this problem, artificial intelligence machines can be used for deep learning to teach intelligent agents to assist in local salt cultural protection and inheritance.[1]

With the continuous development of science and technology, the proportion of artificial intelligence in the protection and inheritance of Yanjing salt cultural heritage will gradually increase. Its application is not to deny the traditional technology, but to better meet the needs of various aspects, and give full play to the complementary advantages of the two, so that the ancient local culture can be revived.

c. The Application of Biotechnology

As one of the emerging technologies that has developed rapidly in recent years, biotechnology by biological principles is not only limited to the biological field, but may also be applied to cultural heritage protection and restoration. With many

1 Ibid.

advantages such as green environmental protection, high compatibility, and easy promotion, etc., it has become a new development direction for cultural heritage protection in many countries.

The cultural relics related to the production and living activities of the salt industry in Yanjing area, with a long history, have experienced the impact of external and internal factors for a long time, resulting in varying degrees of weathering. For example, mildew, cracking, and degradation, etc. may appear in organic cultural relics, stone cultural relics, bamboo and wood lacquerware, and other cultural relics, and some of them may even deteriorate or be on the verge of disappearing. Proper use of biotechnological means can improve the efficiency of cultural relics protection, reduce the loss of cultural relics, and thus prolong their "lifespan" to a certain extent. The application of biotechnology in the protection of Yanjing salt cultural relics mainly includes the following aspects: Firstly, biological detection. Some damages of salt-making cultural relics such as colored spots, degradation, corrosion, holes, etc., can be detected by biological methods, including microscopy, molecular biology, biochemistry, and so on. With these new technologies, we can promptly ascertain the source of cultural relic damages; if it is biological, we may first identify the harmful microorganisms, understand their physicochemical property, and then choose appropriate methods to remove them.[1] Secondly, cleaning protection. Unlike mechanical, physical (laser, steam, etc.) or chemical (cleaning agent, water, etc.) methods to remove the surface pollutants of salt-making cultural relics, which will either cause damage to the body of the cultural relics, or have potential toxic hazards and pollute the environment, biological methods, including microbial cleaning technology, microbial desalination cleaning technology, and biological enzyme cleaning technology, will not affect the surface color of the cultural relics while achieving very good cleaning results, and will be safer for cultural relics, the environment, and operators. Thirdly, reinforcement protection. Effectively utilizing biological principles, such as biological calcification bacteria (BCB), may strengthen the stone and delay the deterioration of calcium matrix in stone relics such as ancient brine wells and storage tanks, etc. Similarly, mineral precipitates such as oxalates,

1 Feng Zhengteng. Review of Biotechnology Applied to Cultural Relics Protection [J]. *Hakka Cultural Heritage Vision*, 2020(2).

carbonates, and oxides formed in microbial activities can also improve the durability of cultural relics.[1] After all, if the wooden cultural relics such as salt shelves and ancient folk houses are not preserved properly, especially the former, water-saturated for a long time, the cellulose that plays the role of the supporting structure is easy to be degraded and its strength will be reduced and decayed.

Yet, as a new cultural relics protection technology, biotechnology is not very mature and has limitations, so most of this relic preservation methods are still in the experimental stage. As the research work is continuously carried out, the technical system of "biotechnology--cultural relics protection" will play an increasingly important role in the scientific and technological protection of cultural relics.

The above mentioned are just a few representative new technologies, and there are many more available. With the continuous scientific and technological innovation, the technology used to protect and inherit Yanjing salt cultural heritage is also constantly upgraded and updated. In order to effectively protect Yanjing salt-making cultural heritage and constantly give it a new life, only by applying advanced scientific and technological methods to the protection process can we achieve twice the result with half the effort.

4) Establishing a Joint Protection and Sharing Mechanism, and Promoting Pluralistic Participation to Form Composite Force

The protection of cultural heritage should be in line with the development of the times and the needs of society. The traditional protection method of Yanjing salt cultural heritage always existed in a closed way, with little communication with or acceptance of the outside world, mainly relying on the power of local governments. However, in recent years, with the accelerating process of globalization, the traditional protection mode of Yanjing salt cultural heritage has been unable to meet the needs of society. In order to fully mobilize the enthusiasm and initiative of all parties and attract more people to participate in the inheritance and protection of Yanjing salt cultural heritage, it is necessary to establish a joint protection and sharing mechanism to promote pluralistic participation, so as to comprehensively

---

1 Wu Fasi et al. Advancement of Biotechnology for the Conservation and Restoration of Cultural Heritage Objects [J]. *Sciences of Conservation and Archaeology*, 2022(1).

improve the collaborative innovation ability of local salt cultural relics protection.

a. Establishing an Inter-regional Cooperation and Sharing Mechanism

The protection and inheritance of regional cultural heritage cannot be done without the joint efforts of neighboring provinces and regions. Therefore, it is an important task to enhance cross-regional cooperation between Yanjing and neighboring areas, seek project cooperation, and realize resource sharing, for the purpose of building a joint protection and sharing mechanism of local salt culture.

First of all, in order to broaden the way of thinking in Yanjing salt cultural protection, inter-regional cooperation and exchanges should be actively strengthened. On the one hand, policy communication should be strengthened between the neighboring areas of Yunnan, Tibet, and Sichuan to establish a cooperative mechanism of mutual trust. Through regular high-level meetings or cultural heritage forums, Yanjing may work with other regional cultural institutions to jointly study and formulate relevant policies, and form a consensus, so as to provide policy support for the protection of Yanjing salt cultural heritage. On the other hand, exchanges and cooperation should be strengthened between experts and scholars in various regions of Yunnan, Tibet, and Sichuan. Through academic seminars, joint archaeological projects, and cultural heritage protection projects, etc., Yanjing cultural heritage protection institutions can participate in joint research and exploration of local well salt cultural heritage protection and inheritance methods, sharing experience and improving the protection level.

Secondly, in order to better realize the joint protection of salt cultural heritage, Yanjing and the surrounding areas may consider establishing an inter-regional cooperation mechanism. On the one hand, a regional cultural heritage protection organization can be established, which is composed of cultural heritage protection institutions in Yanjing and the surrounding areas, and is responsible for the protection and inheritance of regional cultural heritages like Yanjing well salt. The organization may formulate relevant conservation policies and standards, and coordinate cultural cooperation projects in neighboring areas, so as to promote the transmission and sharing of regional cultural heritages. On the other hand, Yanjing and other areas can establish a regional cultural heritage database to integrate and share cultural heritage information from nearby places. Through this database, they can better understand the cultural heritage resources in the neighboring areas, and provide reference and support for conservation and inheritance work.

Thirdly, in view of the fact that the sharing of regional cultural heritages is an important goal of joint protection, a sharing mechanism for regional cultural heritages should also be established. On the one hand, the neighboring areas can display and exchange local cultural relics, including Yanjing salt cultural heritage, by setting up a regional cultural heritage exhibition center. Through such a center, these areas can work together to promote and disseminate their regional cultural heritages, attracting more tourists and culture lovers. On the other hand, the neighboring areas can strengthen the digitization and networking of regional cultural heritages, and share and disseminate their information through new media platforms such as the Internet. In addition, these areas can also carry out joint protection and development projects of regional cultural heritages, and work together to explore their protection and utilization modes, so as to achieve win-win development.

In short, the establishment of a joint protection and sharing mechanism of regional cultural heritages is an important way to promote the protection and inheritance of such regional cultural heritages as Yanjing well salt, etc. By strengthening inter-regional cooperation and exchanges, establishing a cooperation mechanism, enhancing protection and inheritance work, and realizing the sharing of regional cultural heritages, we can better protect and pass on Yanjing salt cultural relics and promote the development of cultural diversity.

b. Establishing a Coordinated Development Mechanism between Traditional Protection and Social Management

Due to its wide range, long age, lack of inheritors, and other reasons, the discovery, rescue, and protection work of Yanjing salt cultural heritage is extremely arduous, and far from enough to rely solely on the strength of the government or cultural departments, so it's necessary to strengthen the cooperation and coordination with forces from all sectors of society. The coordinated development mechanism of traditional protection and social management refers to the joint participation of the government, social organizations, professional institutions, and the public in the protection of Yanjing traditional mode, and the protection, inheritance, and development of the well salt cultural heritage can be achieved through consultation and cooperation, division of responsibilities, and resource integration, etc.

First of all, in order to further carry out the rescue, preservation, excavation, maintenance, sorting and utilization of Yanjing salt cultural heritage, a large amount of capital investment is required. If we stick to the traditional management concept, only relying on national and local financial investments, facts have proved that it can not meet the practical needs. Therefore, it is necessary to build a multi-faceted effective model of central, local and private funds by opening up diversified funding channels that combine government allocations, social fundraising, and acceptance of domestic and foreign donations, etc., and reasonably guide non-government funds into the protection and development of Yanjing salt cultural heritage, so as to ensure adequate financial support in its preservation and daily maintenance. At the same time, the basic living of salt makers and folk artists should be guaranteed, lest the local salt cultural heritage go to extinction before we know it.

Secondly, in order to promote the professional, systematic, and sustainable development of Yanjing salt cultural heritage protection, the participation of scientific research institutions is indispensable. The government may carry out cooperative archaeological work centering around the theme of well salt culture, excavate the connotation of salt culture, and organize academic exchanges on the protection practice of well salt cultural heritage with them. For example, with the support of local government, scientific research institutions organize seminars to bring together salt culture research experts as well as relevant researchers to discuss the protection and inheritance of traditional well salt culture, and develop the inheritance and promotion of Yanjing salt culture into a conscious behavior of experts and scholars in their study of local culture.[1] The two sides may also jointly carry out protection and restoration of important monuments and sites, as well as rescue and renovation of relics after disasters, and launch cultural heritage protection projects with regional characteristics, such as joint research on its protection technologies and methods, etc.

Thirdly, public participation is also an indispensable part of the coordination and cooperation mechanism. The government should, by introducing relevant supporting policies, encourage enterprises, especially private enterprises and individuals, to effectively protect the cultural heritage of the salt industry in compliance with laws and regulations. For example, a sound living heritage protection management

---

1 Gong Jian, Mao Bo. Research on the Historical Cultural Value and Protection of "Satern Music" in Southern Sichua [J]. *Salt Industry History Research*, 2016(3).

system can be established, with public enterprises for heritage management. After all, the direct management of local government is prone to bring about the tragedy of the commons, the dissipation of rent value, high management costs and other negative effect. But by entrusting management to public enterprises, it can greatly save costs and effectively avoid the problem of overutilization that may be caused by the operation of private enterprises. Surely, these enterprises can also play to their strengths and enhance their images by sponsoring cultural heritage restoration. In addition, universities can organize their students to participate in the restoration and protection of Yanjing salt cultural heritage, and develop their passion and sense of responsibility for the cultural heritage through practice.

The protection of Yanjing salt cultural heritage requires the participation and support of the whole society. Various stakeholders should create synergy and establish a collaborative working mechanism to jointly promote the integration of local salt cultural heritage protection and regional development.

c. Attaching Importance to International Experience and Cooperation in Salt Cultural Heritage Protection

International experience and cooperation also play an important role in the protection of Yanjing salt cultural heritage. In practice, by learning from the successful cases of other countries and combining the local conditions, we can explore a coordinated development mechanism suitable for the protection of traditional well salt cultural heritage and social management, so as to achieve social development characterized by culturally prosperity.

First of all, by learning from the experience of other countries, Yanjing local government and cultural departments can draw useful inspirations and lessons to improve their own protection level. For example, the "Machiya Inheritance Program" in Wakayama, Japan was a successful practice. The Wakayama municipal government, in cooperation with local residents and businesses, has preserved and restored a large number of traditional Japanese-style machiya buildings, and put them to use. This initiative has not only promoted the preservation and innovative use of traditional cultural heritage, but also created economic value and accelerated local urban development. In addition, the Korean "cultural heritage management series of activities" was also a case worth learning from. Through systematic training and research conducted in cooperation with universities and research institutes, the

Korean Government has improved the quality and competence of cultural heritage management professionals. At the same time, the government has also increased the public attention and participation in the protection of cultural heritage through community education and tourism promotion, etc. This series of activities has effectively raised the level of coordinated development of traditional cultural heritage protection and social management. Of course, Yanjing can also actively share and promote its achievements in salt cultural heritage protection through joining international organizations, attending international conferences, and other channels. Through exchanges and cooperation, we will help to promote the overall progress of global cultural heritage protection.

Secondly, by sharing technology and resources through international cooperation, the efficiency and quality of Yanjing salt cultural heritage protection can be improved. In order to better protect the shared cultural heritage, many countries are generally aware of the importance of transnational cooperation. Through the establishment of cooperation mechanisms, it is possible to jointly formulate standards and principles for cultural heritage protection, and strengthen consultation and cooperation on transnational issues. For example, the World Heritage Convention formulated by UNESCO is an important mechanism for such transnational cooperation. Yanjing can also strengthen cooperation with other countries by jointly carrying out cultural heritage protection projects, which may include co-restoration, shared research, and co-promotion. Through these projects, local experts in Yanjing can work together with foreign experts to protect and pass on each other's cultural heritages. In addition, the protection of salt cultural heritage requires the utilization of a wide range of expertise and resources. For example, some countries have advanced restoration technologies and equipment, which can be shared with other countries to help local cultural heritage restoration and protection. At the same time, some conservation resources, such as funds and talents, etc., can also be effectively allocated and utilized through international cooperation.

Yanjing local government and cultural departments should actively participate in the sharing of experience while learning from the successful practices of other countries, strengthen the establishment of transnational cooperation mechanisms and projects, and share technology and resources. Only through the sharing of international experience and cooperation can we better inherit and protect the precious local salt cultural heritage. Meanwhile, the protection of cultural heritage should be combined with sustainable development, and joint efforts should be made to protect the common cultural heritage of mankind.

Of course, there are far more effective methods and strategies for the protection of Yanjing salt cultural heritage than those mentioned above. In view of the fact that professionals can not only provide intellectual support for the protection and utilization of local salt cultural heritage, but also lead the progress and innovation of science and technology, to innovate the protection and utilization of salt cultural heritage, it is also necessary to build a diversified cultural heritage talent training model for educating and cultivating talents who can adapt to social development and environmental changes. Moreover, efforts should be made to include the ancient salterns as a living heritage site in the World (Cultural) Heritage List as soon as possible, so as to gain wider attention, and strive for more policy, financial, and technical support, for better protection and research of this living heritage site. Positive and constructive ways of protecting it, as such, will make it more vigorous, and thus escape the path of decay that it may embark on.

3. Several Issues Worth Noting

In recent years, the salt cultural heritage protection in Yanjing, Markam has just started, basically in the stage of concept indoctrination, knowledge dissemination, and practical exploration. From a theoretical point of view, its protection concept is shifting from passive to active, and from negative to positive. From a practical point of view, its protection work is moving from normative to scientific, from extended protection to connotative protection, and from extensive protection to intensive protection.[1] Although remarkable achievements have been made in the inheritance and protection of Yanjing salt cultural heritage with the great attention of the local government and the extensive participation of the society, some problems are still unavoidable due to such conditions as incomplete laws, regulations, and rules, etc. Of these confusions, some are unique to Yanjing, while others are common throughout

---

1 Zhao Lili, Nan Jianfei. Research on Status quo, Problems and Countermeasures of the Conservation of Zigong Salt Cultural Heritage [J]. *Salt Industry History Research*, 2010(1).

the country. On the whole, the following aspects are worth noting.

1) Classified Protection with Focuses

As Yanjing salt cultural heritage has a wide range of contents and various categories, it's very difficult to inherit and protect such a large number of cultural heritage resources completely. In order to implement effective protection, instead of simply applying the "one-size-fits-all approach", we must correctly handle the relationship between economic and social development and cultural heritage protection, adopt corresponding protection methods and countermeasures according to different types of cultural objects, make overall planning, classify management, highlight key points, and implement them step by step.

a. Clarifying the Distinction between Material and Non-Material Cultural Heritage

Material cultural heritage, also known as "tangible cultural heritage"; according to the definition of the Convention Concerning the Protection of the World Cultural and Natural Heritage introduced by UNESCO, anything belongs to one of the following categories is of this type: (1) cultural relics: buildings, carvings, and paintings of outstanding and universal values from a historical, artistic or scientific point of view, as well as elements or structures of archaeological significance, and inscriptions, caves, settlements, and complexes of various cultural objects; (2) architectural complex: a single or interrelated group of buildings of outstanding, universal values from a historical, artistic or scientific point of view because of their architectural form, identity, and place in the landscape; (3) historic sites: man-made works, joint masterpieces of man and nature, or archaeological sites of outstanding universal value from a historical, aesthetic, ethnographic, or anthropological point of view. Intangible cultural heritage is a concept that has been developed in relation to tangible cultural heritage, and the United Nations made at least five revisions before it came into being.[1] According to the Interim Measures for the Application and Evaluation of Representatives of National Intangible Cultural Heritage promulgated by the Chinese government, intangible cultural heritage is defined as: "the various

1 Wang Wenzhang. Introduction to Intangible Cultural Heritage [M]. Beijing: Culture and Art Publishing House, 2006:2-7.

traditional cultural expressions (such as folk activities, performing arts, traditional knowledge and skills, as well as related utensils, objects, handicrafts, etc.) and cultural spaces that have been passed down from generation to generation by people of all ethnic groups and are closely related to people's lives. Its scope includes: (1) oral traditions, including including languages as cultural carriers; (2) traditional performing arts; (3) folk activities, rituals, and festivals; (4) traditional folk knowledge and practices related to nature and the universe; (5) traditional handicraft skills; (6) cultural spaces associated with the above expressions."[1]

Yanjing salt cultural heritage also consists of tangible and intangible types. The former includes immovable salt cultural relics such as ancient ruins, ancient tombs, ancient buildings, cave temples, stone carvings, murals, etc., movable salt cultural relics such as important physical objects, works of art, documents, manuscripts, books, etc., of various eras in history, as well as famous historical and cultural streets and villages with outstanding universal value in terms of architectural style, uniform distribution, or combination with environmental scenery, etc. And the latter includes oral traditions, traditional performing arts, folk activities, rituals and festivals, traditional folk knowledge and practice related to local salt-making activites, traditional handicraft skills, as well as cultural space associated with traditional salt cultural expression forms.[2] In December 2005, The State Council issued the Notice on Strengthening the Protection of Cultural Heritage, which clearly stated that intangible cultural heritage and tangible cultural heritage are two major components of human cultural heritage, thus the intangible cultural heritage had achieved the same status as tangible cultural heritage in Yanjing.

However, from the perspective of conceptual definition, we believe that there are the following significant differences between the two: First of all, the morphological differences are the most striking. The tangible cultural heritage is material, visible, touchable, and can be sensed by touch; whereas the intangible cultural heritage, which cannot be touched or perceived by the sense of touch, exists

1 China Intangible Cultural Heritage Protection Center, under the Chinese National Academy of Arts (ed.) Handbook for the Census of China's Intangible Cultural Heritage [M]. Beijing: Culture and Art Publishing House, 2007:23.

2 Wang Jushan, Xia Xiaochen. The Integrity Principle and the Protection of Intangible Cultural Heritage [J]. *Ethnic Art Studies*, 2011(3).

in the spiritual realm. Secondly, there are differences in the subjects of protection. Of each tangible cultural heritage item, its heritage subject and protection subject are one and the same; while the intangible cultural heritage, which is invisible, and can not be touched, its inheritance and continuation must rely on the "relay" of socialized individuals to realize the transmission of living culture from generation to generation. Therefore, there's dislocation between the heritage subject and the protection subject of an intangible cultural heritage item, and its protection work is to a large extent to protect its carrier--the socialized beings. Thirdly, the tangible cultural heritage is non-renewable, and once it disappears, it will disappear forever; while the intangible cultural heritage can be restored through the records of documents and the reference of relevant materials, and it has a certain degree of recoverability. Fourthly, once formed, the tangible cultural heritage has fixed spatio-temporal characteristics, and its original functions will be gradually lost or weakened over time; while the intangible cultural heritage, constantly changing in the process of its existence, its manifestations and characteristics are constantly changing with the changes of time and space, and its cultural functions will not be lost although they may be changed in the process of inheritance. As can be seen from the above, there is a huge difference between the tangible and intangible cultural heritages related to Yanjing well salt. In particular, the intangible cultural heritage, with human beings as the protection subject, is a "living" culture constantly developing with time and space, and is an integral part of the regional cultural ecosystem. Obviously, the "content view" and "environmental view" of the integrity protection principle cannot meet the requirements of the holistic protection of intangible cultural heritage, and the integrity principle in the context of intangible cultural heritage should have its own unique connotation as well as different content and requirements.[1]

Due to different forms of expression of the two, their protection methods will also be different. In terms of the policy "prior protection, rescue in the first place, rational utilization, and strengthening management", for the tangible cultural heritage we mainly focus on the preservation and maintenance of cultural relics, and try to leave them as intact or repair as possible, so as to prevent them from being damaged and lost. But, for the protection of intangible cultural heritage, more attention is

---

1 Ibid.

paid to its inheritance and development. Due to the high fluidity of its inheritance, the inheritors play a key role in this process.[1] Therefore, it is necessary to follow the vitality of intangible cultural heritage, and through the training of inheritors as well as the publicity and education of traditional culture, so that the heritage can be passed on from generation to generation, and will not be turned into museum art.

In short, the protection of tangible cultural heritage takes objects as the target, while that of intangible cultural heritage takes human beings as the target; the former focuses on maintenance, while the latter on inheritance; the former can focus only on the objects, while the latter must relate to the consuming body.

b. Protecting Different Cultural Heritage Items in a Classified Manner

The richness and complexity of the existing forms of Yanjing salt cultural heritage also determine the corresponding diversity of its protection methods. In order to rescue and protect the existing well salt cultural heritage in a comprehensive and systematic way, it is necessary to establish a classification system with scientific generalizability and in line with China's census and protection application practice according to the different survival status and development status of the local salt cultural heritage, and by drawing on international and domestic norms.

About Tangible Cultural Heritage

For the tangible cultural heritage items still in operation, the protection should be strengthened during their operation process; for those that no longer produce salt but the relics still exist, various special technologies should be used to properly protect them, highlighting their cognitive functions, so that they may continue to serve the lives of the local people. The focus is on ensuring its integrity and durability through enhanced preventive protection, routine maintenance, and conservation repairs.

- Maintenance

Maintenance is a major means of daily management, especially for non-architectural salt culture sites, such as brine well sites, brine pools, roads, etc.; meanwhile, salt streets, ancient villages, and other important sites should also be maintained frequently, timely eliminating natural erosion without changing their

---

1 Yi Shanbing. Application of Artificial Intelligence in the Protection of Intangible Cultural Heritage [J]. *Science· Economy· Society*, 2020(1).

original state. Maintenance work, running through the daily maintenance of cultural relics and monuments, requires perseverance. In addition, the ethnic cultural heritage of a single physical type, such as production tools, living utensils, handmade works, musical instruments, costumes, etc., should be kept in their original states.

- Preservation

On the one hand, it refers to the protection of cultural relics and monuments such as salt pans, brine wells, guild halls, salt roads, salt production and living utensils, etc., under the current state of preservation of cultural heritage through museums and cultural institutions; on the other hand, it refers to the preservation of cultural relics and historic sites, also known as protective reinforcement. It is mainly to adopt additional protective measures for Yanjing salt cultural relics, so as to prevent or delay the damage caused by natural forces. In terms of technology, it is mainly to use modern protective materials to protect rock paintings, stone carvings, colored drawings, etc., as well as cultural relics and monuments such as ancient salt buildings and salt merchant guild halls, etc., so as to strengthen them without changing their external patterns.

- Renovation

It means that once damage occurs or is about to occur, we should, according to its actual situation, remove the parts added by the descendants, restore the broken parts, and repair the incomplete parts, under the principle of "restoring the old as the old", in order to maintain the absolute existence of its entity. For example, salt roads should be kept in their original state as much as possible and repaired regularly. The core significance of the millennium Yanjing salt cultural heritage is that it represents a period; so, what the cultural heritage should display and promote is its core cultural connotation, and the physical substances that have nothing to do with it and were simply added by later generations for restoration should be removed to maintain its original purity.

- Reconstruction

It refers to the renewal of the missing or damaged parts of Yanjing salt cultural heritage, such as cultural relics and monuments, under the condition of fully grasping their original state; for example, the renewal of salt industry sites such as streets, shops, and dwellings related to local salt culture. There is a big difference between reconstruction and restoration, for in reconstruction, there must be much and relatively complete remains, and the amount of renewal accounts for only a small part of the stock. However, reconstruction must be conducted cautiously and justified beforehand by experts in the relevant sectors, mainly on the basis of the residual construction. For example, the ancient ruins of the salt cultural heritage that have been severely damaged can be considered for reconstruction.

- Remodeling

Remodeling is also known as functional update. It mainly refers to the moderate transformation of the original Yanjing salt cultural heritage, such as cultural relics, in order to adapt to the current use function, while retaining some parts with historical and cultural characteristics of local salt industry. The act of remodeling and its extent are directly related to the compatible use of the cultural objects and historic monuments. The cultural relics and monuments of Yanjing Salt Cultural Heritage should be utilized reasonably, but such utilization must be compatible with the original, such as the protection and improvement of the Yanjing Old Street Historical and Cultural District.[1]

About Intangible Cultural Heritage

Intangible cultural heritage is the invisible heritage in human history and culture, whose protection needs to follow specific traditional ways and methods, mainly in the following three aspects.

- Rescuing Conservation

The so-called "rescuing conservation" refers to the collection, recording, sorting, filing, preservation, and research of the intangible cultural heritage scattered among the people, which is the basic work in the protection of the intangible cultural heritage in Yanjing area. For such representative items of Yanjing intangible cultural heritage as literature, art, writing, language, classics, etc, including oral traditions, ballads, and folk songs, their contents, forms of expression, and technical processes shall be recorded and sorted out, compiling and printing books, producing audio-visual materials, and establishing archives; at the same time, rescue and protection plans shall be formulated, arranging protection funds as well as venues for transmission, display, and exhibition; appropriate subsidies for regular students shall

1 Zhao Lili, Nan Jianfei. Research on Status quo, Problems and Countermeasures of the Conservation of Zigong Salt Cultural Heritage [J]. *Salt Industry History Research*, 2010(1).

be provided to strengthen the inheritance and protection of traditional knowledge.

- Memory Protection

For the traditional salt-making method and process, it is necessary to select right persons to inherit them, coupled with text and pictures for preservation. For intangible cultural heritage items that have lost their inheritors or whose objective conditions for existence have disappeared or have basically disappeared, the local people's government should establish a memory list and implement memory protection. For example, in terms of institutional salt culture, it is mainly necessary to protect its carriers, such as historical documents, books, inscriptions, etc., record the salt law, salt administration, salt industry system, taxation, etc., of local well salt, which can be systematically sorted out and published for preservation and circulation. The competent departments of culture and tourism should also organize relevant units to carry out investigations on the intangible cultural heritage items included in the memory list, collect relevant materials and physical objects, such as oral traditions that need to be recorded in audio and written forms, establish lists of representative items as well as project memories, etc., and establish databases, archives, and digital protection platforms.

- Productive Protection

As early as August 2009, the Ministry of Culture and the National Tourism Administration jointly issued the Guiding Opinions on Promoting the Combined Development of Culture and Tourism, which stated that "the intangible cultural heritage of the traditional skill category should be rationally utilized through productive protection..." For salt making skills and folk culture resources in Yanjing, they should be explored and utilized so as to be given full play in today's cultural construction. Similarly, representative items of intangible cultural heritage with productive skills and social needs in Yanjing, which can be transformed into commodities or services by means of production, circulation, sales, etc., such as cultural tourism, cultural creativity, cultural sports, etc., should also be protected through production. Relevant departments should provide support in terms of venues, funds, infrastructure construction, etc., strengthen industrial management, and improve the quality of the industry, etc.; they should maintain the integrity of the traditional process and the authenticity of the core skills, and shouldn't change its core skills.[1]

In short, the classification of Yanjing salt cultural heritage protection is to better protect the diversity of local history and culture. We need to take appropriate measures actively to ensure that different types of cultural heritage are properly protected and passed on. It is believed that with the gradual deepening of the protection of Yanjing salt cultural heritage, its scientific and precise level will continue to improve.

c. Defining the Protection Objects and Levels

In the long history, Yanjing salt cultural heritage has a profound accumulation and rich connotation. However, due to the different forms of existence, modes of value manifestation, and environment faced by various cultural heritage items, the intact degree of cultural relics and monuments of different times also varies, which requires differential treatment. At present, what salt cultural heritage resources need to be protected? How are these salt cultural heritage resources classified? There are different views in the academic circles. Therefore, on the basis of establishing the scope of their protection through research, they should be categorized, and divided into different levels according to the degree of importance.

With reference to the protection scope of salt cultural heritages in other regions, they may generally include production remains, management remains, transportation and marketing remains, living facilities remains, sacrificial cultural remains, intangible cultural remains, etc. Specific to the material forms of cultural heritage related to local salt industry, in terms of type and composition, there are mainly ancient salt road ruins, gallery road ruins, courier station ruins, salt shop ruins, salt warehouse ruins, guild halls, temples, ancestral halls, memorial archways, water transport docks, ancient bridges, ancient streets, ancient villages, stone inscriptions, cliff stone carvings, and various salt transport equipment of water and land transport system, etc., as well as intangible cultural heritage related to salt industry activities, such as diet, folk customs, labor songs, etc.[2] According to the current situation of salt

---

1 "Grading and Classification Protection: Weihai Intangible Cultural Heritage Protection Has a Rigid Standard!" https://m.thepaper.cn/baijiahao_5082426.

2 Deng Jun. The Present Situation and Protection of the Cultural Heritage--the Ancient Road for the Transportation of Sichuan Salt[J]. *Journal of Sichuan University of Science & Engineering (Social Sciences Edition)*, 2015(5).

cultural landscape and relics in Yanjing area, the local people's government should conscientiously formulate a protection plan and strictly implement it. By breaking through the existing point-like distribution and achieving "connected points intersect at the line, and the line intersects at the surface, covering the whole", its protection planning should connect the local salt heritage resources into a coordinated and unified complete system, so as to realize the combination of static protection and dynamic protection, as well as the combination of closed protection and development protection. At present, the tangible and intangible salt cultural heritage such as production tools, utensils, customs, ruins, sculptures, cultural classics, etc., should be mainly included in the scope of protection work, and special attention should be paid to the preservation of ethnic characters, ethnic languages, ethnic costumes, ethnic customs, and various cultural relics and monuments, so as to maintain their unique characteristics.

The protection and utilization of cultural relics are treated differently in foreign countries. Taking building relics as an example, Italy divided them into four levels, and the protection requirements are very strict for architectural works of great historical value. The appearance of a lower level cannot be changed, but their structures might be updated, and the interior of an even lower level can be changed to provide convenience for reasonable use. The United Kingdom divided the "listed building" into three grades, with the first accounting for 2%, the second 4%, and the third 94%; the protection requirements for the first and second grades are quite strict, while for the third, internal alterations might be made. Although only 6% are strictly protected, which is not a large proportion, the absolute number is still as high as 30,000. According to the classification standard of China's cultural heritage protection, from high to low, they are of the first, second, third, and fourth grades respectively: (1) The first grade refers to those with a very high historical, artistic, scientific, or cultural value, representing and reflecting the uniqueness of the country, nation, and region. The protection scope of such cultural heritage should be comprehensive and strict, and no form of destruction or alteration should be permitted. (2) The second grade refers to those with relatively high historical, artistic, scientific, or cultural value, representing and reflecting the significance of the country, nation, and region. The protection scope of such cultural heritage should also be comprehensive, but appropriate alterations and restorations can be made over time. (3) The third grade refers to those with certain historical, artistic, scientific, or cultural value, which needs to be preserved and repaired, representing and reflecting the common quality of the country, nation, and region. The protection scope of such cultural heritage should be properly maintained and preserved, but it can be achieved with some reasonable alterations and repairs. (4) The fourth grade refers to those with certain historical, artistic, scientific, or cultural value, which needs to be maintained and preserved, representing and reflecting the common quality of the region. The protection scope of such cultural heritage should be kept to a low level, and necessary repairs can be carried out. According to the actual situation of Yanjing, we should also explore the salt cultural heritage from such different perspectives as the historical background, constitutive characteristics (including structure, materials, technology, artistry, etc.), cultural connotation (including cultural value, history, cultural inheritance, etc.), etc., evaluate its existence and establish local directory classification and grading standards, and form protection and restoration strategies with different priorities. The following two points need to be focused on:

Priority to endangered heritage items. Some need to be preserved and rescued, such as salt-making tools, which are the best materials directly reflecting the sun-drying process, but with the transition to modernization, traditional handmade tools have been disappearing. In addition, the rich and tough horse caravan culture related to salt exchange, with horses gradually disappearing, will also become difficult to be traced. The same is true of the life experience of salt makers. With fewer and fewer families engaged in the livelihood of salt production, this experience will also be erased from the memory with the development of the times, and it will be difficult to present it to society and future generations if it is not preserved and rescued urgently. To avoid the loss of culture, there must be a distinction of order.

Adhering to key protections. In view of the large amount of Yanjing salt cultural heritage, the limited manpower and material resources should be effectively used for the key conservation projects. Firstly, it is necessary to focus on protecting the cultural heritage types with relatively high value. Generally speaking, the expression forms of architectural landscape heritage value, which is more complex, with greater influence on contemporary economic, social, and cultural development, have the most comprehensive utilization value. Secondly, the means and methods for the protection of each type of cultural heritage, which are multi-gradient and multi-level, should highlight the most critical parts to achieve the goal of protection in a domino effect. Protection zones or necessary construction control zones can be established to provide a suitable space for the survival and development of key protection projects, so as to mitigate the impact of the industrialization process on their survival foundations and the impact of the mainstream culture of modern society on them.

In short, the classification criteria for the protection grades of Yanjing salt cultural heritage are the cornerstone of its protection work. Clarifying the content and corresponding grades of salt cultural heritage resources can provide guidance for their protection work, so as to formulate comprehensive and operational protection laws and regulations, which are conducive to necessary restoration and protection.

2) Taking into Account both Protection and Utilization

The Central government of China attaches great importance to the protection and utilization of historical and cultural heritage, and by writing "strengthening the protection and utilization of cultural relics and the protection and inheritance of cultural heritage" into the report of the 19th CPC National Congress, emphasized that the protection and utilization of historical and cultural heritage should be placed in a more prominent position. As far as Yanjing salt cultural heritage is concerned, it is our duty and lofty mission to protect and make good use of it, inherit and carry it forward, and further promote local economic development. But on the other hand, how to effectively protect and rationally use it is also a problem that has troubled us for a long time.

Good work of cultural heritage protection includes early restoration and later utilization, both of which are indispensable. If Yanjing salt cultural heritage is clozed for conservation, it will lose the value of making it "live". So, protection should not be set against development and utilization; instead, we must balance the relationship between them. And we should clearly recognize that the protection of cultural heritage is constantly evolving, whose ultimate effect is bound to benefit the people. Therefore, attention should not only be paid to the historical value brought to us by the local salt cultural heritage, regarding the historical and cultural heritage as only historical relics, and only blindly protecting and inheriting; more inportantly, its practical value must be stimulated to "come alive". The local cultural relics resources should be protected and inherited while being innovatively developed and rationally utilized.[1]

a. Protection and Utilization of Salt-making Villages

At present, the utilization and protection of traditional villages have been continuously emphasized. As a special type, salt-making villages deserve more attention, too. The salt-making villages in Yanjing, Markam County, which have inherited the salt production technology, formed cultural landscapes such as salt pans, brine wells, etc., and demonstrated the influence of salt as a trade and transportation product on the integration of ethnic groups in various villages, are of high cultural value. As one of the first batch of villages with important conservation value, Upper Yanjing Village has even been included in the China traditional villages list. Therefore, it is of great practical significance to protect such villages by making full use of them.

To Create Tourist Attractions with the Existing Natural Landscape

The ancient salt pans have immeasurable tourism value. On the premise of not destroying or reducing the cultural value, make full use of them, and appropriately collect admission fees through the development of tourism, so as to maintain the sustainable operation and management of the scenic spots. Meanwhile, tourist souvenirs such as salt sculptures, salt lamps, salt-making tools, etc., should be developed to increase the added value of salt, so that salt makers can profit from it and have the motivation to engage in salt production, restoring the vitality and vigor of the salt industry. In addition, salt drying activities may be displayed on site in the scenic area, so as to appropriately create opportunities for tourists to participate in production and labor, experience the salt-making process, taste the local food culture, and truly understand the production and lifestyle of salt makers in the past.[2]

To Promote Salt Culture with Related Myths and Legends

There are abundant myths and legends in the local area, which is very helpful to understand the importance of salt in different stages of human history. Therefore, the collection and sorting of myths and legends, and the integration of them into the explanation of the scenic spots, are conducive to enriching people's understanding of

---

1 Jin Yantong. Research on Strengthening Protection and Utilization of Traditional Cultural Heritage in Jilin City [J]. *Journal of Jilin Radio and TV University*, 2023(2).

2 Gao Yue, Zhao Shubin. The Current Problems and Approaches of Yangpu Ancient Saltern Cultural Landscape Heritage [J]. *Huazhong Architecture*, 2016(9).

salt culture itself. For example, as to how the brine was formed, there have long been multiple versions among the ordinary people in the Yanjing area, including legends of the golden rooster and the silver rooster, as well as a legend of the phoenix, etc. At least three involve golden and silver roosters. These legends, between beliefs and fantasies, all reflect the strong flavor of "salt brine", highlighting the unique charm of the local well salt culture.

To Explore the Food Culture through Local Salt

Although the local salt produced by traditional production methods cannot be compared with industrially produced salt, peoples who have consumed it for a long time have already formed a dependency, and often cannot do without the addition of it in their diets. For example, when they make yak butter tea, local people must use Tibetan red salt to produce the taste they like. And for another example, Yanjing Jiajia noodles are also inseparable from local salt, etc. All reflect that the locally produced salt has been deeply rooted in people's hearts and has become an indispensable material for their production and living.[1]

Relying on the natural landscape of villages formed by traditional salt production, Yanjing has achieved some results on the way of utilization and development. The value of traditional villages will be more and more appreciated by the government and accepted by the society. Especially under the impact of industrial salt production on the traditional salt industry, the protection and utilization of traditional villages has become an advantage for its own existence.

b. Protection and Utilization of Craft Heritage

Craft protection is the focus of technical intangible cultural heritage protection. The preservation and reuse of the traditional salt-making process in Yanjing mainly includes two aspects: the first is to study the scientific recording and preservation of the salt-making heritage, including museum preservation and display; the second is to carry out research and development of derivative technologies and products for the traditional salt-making process that is on the verge of disappearing.

The scientific recording and preservation is an important prerequisite for its

---

1 Li Hechun, Huang Minling. Discussion on Traditional Salt-making Villages and Their Cultural Characteristics in Qinghai-Tibet Region: Also on the Utilization and Protection of Traditional Salt-making Villages [J]. *Tibetan Plateau Forum*, 2019(1).

reuse. Taking the Yanjing Historical and Cultural Exhibition Hall as an example, its preservation of the salt craft heritage is not only to display the so-called "traditional craft heritage", but through display to "revitalize" the traditional craft that has been permanently broken, so that the traditional craft heritage of the salt industry that has passed away or is about to pass away may return to people's daily life in another way, playing a positive role in influencing people's production and living.[1] The preservation of the salt craft heritage in the Yanjing Historical and Cultural Exhibition Hall,which is fragmented, extracts, selects, transforms and displays technical and cultural details from the millennium heritage repository, and reconfigures and combines them, forming a "Chimera" reconstruction in the exhibition of the history of salt technology. Such "museum-style" reconstruction is obviously different from the real production. It expresses local people's self-cultural imagination, and at the same time, as a practical activity closely related to the reconstruction of local economy and culture, it is more inclined to the reuse of traditional craft heritage. By integrating the traditional salt craft heritage with local cultural resources, it has formed a complete cultural space, and turned it into tourism resources, so that it may conform to the actual interests of local people, achieving the purposes of communicating with foreign tourists and reviving the local culture and economy.

On the other hand, although the changes in social form and living space have led to the transformation of people's traditional living habits, aesthetic concepts, and interactive ways, and also posed severe challenges to the well salt handicraft skills originating from traditional culture, yet by applying modern design concepts and technologies, the production of local handicrafts such as edible salt, agricultural and animal husbandry salt, salt for breeding and processing, salt for hot compresses, beauty salt, etc., have become not only more accurate and efficient, but also more personalized and innovative, showing more fashion sense and commercial value. This combination of commercialization and modern design, while satisfying consumers' needs for individuation and uniqueness, has revitalized salt-making handicrafts in

---

1 Zhu Xia. Preservation and Reuse of Traditional Craft Heritage: A Case Study of Craft Display at Salt Culture Museum in Shiyang Town, Yunnan Province [J]. *Social Sciences in Yunan*, 2018(1).

modern society and become a hot consumer trend, thus promoting the prosperity and development of local salt industry. It is undeniable that with the commercialization of salt-making handicrafts, some people have begun to question their commercial value and the issue of modern design, but the combination of commercialization and modern design of salt handicrafts still provides consumers with more choices and opportunities, gaining more attention and market recognition.

Of course, we also need to pay attention to the uniqueness and cultural connotation of salt-making handicrafts, and through reasonable commercialization and the use of modern design, enable them to keep up with the times while maintaining tradition. Only in this way can Yanjing salt handicrafts achieve better development and circulation under the promotion of commercialization and modern design, making greater contributions to the process of human civilization.

c. Protection and Utilization of Folk Custom Relics

At present, in the face of the extremely serious destruction and loss of folk custom relics in our country, the museum protection mode has been adopted in most areas to preserve and utilize them. However, due to many factors, coupled with the fact that museums restrict cultural relics to fixed exhibition halls, the efficiency of their protection and utilization is quite low, which may hinder the inheritance and promotion of history and culture, and is not conducive to the continuation of traditional folk customs in China. As far as the various cultural museums or exhibition halls in Yanjing are concerned, in order to solve the common problem of attaching importance to protection while ignoring development and utilization, it is necessary to do the following:

To Research on the Essential Attributes of Folk Custom Relics

Research of the essential attributes is the basis for the utilization of folk custom relics in Yanjing museums. From the perspective of protection and utilization, salt culture museums in Yanjing, with a high degree of professionalism, can not only conduct scientific research on the collected salt cultural relics and explore their connotations, but also achieve further excavation and protection of other related cultural relics. Specifically, the professionals of these museums, with first-hand materials of salt cultural relics resources, when a certain amount has been collected, they may obtain more effective research information and in turn more valuable research results. In addition, in the study of cultural relics attributes, the professional level improvement of cultural relic investigation and collection personnel, managerial staff, researchers, etc., can also promote the overall development of museum undertakings.[1]

To Lay Emphasis on the Social Value Development of Folk Custom Relics

Exhibitions in museums have lots of functions, and the social function is an important one. The exhibitions of Yanjing folk custom relics can make use of cultural relics to do a good job in folk culture education, realizing the exchange and publicity of local folk culture. Specifically, these museums may record the appearance characteristics of folk custom relics with a variety of exhibition methods, such as audio and video, etc., accompanied by vivid explanations, so that viewers can intuitively understand the cultural relics and the folk culture hidden behind them.[2] In the process of observation, viewers use their ears, eyes, noses, and brains simultaneously to feel and think about the rich connotation and true meaning of salt cultural heritage, which may help to enhance their ethnic self-confidence and cultural self-confidence, realizing the effective embodiment of its social function.

To Achieve Commodity Development of Folk Custom Relics

The development of similar commodities is an important utilization form of folk custom relics in Yanjing museums. Under the new economic formation, China's influence in the world is gradually expanding, and many foreigners, with more interest in our history, culture, and folk crafts, etc., are willing to pay visits to our famous mountains and rivers and museums. Therefore, some traditional workshops or workshops specializing in the production of simulated folk custom relics as well as folk cultural and creative products can be built next to various Yanjing museums for tourists to observe the production process and production techniques of well salt, or buy finished products. This may not only increase the income of museums, contribute to the development of Yanjing regional economy, but also have a profound impact on the international dissemination of Chinese folk culture.

To Highlight the Educational Function of Folk Custom Relics

Relying on museums to educate students is an important way for primary and

---

1 Cai Yunhua. Research on the Protection and Utilization Approaches of Museums and Folk Custom Relics [J]. *China National Exhibition*, 2019(9).

2 Wan Yousheng. A Preliminary Study on the Protection and Utilization of Museums and Folk Custom Relics [J]. *Identification and Appreciation to Cultural Relics*, 2019(1).

secondary school students to carry out social practice in China.[1] In the utilization of folk custom relics in Yanjing museums, it is also necessary to contact local primary and secondary schools to open special exhibitions for them, and then under the explanation of museum professionals, the folk history will be explained and the ethnic spirit indoctrinated to them. Moreover, the students will be encouraged to observe folk custom relics and make some simple handicrafts. From the perspective of the utilization process, the function of Yanjing museums in the social and public attributes is maximized; while expanding the influence of the museums, it may realize the full education of students, and ensure that they have a strong sense of ethnic pride and patriotism.

In short, to achieve effective protection of salt cultural resources and cultural ecology in Yanjing, Markam County, we must combine protection and utilization, and realize protection in the process of utilization. However, as the stakes of different ways of utilizing cultural heritage are different, they must be analysed concretely and treated with discretion. In the process of utilization, attention should be paid to the limits, and it is necessary to make rational use of cultural resources, so that they may continue to come into play; we must not kill the goose that lays the golden eggs. Paying equal attention to "effective protection" and "rational utilization" is the best way for their survival and development.

3) Balancing and Coordinating Interests of All Parties

In recent years, the preservation of salt cultural heritage in Yanjing, Markam County has achieved fruitful results, which is a fact for all to see. Yet these results have been achieved through the joint participation of various social classes, institutions, and organizations in the local society. As participants in the protection of Yanjing well salt cultural heritage, governments (central and local governments), common people (local people as well as people from outside), and organizations (economic groups, social institutions, etc.) have not only made their own contributions to the protection of local cultural heritage, but also different interest demands. And the choice of these demands will also have a direct impact on the

---

1 Wu Haitao. A Brief Discussion on the Protection and Utilization of Folk Custom Relics in Museums [J]. *China National Exhibition*, 2017(24).

effectiveness and quality of cultural heritage protection and inheritance. Hence, how to adjust the interests of all parties and effectively mobilize the enthusiasm of them is an important condition for promoting the healthy development of the protection of Yanjing salt cultural heritage.[1]

a. Overall Planning for the Interests of Local Governments

In the protection of Yanjing salt cultural heritage, local governments at all levels in Tibet have played a key role in linking the preceding and the following. In practice, they have two types of goals and tasks. On the one hand, they indubitably have to implement the overall goals and tasks of cultural heritage protection as required by the central government, which is is a long-term goal and task. On the other hand, they have to take into account the relevant interests within their jurisdictions: what can be benefited from the protection and inheritance of Yanjing salt cultural heritage? This is a short-term goal and task.

Judging from the actual protection and inheritance work, local governments may have been attaching more importance to short-term goals and tasks, regarding the protection of Yanjing salt cultural heritage as a short-term political and economic behavior. Since the rise of intangible cultural heritage protection projects in the 21st century, they have become an important cultural project in China and received widespread attention. In cultural and even economic work, the protection of intangible cultural heritage has become a very explicit part. As it can better reflect work “achievements”, “the intangible cultural heritage may be targeted for economic interests and political performance”[2]. For local governments at all levels in Tibet, the inclusion of some intangible cultural heritage items related to salt in the United Nations, national, and provincial representative lists can not only obtain the praise and political achievements of “doing well” in the protection of cultural heritage, but also stimulate local economic development by making industrial use of them.

For the sake of political action, local governments at all levels, including Yanjing Township, may only pay attention to the “image” of cultural heritage protection, which has led to many “image projects”, without giving much thought

---

1 Tan Hong. Protection of Intangible Cultural Heritage Based on Stakeholders[J]. *Journal of Original Ecological National Culture*, 2017(1).

2 Liu Qiuzhi, Liu Yan. The Effective Role-playing of Local Governments in the Protection of Intangible Cultural Heritage [J]. *Journal of Lanzhou Jiaotong University*, 2013(2).

to whether the protection has achieved practical results. This phenomenon is very common throughout the country. For example, in the process of establishing national cultural ecological reserves, various regions may compete for "quotas" from the central government, for as long as a quota is obtained, they will have an "image", and in turn have "data" when making summaries and reports. However, after truly obtaining it, it's difficult for them to achieve the goals, tasks, and requirements of the Guiding Opinions of the Ministry of Culture on Strengthening the Construction of National Cultural Ecological Reserves. This is because some local governments are unwilling to adjust some plans and projects that may bring economic benefits in the near future for the sake of long-term goals, or to temporarily sacrifice local and partial interests for the sake of the long-term and overall interests of the nation and the country. For the sake of economic interests, various regions have put forward the strategy "culture as a platform for economy". Thus, in the process of protection, intangible cultural heritage has changed from a "protagonist" to a "supporting role", or a "cultural green leaf" to accompany the "economic red flower", and has become an advertising slogan and signboard for attracting investment. More seriously, only the intangible cultural heritage items that may contribute to the economy may have the "privilege", and those that are not related to the economy are not even qualified to be "green leaves". So, this has resulted in an imbalance in the protection of intangible cultural heritage: The intangible cultural heritage items that are fortunate enough to be "green leaves" may receive more attention and protection funds, and their survival conditions will be better; whereas those that have failed to become "green leaves" may be ignored, and their protection funds are insufficient, unable to provide necessary guarantee for their survival. In fact, although we advocate the productive protection of Yanjing intangible cultural heritage, it is certainly inappropriate to include all its protection work into the cultural industry, because on the whole, the protection of intangible cultural heritage belongs to cultural undertakings, and it is impossible to fully develop, utilize, and manage all of them in a market-oriented and industrialized manner.

As for the intangible cultural heritage protection project, early in the beginning of its protection movement, the central government established the above-mentioned policy "prior protection, rescue in the first place, rational utilization, and strengthening management". This working policy clearly puts "protection" and "rescue" in the most important position, which also shows that the protection of intangible cultural heritage is an important social and cultural work, more reflecting the attributes of cultural undertakings. Since it is a social and cultural undertaking, there will be a tremendous amount of manpower, financial resources, and material resources input, or the so-called social cost. The social responsibility of people's governments at all levels in the protection and inheritance of intangible cultural heritage has been clearly stipulated in the Intangible Cultural Heritage Law of the People's Republic of China. From the perspective of protecting the cultural interests of the whole nation, such efforts and investments are both necessary and worthwhile. "Utilization" is an important method and approach to protect and inherit intangible cultural heritage, which is also advocated by the United Nations' Convention for the Safeguarding of the Intangible Cultural Heritage and stipulated in the Intangible Cultural Heritage Law of the People's Republic of China.

However, it should be rational, not excessively industrialized and marketized; otherwise, it will cause harm and damage to the specific content and style of intangible cultural heritage. Therefore, local governments at all levels in Tibet must make overall plans for the protection of the well salt intangible cultural heritage in Yanjing area, regard it as an important part of local social and cultural undertakings, and invest when necessary, to promote the synchronous development of local economy and society, "local governments should always adhere to the direction of inheriting the excellent culture of the Chinese nation, adhere to the concept of scientific development, harmonious development, and sustainable development, so as to realize the balanced development of social culture and economic values"[1].

b. Safeguarding Interests of the Local People

In the specific natural and social environment, various ethnic groups in Yanjing have created rich and colorful salt cultural heritage itmes, which belong to their own cultural traditions, and the resoinsibility of whose ultimate protection and inheritance is mainly on their shoulders, "only by treating the local community as an important stakeholder can the balance between protection and development be effectively

1 Wang Jun, Zhang Yanguo. On the Role of Local Government in the Protection and Utilization of Intangible Cultural Heritage: A Case Study of Jiangxi Province [J]. *Jianghan Tribune*, 2013(10).

addressed"[1]. Since local people bear the responsibility of protecting and inheriting their own salt cultural heritage, they should also gain benefits in the whole process, so that their enthusiasm and consciousness for them will be raised.

As far as the local well salt intangible cultural heritage site is concerned, the common people's interests involve two aspects: Firstly, compensation for their lost benefits. For local people in Yanjing area, in order to protect the salt intangible cultural heritage, they may have lost many opportunities for developing other industries, which requires a certain amount of financial or other forms of compensation. This is already mentioned in the Intangible Cultural Heritage Law of the People's Republic of China and the intangible cultural heritage protection regulations of governments at all levels, but not perfect and substantial enough. Secondly, sharing of the created benefits. As a type of cultural resource, many local well salt intangible cultural heritage items have been developed and utilized by outsiders, gaining huge economic benefits. So, how the local people can share the benefits is also an important issue that needs to be solved in the "rational utilization" of well salt cultural heritage. It is necessary not only to encourage the participation of various social resources, letting them obtain certain benefits in the protection process, but also to pay attention to the needs of the local people for their own development and let them share appropriate benefits. This can mobilize the enthusiasm and consciousness of the local society or ethnic group to protect their own cultural heritage, and meanwhile be more conducive to the protection of its regionality, integrity, and inheritability, so that Yanjing salt cultural heritage protection can be put into practice, and the quality and level of its protection and inheritance will be truly improved.

In reality, affected by the geographical location of Yanjing, the local economic development level is relatively lagging behind, and people's living standards are relatively low. Hence, the protection and inheritance of well salt cultural heritage may limit the development of some industries in the local area, making them lose some growth opportunities. However, the protection and inheritance of Yanjing salt cultural heritage will only be achieved in a "living state", instead of being mechanically

---

1 Song Rui. On Stakeholders of Ecotourism: With China As an Example [J]. *China Population, Resources and Environment*, 2005(1).

and completely protected and passed on as if it were a "mummy" according to the situation when a certain cultural pattern was produced. After all, in the protection of local heritage, it is difficult to do a good job when remaining in poverty. The evolution and development of culture is inevitable, and the innovation on the basis of protecting the core value of well salt cultural heritage is also required by the inevitability of cultural development; it should be "constantly recreated". The UN Convention on the Protection and Promotion of the Diversity of Cultural Expressions states in its "Principles of complementarity in Economic and Cultural Development": "All individuals and peoples have the right to participate in and benefit from the development of both." Comrade Xiaoping's famous dictum--"Development is the absolute truth" is very convincing here, too. The local society and ethnic groups in Yanjing also have the right to enjoy modern scientific and technological achievements and social development. If they are left unchanged for a long time and protected for the sake of protection, the local people will obviously be unhappy about it, and this should not be the purpose of cultural heritage protection. Protection is for better development. The protection of Yanjing salt cultural heritage is to reflect the diversity of human culture, and make local residents' life more colorful and happier. It is by no means to preserve the content and style of the well salt cultural heritage simply to satisfy some people's "curiosity" or "inquisitiveness", anyway. Therefore, in the process of protecting Yanjing salt cultural heritage, the concept of people-centered protection and development must be reflected, and necessary economic compensation must be made to the local people, so that the they can obtain the same material life rights and development opportunities as other members of the society.

The purpose of returning a certain proportion of the income from the marketization and industrialization of Yanjing salt cultural heritage to the government and people in the protected area is to better protect the intangible cultural heritage and make it sustainable. "The distribution of profits necessary for the preservation and sustainable development of cultural heritage must be guaranteed."[1] In the actual development and utilization of Yanjing salt cultural heritage, the local people are willing to provide some resources and services for developers, hoping to obtain

1 Wang Songhua, Liao Rong. Intangible Cultural Heritage Protection from the Perspective of Industrialization [J]. *Journal of Tongji University (Social Sciences Edition)*, 2008(1).

certain benefits. But in reality, the benefits to local people are far from enough; their efforts are not proportional to their gains, and a lion's share is often taken by outsiders (developers, marketers, travel companies, etc.). This state of affairs has resulted from the capital advantage of outsiders and the government's constant concessions for short-term economic goals. In the "rational utilization" of well salt cultural heritage resources, due to the lack of its original accumulation, Yanjing has to rely on external political institutions and economic groups for operation and investment, and these institutions and groups, after intervening as "poverty alleviators" or "investors", have gradually become the controllers of resources and the decision-makers of management. In order to attract more external resources, the government has to formulate many "preferential policies" and make concessions to external institutions and groups. Practice has proved that in the process of the utilization of well salt cultural heritage, Yanjing residents will show a positive attitude of support when they feel it is profitable; on the contrary, they will resist or react negatively to it. Therefore, in the protection and inheritance of Yanjing salt cultural heritage, whether in terms of protection interests or utilization interests, we must fully consider the benefits of local people, so that they will have a stronger sense of responsibility and consciousness of protection.

Of course, in this regard, we should be vigilant about the impact of excessive "commercialization" on the protection of Yanjing salt cultural heritage, which may result in an "empty-shell phenomenon" in the local area, i.e., residents will "voluntarily cater to such cultural changes" and "voluntarily move away" because of the "real benefits" they have received[1]. If so, the local area will lose the main body of "live transmission", causing fundamental damage to the salt cultural heritage. After all, excessive commercial exploitation will inevitably cause loss and destruction to its authenticity and regionality. Therefore, for the protection and utilization of Yanjing salt cultural heritage, a relatively better way is to adopt the principle "who benefits, who protects", and all beneficiaries should assume the responsibility and obligation of protection.

c. Constraint for the Interests of Economic Organizations

It is beyond reproach that the protection of Yanjing salt cultural heritage is the

---

1 Peng Zhaorong, Lin Yaqiang. Interpretation of Heritage[J]. *Guizhou Social Sciences*, 2008(2).

responsibility and obligation of every local social member and social organization. With liveliness and inheritability as its most important characteristics, it can only be better protected and inherited in actual production and living practice. Precisely for this reason, Article 37 of the Intangible Cultural Heritage Law of the People's Republic of China stipulates that: "On the basis of effective protection, representative projects of intangible cultural heritage should be rationally utilized to develop cultural products and services with local and ethnic characteristics and market potential."

From the perspective of resources, Yanjing well salt intangible cultural heritage is a special type with unique economic value, which is of great commercial appeal to social organizations and groups, and will induce them to actively participate in its protection and inheritance. As far as specific economic organizations are concerned, there should be two benefits targets when they participate in the protection and inheritance of Yanjing salt cultural heritage: (1) by making good use of the capital power, "reasonably" utilize and develop the specific content and forms of the well salt intangible cultural heritage, and promote its participation in the production practice of creating contemporary social wealth, so that such capital may contribute to the protection and inheritance of Yanjing cultural heritage; (2) by making good use of market forces, the content and forms of the well salt intangible cultural heritage will gain market recognition during its development and utilization, and obtain certain economic benefits. With regard to the first target of the economic organizations, we should actively advocate and offer guidance to increase the "hematopoietic" ability of Yanjing salt intangible cultural heritage in its protection and inheritance, so that it will be always in a virtuous cycle. For the second target, while supporting its development, we should also put in place necessary constraints to prevent the use of protection as a "slogan" as well as the "excessive" development.

In economic development, the infinite expansion of capital and the finiteness of resources are always contradictory; the unlimited capital expansion will inevitably lead to the unlimited plundering and exhaustion of natural and cultural resources that may be used as means of production. As a limited resource, Yanjing salt cultural heritage cannot be excessively overused and exploited, so it is necessary to protect it anytime and anywhere. In terms of resource attributes, it is a special cultural resource with the "nature of quasi-public goods"[1]. Thus, in the market development, various capitals will inevitably compete to join in so as to obtain the profit brought by such capital expansion. As a result, although the protection policy of Yanjing salt cultural heritage requires "rational utilization", yet in reality there will often be the problems of "over-exploitation and over-utilization". Under the strong capital force, the local government keeps making concessions and the public keeps giving way, which have in turn resulted in the "owner absence" in the protection of local well salt cultural heritage, as well as the economic organizations gradually "ascending" as the actual users and operators. At this time, if they ignore the first benefit target, they may continue to magnify the second one, and treat the protection and inheritance of Yanjing salt cultural heritage of Yanjing with the mentality of "knowing that the storm will come one day, everyone hopes that it will fall only on the heads of their neighbors after they have made a large fortune and hid the money"[2]. Eager to recover investment costs and make money, investors are unwilling to stop and plan for the sustainable development of salt cultural heritage, which has led to many new problems that go against the original intention of protection, and the "tragedy of the commons" and "heritage fraud" may occur from time to time.

Therefore, instead of being completely regulated by the "invisible hand" of the market, the participation of economic organizations in the utilization of Yanjing salt cultural heritage must be restrained by the power of institutions and policies to ensure the regionality, integrity, inheritability, and authenticity of Yanjing salt cultural heritage as much as possible. In the Intangible Cultural Heritage Law of the People's Republic of China, special emphasis is placed on the support and preferential treatment for economic organizations to rationally utilize intangible cultural heritage, i.e., they "enjoy tax incentives stipulated by the state in accordance with the law", while ignoring the management and restraints on them. In "Legal Responsibility", Chapter 5 of the Intangible Cultural Heritage Law of the People's Republic of China, punitive expressions are only made on some destructive activities and behaviors, but no explanation is given for the specific harms and damages

1 Li Xin. Research on the Evaluation of Intangible Cultural Heritage Entering the Cultural Industry [J]. *Dongyue Tribune*, 2011(4).

2 Karl Marx. Das Kapital (Vol. I) [M]. Beijing: People's Publishing House, 2004:311.

that occur in the development and utilization. It can be seen that the restraint and punishment of economic organizations in their rational utilization of Yanjing salt cultural heritage resources should be strengthened. A market access system might be adopted to set a "threshold" for what kind of economic organizations can enter; after all, Yanjing salt cultural heritage development and utilization should not be open to all economic organizations and groups, and this requirement is necessary. Restrictive measures can be taken to restrain the rational utilization of well salt cultural heritage in terms of scale, quality, and profit in accordance with the local cultural heritage protection plan. For the development and utilization of Yanjing salt cultural heritage, it is not the larger the scale, the better, or the higher the degree of industrialization, the better, which is determined by the characteristics of cultural heritage resources. The supervision mechanism can be strictly enforced, and effective supervision and management is an important guarantee for the "rational utilization" of Yanjing salt cultural heritage. In this way, it may prevent the excessive entry of economic organizations, improve the quality and level of their development and utilization, and avoid the over-exploitation and over-utilization of Yanjing salt cultural heritage resources, lest the exhaustion of its resources should occur. As far as the participation of economic organizations in the protection of Yanjing salt cultural heritage is concerned, we should advocate their social responsibilities, "including legal as well as moral social responsibilities"[1]. Judging from the actual effect of the protection and inheritance of Yanjing salt cultural heritage, the social responsibilities of economic organizations are emphasized in the process of "developing" and "utilizing" it, which does not require them to sacrifice their own "economic interests", but to reduce the hazards and risks caused by excessive exploitation and utilization from a longer-term goal, so that its development and utilization can be orderly and sustainable. In the end, the benefits of economic organizations and the protection of local well salt cultural heritage will be both ensured.

In a word, the protection and inheritance of Yanjing salt cultural heritage is a task that requires the joint participation of social organizations and social members, and different participants may have different interest demands and constraints. So,

---

1 Zhang Zhaoguo, Liang Zhigang, Yin Kaiguo. Research on Corporate Social Responsibility from the Stakeholder Perspective [J]. *China Soft Science*, 2012(2).

it is very necessary to clarify and standardize the roles of various stakeholders and establish an effective management mechanism for the participation and coordination of all parties. We should, by taking the grand goal of inheriting excellent cultural traditions, ensuring national cultural security, and improving national self-confidence and cultural consciousness as the premise, subordinate local interests to the overall interests, and short-term interests to long-term interests, so that the protection and inheritance of Yanjing salt cultural heritage will achieve practical results and be passed on from generation to generation.

CHAPTER EIGHT

Innovative Development and Utilization of Yanjing Salt Cultural Heritage

Since ancient times, Yanjing in Markam County, Tibet has been one of the important well salt producing areas in China. Due to the superior brine resource conditions, the local well salt occupies an important position in the history of the development of China's salt industry, and the local residents have created a rich and precious salt cultural heritage in the long-term historical development process, which is of great value for development and utilization.

In recent years, the Party committees and governments at all levels in the Tibet Autonomous Region have attached great importance to the protection, inheritance, and promotion of the Yanjing salt cultural heritage, and insisted on combining protection with development and utilization, achieving remarkable results with the joint efforts of all sectors of society. However, it's undeniable that the development and utilization of Yanjing salt cultural heritage is still facing some difficulties and problems, such as the low utilization rate of some cultural heritage sites, the overall lack of competitiveness of the relevant cultural industry system, and the significant reduction in tourism revenue affected by the epidemic. How to make Yanjing salt cultural heritage "come alive", realize creative transformation and innovative development, and revitalize, which is related to the historical inheritance and sustainable development of Yanjing area, is of extraordinary importance.

1. The Basic Principles for Its Development and Utilization

Regarding the rational development and utilization of salt cultural heritage, scholars, based on a specific ancient saltern, usually investigate it at the micro level. Objectively speaking, such investigation may be helpful to the quantification and deepening of the research, but there are also some problems such as the lack of macro guiding significance of the research results, which is thus not conducive to the formulation of the development and utilization strategy of salt cultural heritage. The author believes that Marxism provides a unique perspective for the development and utilization of salt cultural heritage in Yanjing, Markam County. Guided by Marxism, we may establish basic methods and ideas for the development and utilization of Yanjing salt cultural heritage from a macro perspective, and then formulate development and utilization strategy of local well salt cultural heritage as a whole.[1]

1) To Follow the People-oriented Thought and Put People First

Marxism contains a wealth of people-oriented ideas, the core of which is to put people first. The development and utilization of Yanjing salt cultural heritage should follow the people-oriented thought of Marxism and take it as the basic principle. In the history of mankind, the people-oriented thought is of a long long-standing and well-established. In the discourse systems of different historical periods, different classes, and different schools of thought, the connotations of the people-oriented thought differ in thousands of ways. After the birth of Marxism, the people-oriented thought, as a new ideological system, has become an important part of it. However, there is an essential difference between Marxist people-oriented thought and humanism of other schools in history. "The people-oriented thought in Marxism is no longer the original abstract humanism, but a humanism has been transformed and guided by Marxism."[2] In the Marxist ideological system, the people-oriented thought is embodied through two dimensions: ontology and axiology. These two dimensions constitute the two aspects of adhering to the people-oriented thought in the development and utilization of Yanjing salt cultural heritage.

---

1 Li Xuhua, Yan Xiaorong. Development and Utilization of Ancient Town Cultural Heritages from the Perspective of Marxism[J]. *Journal of Xiamen Radio & Television University*, 2020(1).

2 Huang Nansen. Marxism and "Humanism": Answering Some Questions in the Study of Humanism [J]. *China Higher Education Research*, 2004(4).

a. Highlighting the Practice Subject Status of Local Residents in the Ontological Dimension

In Marxist philosophy, humanism in the ontological sense is more fundamental; without it, there will be no humanism in the axiological sense. In the ontological sense, "'humanism' is not to explain that man is the origin of the world and the first nature from which other material forms are derived, but that man is the most fundamental and important matter in the material world"[1]. Specifically, man is the subject of practice, and he should fully display his essential power in practice. Practice is an objectification activity unique to human beings, and people must give full play to their subjective initiative in order to achieve the expected practical purpose. It is precisely because of this that in practice, people can not only understand the objective laws, but also use them to objectify the essential forces of human beings, such as their goals, ideals, knowledge, abilities, etc. into objective reality, creating a human object world. "By creating the object world and transforming the inorganic world through practice, man proves himself to be a conscious species-being"[2], so that the relationship between man and things has changed from the latter dominating the former to the former dominating the latter, thus demonstrating man's subject status in practice.

Accumulated by the local residents in their long-term practical activities, Yanjing salt cultural heritage is the product of their practical activities combined with local natural and innate genetic factors. Through their own practical activities, local residents have realized the transcendence of animal instinct and biological nature, and established a "second nature" as their own unique survival system, which is exactly Yanjing well salt culture. It must be pointed out that local residents "are not people in some illusory isolated and fixed state"[3], but are a combination of living individuals engaged in specific salt production and practice activities, i.e., living and realistic people with regional characteristics. It is these practice subjects with

---

1 Zhou Jiarong. From Subjectivity to Inter-subjectivity of Human-Oriented: On Marxism Human-oriented Theory Leading to Inter-subjectivity Philoso[J]. *Journal of Social Sciences of Hunan Normal University*, 2008(5).

2 *Collected Works of Marx and Engels:Volume 1* [M]. Beijing: People's Publishing House, 2009: 162.

3 Ibid., P. 252.

distinctive characteristics that have created the unique tradition and characteristics of Yanjing salt culture, and fundamentally define the Yanjing salt cultural heritage as the dominant way of living in a specific era, a specific region, a specific nation, or a group of people. Thus it can be seen that in the formation process of Yanjing salt cultural heritage, local residents have shown strong autonomy and creativity, demonstrated their essential power, and reflected their subject status in practice.

Local residents are not only the creators of Yanjing salt culture, but also the inheritors of this cultural heritage. In the development and utilization of Yanjing salt cultural heritage, it is necessary to firmly rely on the local people and continue to highlight their subject status in practice. Local residents "are bearers of the folk culture, as well as subjects of the folk tourism exhibitions and performances, and they know best what is the real and valuable part of the ethnic minorities here. They modify the display and development of the intangible cultural resources according to their own wishes and understandings"[1]. The characteristics of Yanjing salt cultural heritage, derived from the characteristics of local residents' production and living practice, are a concentrated reflection of the local practice situation and the spiritual temperament of the people, which have been accumulated in long-term practice and are still being continuously generated. Adhering to the people-oriented starting point, it's necessary to earnestly respect the initiative of local residents. We must, by starting from the "realistic people", fully respect the ownership status and innovative subject consciousness of each individual in Yanjing, respect the personality and characteristics of different people, and respect their free exploration and initiative spirit. In addition, we must extensively mobilize local residents to actively participate in the development and utilization of well salt cultural heritage, and respect and carefully listen to their opinions and suggestions, creating a good atmosphere for their active and conscious participattion.

b. Fully Respecting and Caring for the Vital Interests of Local Residents in the Axiological Dimension

In the final analysis, all practical activities of human beings are aimed at creating values and satisfying their own needs. According to dialectical materialism,

---

1 Chen Tianpei, Tan Hong. Research on Economic Development Strategy of West-chongqing Economic Corridor [M]. Chengdu: Southwest Jiaotong University Press, 2006: 274.

truth and value are the two basic yardsticks that people must follow in their practical activities. The value yardstick of practice refers to the people-oriented nature in practical activities, i.e., to meet the survival and development of human beings as the basic value orientation. In another word, all practical activities are fundamentally for the sake of human beings, and the satisfaction of human needs is a specific content. In Marx's view, the fruits of practical activities naturally belong to the practice subjects. He said: "When man is increasingly subjecting nature to his domination by his own labour, and when the miracles of industry have made the miracles of God more and more superfluous, would it not be a great contradiction if man should give up the pleasure of production and the enjoyment of products in order to please these forces...The existence for which labour serves and the product of labour is for whose enjoyment can only be man himself."[1] Yanjing salt cultural heritage is the outcome of local people's production and living practice; as the creators, their vital interests should naturally be respected and taken into account in its development and utilization.

On the one hand, for the protection and development of Yanjing salt cultural heritage, it's necessary to fully respect and take into account the material interests of local residents. Material interests are not only the foundation of human survival and development, but also the most basic and realistic interest demands. Marx said: "Everything that people fight for is related to their interests."[2] Yanjing people are inheritors of local traditional folk customs, as well as owners of residential buildings and users of saltern facilities. The good soft and hard environments of Yanjing must depend on local residents. In the development and utilization of Yanjing salt cultural heritage, no matter what model is adopted, it is necessary to consider what kind of economic benefits it will bring to the local area, how much income it can increase for local residents, and how large the scope of benefits will be. There are realistic conflicts of interests between the government, developers, local residents, other stakeholders, etc., so it is necessary to coordinate them well, especially attaching great importance to the interests of local residents. If the distribution of material

---

1 *Collected Works of Marx and Engels:Volume 1* [M]. Beijing: People's Publishing House, 2009: 164-165.

2 *Collected Works of Marx and Engels: Volume1* [M]. Beijing: People's Publishing House, 1995: 187.

interests is not handled properly, or the interests of local residents are not given enough consideration to, it will cause their dissatisfaction, and in turn seriously affect the effect of development and utilization. Only when the material interests of local residents are fully respected and taken into account, and their living standards are improved and enhanced with the development of salt cultural heritage, can they truly realize that the development of Yanjing salt cultural heritage is closely related to them, and then the development of these cultural heritage can achieve real results.

On the other hand, for the protection and development of Yanjing salt cultural heritage, it's necessary to fully respect and take into account the spiritual needs of local people. Spiritual needs are higher demands than material benefits. According to the report of the 19th National Congress of the Communist Party of China: "The principal contradiction in Chinese society has been that between people's ever-growing needs for a better life and the unbalanced and inadequate development."[1] With the improvement of material conditions, people have even higher aspirations for a better life, not only with higher requirements for material life, but also with stronger demands on the spiritual level. It is undeniable that the development and utilization of some ethnic cultural heritages tend to emphasize the respect and care for the material interests of local residents, but often ignore their spiritual and cultural demands. For example, the excessive commercial development of some ancient towns has led to an overly strong commercial atmosphere, simply imitating "model ancient towns", and they have become all the same and lost their own original cultural characteristics. Local residents find it difficult to identify with such development and utilization emotionally, and thus lose their spiritual belonging to the ancient towns, which is extremely harmful to the sustainable development of the ancient town culture. It can be seen that for the development and utilization of Yanjing salt cultural heritage, we must grasp its unique spiritual connotation and tradition, rely on local people to carry forward its characteristics, and fully respect and care for the spiritual needs of local residents.

---

1 Xi Jinping. Decisive Victory in Building a Moderately Prosperous Society in All Respects and Striving for the Great Victory of Socialism with Chinese Characteristics in a New Era: Report to the 19th National Congress of the Communist Party of China [M]. Beijing: People's Publishing House, 2017:11.

2) To Take Materialist Dialectics as the Fundamental Method

Materialist dialectics, as a science of the general law of nature, society, and thinking development, is the fundamental method for us to understand and transform the world, which has the characteristics of transcending time and space. Mao Zedong pointed out: "The dialectical view of the universe is mainly to teach people to be good at observing and analyzing the contradictory movements of various things, and to show ways to resolve the contradictions according to this analysis."[1] The development and utilization of Yanjing salt cultural heritage should, by following the materialist dialectics, dialectically deal with the relationships between protection and development, the whole and parts, the present and the long term.

a. Dialectically Dealing with the Relationship between Protection and Development

The two sides of a contradiction are opposite as well as united. So we must dialectically deal with the relationship between protection and development in the process of developing and utilizing Yanjing salt cultural heritage. Any attempt to set protection against development will have disastrous consequences. One-sided emphasis on protection and exclusion of development will inevitably lead to the decline of Yanjing salt culture, which will become lifeless and whose heritage will lose its value. Blind development and neglect of the effective protection of Yanjing ancient architecture, traditional crafts, folk customs, festivals, ecology, etiquette, temple fairs, etc. will inevitably result in the fact that salt cultural heritage is completely unrecognizable and its characteristics are gone. To deal with the relationship between protection and development dialectically, it is necessary to adhere to the principle "development on the basis of protection and protection in the course of development". The key is to grasp the degree of protection and development, so as to achieve benign interaction between the two.

Firstly, we should carry out "reasonable protection" of Yanjing salt cultural heritage, laying the foundation for its development. Marxism holds that the development of things is a process of self-denial. Dialectical negation is the link of connection as well as that of development of things, and its essence is to develop what is useful and discard what is useless. The so-called "to develop what is useful

---

1 *Selected Works of Mao Zedong: Volume 1* [M]. Beijing: People's Publishing House, 2009: 304.

and discard what is useless" means that "a new thing criticizes while inheriting the old, and overcomes their negative elements while retaining their positive ones"[1]. The protection of Yanjing salt cultural heritage is exactly to affirm its reasonable elements. As a precious ethnic cultural heritage, Yanjing salt culture is the crystallization of human civilization and an important part of Chinese traditional culture, so it is natural for us to devote to its protection. But in practice, the protection of ethnic cultural heritage is prone to "going beyond" and "falling short" tendencies. "Going beyond" refers to the phenomenon of "overall protection", i.e., accepting the cultural heritage formed in history completely and uncritically, unwilling to move forward and hanging on to the outmoded, plus opposing the development and utilization of cultural heritage as well as bringing forth the new. On the contrary, "falling short" means that the protection of cultural heritage is insufficient, causing damage to it. For example, according to relevant statistics, there are 2,800 ancient towns with conservation value in China, of which no less than 2,000 have been damaged to varying degrees due to lack of effective protection[2]. Therefore, reasonable protection of Yanjing salt cultural heritage is the basis of development. General Secretary Xi Jinping said: "Only by not forgetting history can we open up the future, and only by being good at inheritance can we be good at innovation."[3] The "reasonable protection" of ancient town cultural heritage is to inherit it critically, and to prevent the occurrence of "going beyond" and "falling short" in its protection, so as to lay the foundation for its development.

Secondly, we should realize protection in the process of development, adhering to the policy that the development must be conducive to the sustainable development of Yanjing salt cultural heritage. Reasonable protection of Yanjing salt cultural heritage is the basis for its development, and its effective protection also depends on its successful development. The key to the development of Yanjing salt cultural heritage is to excavate the cultural value of well salt, activate the well salt culture,

---

1 Writing Group of this Book. Introduction to the Basic Principles of Marxism [M]. Beijing: Higher Education Press, 2018: 41.

2 Wang Dasan. "Heshun Mode": The Crystallization of Ancient Town Development and Protection [J]. Socialist Forum, 2006(12): 35-36.

3 Xi Jinping. *Xi Jinping's Discussion on Governance and Administration of State Affairs: Vol. 2* [M]. Beijing: Foreign Languages Press, 2017: 313.

and promote its creative transformation and innovative development, with the aim of realizing the sustainable development of local salt cultural heritage and deriving due economic benefits. Thus it can be seen that sustainable development is the best protection of Yanjing salt cultural heritage, which can only be truly achieved in the development and utilization. However, the development of Yanjing salt cultural heritage should be "moderate", and excessive commercial exploitation must be avoided. For excessive commercial exploitation may not only fail to achieve the desired effect, but also cause serious damage to the cultural characteristics on which Yanjing well salt depends for survival. Cultural characteristics are the basis for the sustainable development of Yanjing well salt; once it is lost, the sustainable development will have no foundation. Therefore, in the development and utilization of Yanjing salt cultural heritage, we should strive to be prudent, formulating a thorough development plan and maintaining its cultural characteristics as the bottom line.

b. Dialectically Dealing with the Relationship between the Whole and Parts

The universality and particularity of contradictions, i.e., the commonality and individuality, are dialectically unified. On the one hand, commonality and individuality are opposed to each other; commonality is the common essence of similar things, but it cannot include all the content and all aspects of individuality, for individuality of things is always richer than commonality. On the other hand, commonality and individuality are unified, and the two are interconnected and interdependent; individuality contains commonality, and commonality resides in individuality and is expressed through individuality. In the development and utilization of Yanjing salt cultural heritage, we should not only see the universality of contradictions, but also pay attention to their particularity, organically unifying the two and dialectically dealing with the relationship between the whole and parts. Specifically, it is to properly deal with the relationships between the top-level design of the government and specific practice plans, comprehensive benefit and economic benefit, collective interests and individual interests, systematic environment and the development of salt cultural heritage, etc.

First of all, we should properly deal with the relationship between the top-level design of the government and specific practical plans for the development and utilization of Yanjing salt cultural heritage. For example, on April 2, 2008, the

State Council promulgated the Regulations on the Protection of Famous Historical and Cultural Cities, Towns and Villages in order to regulate the development and utilization of cultural heritage in famous towns and villages across the country and to make principled provisions. Such regulations have important guiding and normative significance for the protection and development of the salt cultural heritage in Upper Yanjing Village, which has already been included in the list of Chinese traditional villages. According to the local reality and characteristics, Tibetan governments have also successively issued local rules and regulations aimed at achieving a "win-win" situation between protection and development, such as the Regulations on the Protection of Yanjing Ancient Salt Pans in Markam County, Qamdo City, etc. The regulations promulgated by the central government and the relevant rules and regulations formulated by local governments at all levels can be regarded as the top-level design for the development and utilization of Yanjing salt cultural heritage, which have general guiding significance. On the basis of it, we should formulate specific practical plans according to the local reality. Such plans must not only highlight the general principles of top-level design, but also combine with the local reality, giving prominence to the local characteristics.

Secondly, we should properly deal with the relationship between comprehensive benefit and economic benefit of Yanjing salt cultural heritage development. Comprehensive benefit refers to the total sum of benefit obtained within a certain period of time, which can comprehensively reflect various aspects such as economic, social, and environmental benefits, etc. Economic benefit should be an important part of the comprehensive benefit, but not the whole. We cannot one-sidedly emphasize economic benefit, still less can we replace comprehensive benefit with economic benefit.

Thirdly, we should properly deal with the relationship between collective interests and individual interests. In the development and utilization of Yanjing salt cultural heritage, there are not only national and collective interests, but also individual interest demands which formed on the basis of multiple stakeholders. We must give priority to national and collective interests, and meanwhile comprehensively consider the interests of indigenous residents, developers, and other individuals, coordinating interests of all parties and balancing the relationships between them.

Lastly, we should properly deal with the relationship between the system environment and the development of Yanjing salt cultural heritage. Yanjing salt cultural heritage is closely related to the surrounding geographical and ecological environments, constituting a systematic environment. Its development and utilization should not only highlight the cultural characteristics, but also maintain the harmony with the whole system environment. Never can it be achieved at the cost of sacrificing the system environment.

c. Dialectically Dealing with the Relationship between the Present and the Long Term

The fundamental requirement of dialectical materialist methodology is to focus on the present while taking a long-term view. Materialist dialectics holds that any practical activity must not only focus on the present and be down-to-earth, but also take a long-term view and carry out advanced planning. Only by focusing on the present and making long-term plans, can we firmly grasp the initiative. In the development and utilization of Yanjing salt cultural heritage, we must dialectically deal with the relationship between the present and the long term; specifically, it is necessary to deal with the relationships between short-term and long-term goals, as well as the current interests and the long-term benefits.

First of all, we must dialectically deal with the relationship between short-term and long-term goals. Practice is purposeful, and people need to establish certain goals when engaging in practical activities. As the old saying goes, preparedness ensures success and unpreparedness spells failure. There should be both short-term goals and long-term goals. Short-term goals, which are likely to be achieved in the short term, are the direct motivation for people to engage in practical activities. Long-term goals, in contrast, as a kind of strategic vision for the future in practice, which reflect people's strategic thinking ability, is a powerful spiritual driving force for their activities. The short-term goal for Yanjing salt cultural heritage is to promote the protection of salt cultural projects and boost the local economic growth, while the long-term goal is to realize the sustainable development of the well salt culture as well as its development and utilization. In this process, there may be a phenomenon that the short-term and long-term goals cannot be organically unified, or that in order to achieve the short-term goal, the long-term one is not paid enough attention to, resulting in its improper exploitation or or over-exploitation. To deal with the relationship between the short-term and long-term goals in the development and utilization of Yanjing salt cultural heritage dialectically, we should not only focus on the present, facilitating the completion of the short-term goal, but also by taking a long-term view, with the long-term goal as the traction, adjust the deviation of the short-term goal, so that the two may achieve dialectical unity.

Secondly, we must dialectically deal with the relationship between immediate interests and long-term benefits. In the development and utilization of Yanjing salt cultural heritage, there will be not only conflicts between the interests of different stakeholders, but also entanglements between immediate and long-term interests. When dealing with the relationship between immediate and long-term interests, it is necessary to take into consideration in the interests of all concerned and unify the two. We should not only consider the current development needs of Yanjing, but also make planning for its future development needs, i.e., take into account the long-term interests while realizing immediate interests. We should not ignore the current interests of Yanjing residents, but also prevent short-sighted behaviors that are blind to the long-term interests of Yanjing development and seek quick success and instant benefits. Publicity and education should be strengthened, and the main front of education and propaganda where immediate and long-term interests are unified should be built, so as to promote the generation of a deep-seated consciousness of the unity of immediate and long-term interests, so that stakeholders in the development and utilization of Yanjing salt cultural heritage may deeply realize that while obtaining the immediate benefits, they must also take into account the long-term interests of the development of local salt culture.

3) To Be Based on Practice and Serve Practical Needs under the Guidance of the Concept of Practice

Marxism holds that things are continuously evolving from the lower level to the higher level, and in this process, although there will be various setbacks and even temporary retrogressions, the general trend is a kind of forward and upward movement, which is the general law of the development of things. By its very nature, culture is the result of certain social practices, especially the creative labor of human beings, and the product of a certain historical era. Therefore, culture must keep pace with the development of practice to achieve innovation, which is the key to cultural development. Any culture, if only satisfied with the status quo and remain stagnant, will inevitably lag behind the times, and eventually lose its vigor and vitality; this is the inexorable law of cultural development. The development and utilization of Yanjing salt cultural heritage is both a process of cultural inheritance and a practical activity that constantly activates local well salt culture, promotes its creative transformation and innovative development, and forms comprehensive benefits. Its essence is cultural innovation, or to achieve its sustainable development fundamentally. For this reason, it should follow the general idea of basing on and serving the local socialist practice with Chinese characteristics.

Firstly, based on the local socialist practice with Chinese characteristics, we should activate Yanjing salt cultural heritage and promote the innovation of salt culture. Marxism points out that practice is the source of knowledge, and all true knowledge comes from practice in the final analysis. Culture, born in practice, must be developed in practice, too. It can be seen that social practice is the foothold of cultural innovation. To promote the innovation of Yanjing salt culture, we must be based on the life and struggle of Yanjing people today, and on the local practice of socialist construction with Chinese characteristics in Yanjing. It is necessary to closely combine the development and utilization of Yanjing salt cultural heritage with the specific practice of local reform and opening-up as well as socialist modernization, summarize and refine the new understanding, new experience, and spirit of the times generated in practice in a timely manner, and organically integrate it into the salt cultural heritage, giving it a new connotation of the times. Meanwhile, according to the practical characteristics and the requirements of the times of local reform and opening-up and socialist modernization construction, Yanjing salt cultural heritage should be correctly selected, adhering to the principle that the past is used for the present, and the old is brought forth to bring forth the new, "striving to realize the creative transformation and innovative development of traditional culture, so that it can be integrated with the real culture, and jointly serve the tasks of the times of the cultural people"[1]. On this basis, an advanced culture will take shape that reflects the contemporary life of Yanjing, embodies the contemporary spirit here, gladdens

1 Xi Jinping. *Xi Jinping's Discussion on Governance and Administration of State Affairs: Vol.2* [M]. Beijing: Foreign Languages Press, 2017: 313.

the people's hearts, and strives to make progress without losing its traditional characteristics. In short, practice is endless, and the innovation of Yanjing salt culture should never stop. Only by basing ourselves on practice, innovating the local salt culture with the times, and constantly reflecting the practical creativity of Yanjing people, can we realize the sustainable development of the salt cultural heritage.

Secondly, the innovative achievements of Yanjing salt culture must serve the local socialist practice with Chinese characteristics. Marxism holds that all theoretical and cognitive achievements serve practice, and without practice as the ultimate goal, knowledge will be meaningless. Even if it is a correct theory, it would be meaningless to just put it on the shelf, prate about it for a while, and do not apply it to practice. The development and innovation of Yanjing salt culture must be aimed at serving the local socialist practice with Chinese characteristics. The specific requirements are as follows: (1) Directly apply the achievements of Yanjing salt cultural innovation to the practice of Yanjing cultural heritage development and utilization, so as to achieve its better protection and inheritance, bringing tangible benefits to Yanjing people. It can be said that Yanjing salt cultural heritage is not only a synthesis of material culture and non-material culture, but also an extremely valuable ethnic traditional cultural resource. The innovation of Yanjing salt culture is helpful to tap its value, and the results will inevitably produce comprehensive benefits after being applied to the practice of the development and utilization of Yanjing salt cultural heritage. (2) Through innovation, stimulate the vitality of Yanjing cultural heritage, strengthen its unique expression, storage, and transmission, and maintain the continuity and sustainability of Yanjing social history. Yanjing salt cultural heritage appears in a complete form of value, "includeing not only material forms such as buildings, numerous cultural relics, etc, but also immaterial forms such as the profound humanistic connotation hidden in it, etc. They contain a variety of meanings such as traditional folk customs, architectural concepts, historical origins, religious beliefs, ethics and morals, value orientations, and aesthetic concepts of people in different areas, and reflect the social and economic development of various regions"[1]. The innovation of Yanjing salt culture must activate these cultural genes, making them

1 Wang Xuemei. On the Tourism Value of Tourism Resources in Ancient Towns of Sichuan and Their Development, Protection, and Utilization[J]. *Forum on Chinese Culture*, 2004(3).

more perfect and more adaptable to the needs of the times, so as to realize the sustainable development of Yanjing cultural heritage. (3) With the achievements of Yanjing salt cultural innovation, better promote the improvement of the educational and moralizing functions of Yanjing cultural heritage, so as to achieve its role of shaping people, and in turn influence the behavior pattern of Yanjing people, creating a harmonious social practice environment. (4) With the achievements of Yanjing salt cultural innovation, provide intellectual impetus and cohesive power for local practice. The cohesion of social forces depends on cultural identity. The achievements of Yanjing salt cultural innovation, presented in a new form and spiritual outlook, is the high integration of traditional charm, the atmosphere of the times, and the humanistic spirit, which is bound to form a strong attraction and affinity for people, thus helping to inspire and stimulate the simple emotions of Yanjing people to build the motherland and hometown, and strengthen their patriotism and love for their hometown. (5) With the achievements of Yanjing salt cultural innovation, provide intellectual support for local practice. The achievements of Yanjing salt cultural innovation are the results of scientific understanding of the latest practice activities in Yanjing, and its application to practice will help improve the quality, management level, and innovation ability of local laborers, and in turn promote the development of local productivity and the overall progress of society.

We should take the Chinized and modernized Marxist theory as a weapon, fully carry forward its spirit, and apply it to the practice of the development and utilization of Yanjing salt cultural heritage, so as to leave a more beautiful, prosperous, and sustainable salt cultural heritage for the future world.

2. The Primary Strategies for Its Development and Utilization

The process of developing and utilizing Yanjing salt cultural heritage is a process of re-examining, recognizing, and improving them, during which we need to sort out many clues and clarify many problems. Therefore, it cannot be achieved overnight. It requires us to constantly explore according to the actual situation and development trend, constantly discover, research, cope with and predict problems, and constantly adjust and integrate in such a cycle, and finally "keep pace with the times" in the process of social and historical development, so that Yanjing salt cultural heritage can maintain a relatively lasting "vitality", and the precious local traditional culture can be protected and continued in the development and utilization.[1]

1) To Highlight the Characteristics of Salt Cultural Heritage, Correctly Interpreting and Paying Attention to Its Connotation

How to dig deep into the ethnic cultural resources and highlight its ethnic and regional characteristics is an issue worthy of in-depth study. In order to better complete the task of developing Yanjing salt cultural heritage, it is necessary to screen a wide range of folk cultural contents, select the most representative cultural connotation suitable for the local area, and conduct in-depth excavation, so as to better reflect its characteristics.

a. The In-situ Concentration Mode

In less developed areas, the in-situ concentration mode provides an effective way to adapt to the changes of cultural market demand, optimize the structure of cultural industry, and improve the benefits of cultural economy. Due to the development of the times, some ethnic minority villages or areas with rich and unique folk culture, such as Yanjing, have been weakened in terms of architecture, costumes, customs, etc., and are no longer typical, or some important activities of folk culture (such as festivals and weddings) are originally presented in a specific period, making it difficult for outsiders to fully understand the essence of local or ethnic folk culture. The local government or investors can find a suitable location in the local area to build a folk culture theme park focusing on the local salt culture, and present its folk essence in a condensed form.

This kind of theme park, based on Yanjing custom culture, highlights its distinctive characteristics and unique individuality, and organically combines the production and life culture of Yanjing with the local landscape, romance, and display as much as possible. Folk culture gives it spiritual value and emotional connotation, so that it has a unique character and vivid image. It should absorb the essence of the culture, comprehend its spirit, integrate the emotions of design and protection of folk culture, further deliberate, study, refine, and abstract the element symbols of

---

1 Huang Jian. Tentative Plans on Protection and Development of Zigong Well Salt Cultural Heritage [J]. *Salt Industry History Research*, 2011(3).

folk culture, and extend its true meaning into contemporary design.[1] For example, folk food culture, dance and opera, arts and crafts, customs, etc., which can be easily activated and interacted with might become the focus of the park development. It is worth noting that the development of the salt culture theme park should be consistent with the socio-economic environment in which they are located. First of all, we should take miniaturization and specialization as the development direction, and then make it bigger and stronger, rather than pursue the large and complete pattern at the beginning.

b. The Protonatural Mode

"Original ethnic culture" refers to the fact that in some remote areas, due to geographical or historical reasons, local folk customs, less affected by alien cultures and modernization, still retain their relatively original and unchanged traditional ethnic elements.[2] Therefore, the protonatural mode refers to the selection of a most typical village with convenient transportation in Yanjing, a region with relatively rich folk culture, to carry out publicity to the outside world, taking the natural life and production of villagers and the natural form of the village as the content of development and utilization, almost without any processing and transformation except the construction of necessary infrastructure.

As a relic of history, Yanjing salt making villages, having solidified the culture, art, and characteristics of the times, are harmoniously integrated with nature and still retaining the original local culture, traditions, and folk customs. They have not only historical and archaeological value, but also architecturally scientific and technological value and ornamental value. Among the cultural heritage items of Yanjing salt industry, there are not a few historical buildings (groups), which are mainly related to ancient salt making, salt transportation, salt marketing, sacrifice, living and other activities, involving the collection of brine, the sun-drying of brine to make salt, the transportation and marketing of salt, the residence of salt

---

1 Chen Jial, Chen Qi. Folk Custom Theme Park landscape Express Inquiry—Taking "One town with Two Gardens" in Fufeng New Tourism and Cultural Industries District as an Example [J]. *Journal of Green Science and Technology*, 2016(7).

2 Luo Yongchang. The Concept, Goals and Countermeasures in Aboriginal Ecological Folklore Tourism Development: A Case Study of Southeastern Guizhou [J]. *Guizhou Ethnic Studies*, 2007(5): 72.

merchants and salt people, etc. It is not an easy task to make it glow with new brilliance and inherit the customs it contains. From the perspective of local economic development, moving out the residents of ancient villages to turn their living places into tourist attractions will not only disturb their normal production and life, but also change the villages with only buildings and without local cultural elements into soulless dolls. At the same time, the local cultural department should organically combine the geographical environment of the site with the local history and culture according to the psychological requirements of modern people who are eager to return to nature. On the basis of respecting history, culture, and nature, the original natural development should be the mainstay, supplemented by upgrading and transformation, and the original feeling should be maintained as much as possible through restoration, reinforcement, reconstruction and retention of handmade traces, reflecting the particularity and non-renewable nature of the well salt landscape resources.

c. The Highlight Collection Mode

The highlight collection mode refers to that the typical folk customs scattered witin a certain geographical range are concentrated in a theme cultural village (park). Different from the above in-situ concentration mode, which pays more attention to refining and seeing big things through small ones, this mode focuses on the development and utilization of its representative and unique cultural projects, mainly in the following aspects: (1) Historical Relics Among the material cultural heritage of ancient salt industry, from a historical and scientific point of view, the relics with outstanding universal value should be ancient salt wells and ancient salt pans. As salt production essentials, they are the physical objects of social production that can reflect the history of the salt industry in various eras and under various social systems. In addition, there are other cultural relics such as historical documents that record the local salt industry, inscriptions on the transportation of the salt industry, tomb portraits reflecting the production scenes of the salt industry, etc. Although these are not large-scale, they can also serve as an effective supplement to the development of historical cultural relics resources in the salt industry. (2) Various Salt-related Folk Customs As an intangible cultural part of Yanjing ancient salt industry heritage, the primary, intuitive, and conducive to development are various salt industry folklore, mainly divided into production folklore, sacrifice folklore, and food folklore, such as "Napa" festival, marriage festival, as well as *pipa* meat, Jiajia noodles, etc., which may form cultural exhibitions and performances, flavor snack streets, etc., giving full play to the packaging and attracting role of folk culture, and creating local cultural cards. (3) Salt-related Production Techniques and Tools Yanjing salt industry has a long history, and the salt-making technology here has struck a line for itself. The production of traditional salt industry, from brine extraction to brine sun-drying, to salt-making, almost all rely on manpower, which is quite rare today. The tools used in this process, such as the brine pails, wooden bats, salt scrapers, etc., are the crystallization of local salt makers' wisdom. Selling such handicrafts in shops and involving tourists in salt-making activities can play a good publicity effect. (4) Salt-related Music, paintings, literature, and Other Works of Art In the process of salt production and transportation, local salt makers have created a lot of salt-related artworks, such as labor songs and poems reflecting hard work in salt production, etc. These resources not only reflect the unique culture and development history of ethnic minorities, but also highlight their characteristic culture and ethnic memory,[1] which can both enrich the experience of tourists and contribute to the local economic development through the establishment of cultural inns, style bars, cultural performances and other forms of business.

The advantage of this mode is that tourists can enjoy folk culture in a short time within a short distance, but its disadvantage is that the information connotation of some part might be lost in the process of copying and processing.

d. The Retro Reproduction Mode

The retro reproduction mode refers to collecting, sorting, constructing, and reproducing information about the disappeared folk culture through artificial landscape, so that tourists may better understand its past. The advantage of this mode is that by excavating the lost traditional cultural themes, it can "turn back the clock" and meet tourists' wishes that cannot be realized before. However, it has the same disadvantages as the highlight collection mode.

Among this kind of resources, those with outstanding universal value are the

---

1 Chen Qian, Wang Quankang. The Heritage of Ancient Salt Culture and the Value of Tourism Development in Bayu Region [J]. *Journal of Sichuan University of Science& Engineering (Social Sciences Edition)*, 2013(1).

ancient saltworks. The ancient salt processing was mainly completed by small saltworks, whose main work, to dry the extracted brine into salt, is the intermediate link of salt production. Through the restoration of the ancient saltworks, the salt-making scene at that time can be reproduced; the original buildings and styles will be developed and utilized, the old buildings will be restored according to the principle of "restoring the old as the old", and the dilapidated houses and those that are not in harmony with the style will be upgraded. The newly built buildings are basically in harmony with the ancient town style. The original structure and pavement of the block remain unchanged, maintaining the sense of historical vicissitudes. In addition to focusing on antique themes in the form of buildings and scenes, etc., in order to allow tourists to fully experience this culture, people's costumes, behaviors, etc., all have to be retro; only in this way can people feel like they are travelling through time. For example, in the "Living Museum" in the United States, the staff, dressed in the costumes of American working people in the 16th and 17th centuries, play the roles of "immigrants" who arrived hundreds of years ago, and perform ancient traditional customs such as carrying water with square poles and farming with primitive farming tools, which attracts a large number of tourists.[1] In this way, visitors are immersed in it, unable to distinguish between reality and time travel.

The traditional Yanjing culture has satisfied the basic needs of local people for a long time, providing an irreplaceable spiritual home for them to settle down. However, under the increasing commercialization of traditional ethnic culture, Tibetan culture, including Yanjing salt culture, has also been impacted, resulting in the trend of Sinicization and Westernization in local minority areas. In order to avoid cultural alienation or even vulgarization, it is necessary to excavate and develop the excellent ethnic traditional culture of Yanjing and highlight its characteristics. After all, no intelligent ethnic minority will sacrifice its traditional culture in the process of modernization.

1 Shengjing Cultural Tourism: A Brief Analysis of the Four Basic Modes of Folk Culture Tourism Development, https://zhuanlan.zhihu.com/p/625114905.

2) To Advance with the Times and Enhance the Modern Value of Traditional Yanjing Culture

The development of cultural heritage in the new era is faced with both new opportunities and many new challenges. We can only keep pace with the times, keep learning, and constantly change with the times. As far as Yanjing salt culture is concerned, it is necessary not only to be good at developing what is useful and discarding what is useless, but also to incorporate things of diverse nature and maintain an open spirit. This spirit of openness has three layers. For the traditional culture of the region, we must correctly develop what is useful and discard what is useless, absorbing the essence and discarding its dross. For alien cultures, we must be good at borrowing all reasonable and favorable elements. Wht is more, on the basis of inheriting, absorbing, borrowing, and utilizing the excellent cultural achievements of alien cultures and their own cultures, we should, according to the requirements of the times of social development, be good at innovation, bold in exploration, and brave in pioneering, so as to promote the globalization process of local salt culture.

a. Developing What Is Useful and Discarding What Is Useless in Local Culture

To develop what is useful and discard what is uselss in the traditional salt culture, we must first distinguish its essence and the dross, no longer making the mistake of "throwing the child with the dirty water". To distinguish the essence from the dross, there is no fixed standard and boundary, which may change with the change of time and space. In general, the theoretical boundary of reserving or abolishing culture is that those healthy, scientific, upward cultures representing the future development direction and promoting social progress are the essence; on the contrary, those decadent, negative, ignorant and backward cultures hindering social progress are the dross. Or to put it simply, to judge the good and bad of Yanjing salt culture through the pros and cons of its effects on society, the most important thing is to take the interests of the people as the standard, distinguish right from wrong, and judge the advantages and disadvantages. Historical experience shows that all things that are true, good, and beautiful are in line with the interests and requirements of the people; all things that are false, evil, and ugly are contrary to the interests and demands of them.

From a macro point of view, our choices of Yanjing salt culture as well as developing what is useful and discarding what is useless are nothing more than three aspects, namely, inheritance, critical inheritance and elimination. First of all, many Yanjing cultural heritage achievements, due to the continuation of their utilization value, are not only progressive in history, but are still closely related to people's lives, playing a positive role today. For example, the astronomical calendar formed in the production and life of the salt industry, Tibetan medicine related to salt, salt handicrafts, and some folk good morality, etc. To this kind of cultural achievements, we should persist in inheriting and carrying them forward. Secondly, although some cultural phenomena were progressive and played a positive role in history, today, with the change of economic basis for producing them, such cultures have lost their value of existence, yet still passed down to the present day, as is the case with the bow and arrow as a sports tool. This kind of cultural heritage, which has neither positive nor negative effects, can also be preserved, but should be injected with new content. Some have played a positive role in history; today, although they have lost their practical values, their forms and some contents are still maintained, and they are playing a positive role in other aspects and have strong vitality after evolution, such as the marriage festival, the archery competition, and other folk festival customs, etc. We should inherit them critically, transforming their outdated and backward parts. Thirdly, some cultural phenomena have played a positive role in history, but are now obsolete and have become backward; some which were not progressive or positive in history, or even backward and decadent, are incompatible with today's social life and ethical trends, such as feudal hierarchical concepts and superstitious ideas, traditional concepts of being satisfied with the status quo and not wanting to make progress, and so on. These harmful cultural heritages should be resolutely eliminated. As far as the specific cultural achievements are concerned, the essence and the dross often coexist and are inseparable. Even the essence of traditional culture often contains negative elements, which requires us to analyze what is the mainstream and what is the tributary. In this regard, social scientists and cultural workers should make a difference.[1]

1 Yangxi. Correctly Understanding and Treating Tibetan Traditional Culture with Marxist Cultural View [J]. *Journal of Tibet University for Nationalities (Philosophy and Social Science Edition)*, 2000(3).

If we do not correctly develop what is useful in Yanjing salt culture and discard what is useless, regarding traditional things as "treasures", without analyzing them or distinguishing between good and bad, and maintaining an attitude of total affirmation, as a result, they will not only become the shadow of their predecessors, but also lag behind others forever. On the other hand, if we regard the traditional culture as outdated, rigid, or even backward and ignorant, negating it completely and attempting to create a new one, the result will not only be impossible in reality, but also we may become a sinner in history.

b. Being Inclusive of Alien Cultures

Inclusiveness is not only a very distinctive feature of the fine traditional Chinese culture, but also a very valuable advantage. The more inclusive a culture, the easier it is to uphold the concept of "harmony in diversity", adhere to the attitude of incorporating things of diverse nature, and promote the exchange and integration between different cultures, so as to realize continuous cultural progress. For Yanjing salt culture, on the basis of retaining and continuously strengthening its original cultural characteristics, it needs to absorb cultural genes that are conducive to the development of local ethnic minorities and the region, so that the regional cultural characteristics are gradually improved, which is also a wise choice for it to revive vigor and vitality in the combination of tradition and modernity.

A culture that lacks innovation has little vitality. Only by grasping the inheritance of the "cultural gene" and "cultural core", on the basis of protecting its original ecology and authenticity to the maximum extent, advancing with the times, learning from others' strengths, making moderate innovations, and giving traditional culture the spirit of the times and advanced forms, can it become a cultural heritage that is compatible with the contemporary society and full of vitality.[1] In today's excellent situation of opening up and enlivening the whole country, especially in a new era of irresistible ideological and cultural exchanges all over the world, Yanjing salt culture can no longer adhere to a closed and conservative attitude. Isolation, arrogance, exclusion, and resistance to alien cultures can only constrain its own progress and development, and even lead to further isolation and backwardness

---

1 Bi Lianget al. Protection and Development of Intangible Cultural Heritage in Yangzhou City in the New Era. *Journal of Southeast University (Philosophy and Social Science)*, 2009(S1).

in the region. “The result of blind exclusion can only be self-isolation, voluntarily giving up the use of many favorable conditions and rights for its own development, tying its wings to take off, and dragging its own legs.” The current era requires that Yanjing must establish an open spirit, confirm the complementary, stimulating, and inspiring effect of alien cultures on its traditional culture with a more open mind, absorb and learn the excellent components of alien cultures through extensive and in-depth cultural exchanges, so as to enhance its own vitality and realize the benign change of ethnic cultures. But, how can we realize the benign change of ethnic cultures? A very important factor is ethnic self-confidence. Ethnic self-confidence, as an important part of ethnic self-consciousness, is based on the positive evaluation of its cultural characteristics. This kind of self-confidence is manifested in the fact that when the contact with alien cultures reveals differences between local and other cultures, we can adopt an analytical attitude towards both local and alien cultures with an inclusive and broad mind, and carry out cultural exchanges in multiple fields, keeping local advantages while choosing others’ merits. Through the introduction of alien cultures, and with the ability to digest and develop them, we may make it part of our own, so that the local culture can absorb rich nutrients and develop healthily.

In short, in order to prpperly deal with the relationship between inheritance and innovation, Yanjing salt culture cannot be isolated, and it needs to draw nutrients from other cultures. It is necessary to accurately grasp the general trend of globalization and China’s ever-expanding opening up to the outside world, treat alien cultures with an open view, a broad mind, and keen eyesight, and boldly absorb and learn from their useful experience and excellent results. That is, to make reasonable choices, follow what is good, improve in an innovative manner, and use them for our own use.

c. Accelerating the Globalization of Ethnic Cultures

World culture has two meanings: cultures that exist in the world and cultures that are worldwide. The former refers to all existent cultures in the world, including all national cultures, and the world culture in this sense has existed since the beginning of culture. The latter refers to a unified world culture that is universally accepted by all nationalities and is formed by the universalization or integration of national cultures, and this culture is shaped along with the process of world history.[1] The vast Tibet region is not an isolated island as one might think. Geographically, located in the hinterland of Asia, it serves as the connecting belt and hub of East Asia, South Asia, and Central Asia. From the perspective of cultural region, it is at the intersection of Oriental culture, South Asian culture, and Central and West Asian culture. Since ancient times, Tibet has had contacts and ties with its neighboring regions.[2] The remains excavated at the Karuo cultural site in Qamdo suggest that the Yanjing area may have been closely linked to the civilization of the Yellow River Valley as far back as the Neolithic Age. And the Yanjing Catholic Church is more a symbol of local culture brought in line with international practices.

The 21st century is an era of multicultural coexistence, when the mutual influences between different civilizations and cultures will be inevitable. The cultural impact brought about by globalization has threatened the inheritance and development of Yanjing salt culture like never before. On the one hand, globalization has brought more and more alien and emerging cultures, gradually weakening people's perception of traditional ethnic cultures; on the other hand, the cultural industry under globalization is also commercializing, standardizing, and popularizing the traditional ethnic cultures, causing the weakened or even lost original cultural connotation and characteristics. However, according to globalization and the principle of bidirectional influence principle of ethnic cultures, Yanjing ethnic groups not simply accept alien cultures passively, but also bring their local cultures to the world through the process of globalization. The higher the degree of globalization of a culture, the more of its content goes to the world, and the more beneficial it will be to the development of the region; the culture will be able to gain a wider living space and more development opportunities.[3] Therefore, in the face of the tremendous pressure of globalization on Yanjing salt culture, we should broaden our horizons,

1 Chen Xinxia. National and Cosmopolitan Character Of Cultture in the View of Historical Materialism[J]. *Hebei Academic Journal*, 2013(4).

2 Chen Liming. On Tibetan Folk Culture[J]. *Journal of Tibet Nationalities Institute (Philosophy and Social Sciences)*, 2002(4).

3 Ma Xiaojing. Commoditization of Ethnic Tourism Culture and the Development of Ethnic Traditional Culture [J]. *Journal of South-Central University for Nationalities (Humanities and Social Science)*, 2002(6).

insist on opening-up, and actively spread Yanjing salt culture to the world. On the other hand, it is also necessary to carry out dialogue and exchanges on the basis of equOn the one hand, we should absorb the essence of other cultures, strive to learn the advanced science and technology management experience in the world, and actively absorb and draw on the excellent ideological and cultural achievements of other countries and nations in connection with modern civilization, so as to promote the local salt cultural progress; on the other hand, we should carry out dialogue and exchanges on the basis of equality, seeking common ground while reserving differences and properly resolving these differences between different cultures.[1] Only by accelerating the globalization process of Yanjing salt culture through various ways and increasing its internationalization degree can it radiate new vigor and vitality in the trend of globalization.

To sum up, in order to truly keep pace with the times, achieving "all-inclusiveness" and "incorporating things of diverse nature", the development and utilization of Yanjing salt cultural heritage should always be in the spirit of openness towards both local and other cultures. It is necessary not only to critically inherit the excellent parts of the local cultures and carry them forward in light of the characteristics of the times, but also to boldly learn, absorb, and draw on the excellent parts of other cultures, transforming them into our own culture to serve the local people.

3) To Achieve Chinese Modernization of Spatial Production in Traditional Salt-making Villages Driven by Cultural Logic

As a "living fossil" of local civilization, Yanjing traditional salt-making villages have recorded the rich survival philosophy of the ancestors, and also contained the long-standing wisdom of grassroots governance, reflecting the cohesion of various ethnic groups and the inheritance of well salt culture. However, under the impact of urbanization and capitalization, these traditional villages have been declining, which is reflected not only in the disappearance of distinctive ethnic buildings,

---

1 Jiang Miao. The Construction of National Culture in China under the Background of Globalization [J]. *Guangxi Social Sciences*, 2005(1).

but also in the fading away of traditional cultures and values.[1] In order to break through the predicament of capitalization of spatial production, traditional salt-making villages should promote the paradigm renewal of spatial production in the integration of capital logic and cultural logic. The cultural logic of these villages is the dominant one used in the selection of concepts, values, behaviors, customs, and ways of living on the basis of geographical features, factors of the times, and ethnic characteristics in the long-term historical process, forming a rich history and culture. Yanjing traditional salt-making villages are not only a realistic carrier of historical and cultural inheritance, but also a vivid scene showing Chinese civilization. Their spatial production, essentially consistent with Chinese modernization, contains its characteristic elements, too.

a. Protecting the Landscape Culture: Renewal of Physical Space in Harmony between Man and Nature

The landscape culture is the integration of physical space and regional culture of Yanjing traditional salt-making villages, including the protection and repair of the natural ecosystem of the ancient villages and ancient architectural relics, as well as the related activities and values of the people, collective psychology, and ideology. The landscape culture in these villages shows the Oriental wisdom of harmony between man and nature. It is both the way in which the villagers have "adapted to the natural process, the land as well as the space and pattern on the land for the sake of life"[2], and the embodiment of historical memory and rural art in the natural and humanistic landscape. The protection and improvement of landscape culture is not only the foundation of the ecological development of traditional salt-making villages, but also the cornerstone of the restoration and inheritance of the village culture. At present, as far as the physical space renewal of the local salt-making villages is concerned, the first thing is to restore and reconstruct the natural landscape system. On the one hand, after the cessation of human destruction and interference, the damaged natural landscape cultural system, by relying on the self-regulation ability

---

1 Lin Li, Miao Yiyang. From Capital Logic to Cultural Logic: the Chinese-style Modernization of Spatial Production in Traditional Villages--Based on the Investigation of W Ancient Village in Suzhou [J]. *Tribune of Study*, 2023(4).

2 Yu Kongjian, Wang Zhifang, Huang Guoping. Discussion about the Vernacular Landscapes and Their implications to Modern Landscape Architecture [J]. *Huazhong Architecture*, 2005(4).

of the ecosystem, supplemented by artificial measures, can be gradually restored and developed in the direction of a virtuous cycle, such as filling in the abandoned brine wells, etc. For the natural landscape that has been damaged, destroyed, and polluted, such as the water environment, it is necessary to eliminate pollution from the source, increase the intensity of treatment, and improve infrastructure construction, so as to provide a good living and tourism environment. On the other hand, in view of the damage to natural landscape and cultural resources such as mountains and rivers that have not yet occurred, it is necessary to caution against small matters, devote greater effort to supervision, strengthen prevention in key areas, and actively do a good job in publicity and education, so as not to harm and endanger the interests of local residents while developing tourism.

Secondly, it is necessary to strengthen the protection and development of the landscape culture in Yanjing traditional salt-making villages. The physical space production of these traditional villages is regional and unique, and its non-reproducibility and non-replication mean that once the physical space is deconstructed, the attached village culture will also disappear with it. To protect and develop the landscape culture of these salt-making villages is to inherit and continue the visible and tangible history, retaining the nostalgic memories of the local area. The protection of the landscape culture in Yanjing salt-making villages, more than enclosure protection and restoring the old as the old, is not to reduce the natural landscape to an exhibition hall or sightseeing spot through overall relocation, but to make the village landscape culture "come live". Firstly, tourism product development and landscape design through the living culture may allow villagers involved to benefit from the protection and utilization of the village landscape culture, enhancing its economic value. Secondly, in the process of construction and transformation of the village landscape culture, it is necessary to consider the production and life needs of local villagers, in addition to the needs of tourists. In view of the homogenization phenomenon in the development of Yanjing traditional village tourism, the local landscape culture should be organically combined with the excellent traditional Chinese culture to innovate the development of cultural tourism economy, and transformed into a characteristic cultural tourism experience project, making it a specific cultural symbol and experiential product easily accepted by consumers. Thirdly, in the protection strategy of the landscape culture of Yanjing salt-making villages, the local government, as the leading force, should introduce targeted policies, regulations, and protection rules, so as to form a government-led and multi-party participation pattern of cultural protection.

b. Stimulating the Living Culture: Renewal of Spiritual Space Based on Local Values

The living culture is the wisdom of life condensed by the collective members of Yanjing salt villages in the long historical years, which "refers to the culture that people experience and learn through daily practice in a specific time and space, and can only be fully understood by those who actually live in this emotional structure"[1]. It is not only the spiritual values and traditional activities related to the tangible heritage, but also is reflected in the intangible heritage to a certain extent. Including a wide range of daily life and traditional customs, such as regional languages, dietary characteristics, traditional social structure, etc., it is especially reflected in the salt-making villages inhabited by ethnic minorities, with the universality and richness of daily routine life[2]. As the external concrete representation of the living culture, local culture, such as folk customs, folk art, oral literature, etc. is in essence a cultural display of production and life, reflecting its symbolic meaning of the practical value. This is not only the characteristic of production and life in Yanjing salt-making villages, but also the foundation of the differentiated development of traditional villages. To stimulate the living culture means to organically combine cultural inheritance and innovative development, or to integrate the values and production and lifestyles in line with the development of the times in the protection and development of local salt-making villages, so as to inherit but not be old-fashioned, and to innovate without forgetting their roots.

First of all, it is necessary to promote the awakening of local cultural value consciousness. We should recognize not only the inheritance of traditional culture by local culture, but also its inherent vitality and innovation. By combining the

---

1 John Storey. Cultural Theory and Popular Culture: An Introduction [M]. London: Routledge, 2012: 46.

2 Zhang Tianxin, Wang Min. On the Possibility of Living Culture Criteria in Chinese Village Heritage Conservation--Comparison between the Criteria of UNESCO Asia-Pacific Awards for Cultural Heritage Conservation and Chinese Traditional Villages [J]. *Chinese Landscape Architecture*, 2015(4).

acquaintance culture in the traditional society with the spirit of contract in the market economy, we may shape the local village culture that can adapt to the modern society.□Secondly, we should sort out the family genealogies and the history of the villages. Historical legends and celebrity stories of the villages can be explored in the way of oral history, and local chronicles museum can be established in villages with favorable conditions, and even special festival and ritual routes can be designed according to different themes, so as to create a spiritual cultural place, and make the cultural brand bigger and stronger, realizing the simultaneous development of material civilization and spiritual civilization. Thirdly, we should establish and raise the cultural self-confidence in Yanjing salt-making villages, strengthen cultural publicity and education, vigorously carry out the action of transforming social traditions, meet the spiritual and cultural needs of the masses with rich cultural products, and intensify the cultural consciousness and subject consciousness of the villagers, so as to enhance the internal driving force of culture.

In addition to the concrete living cultural form, the reproduction of spiritual space in Yanjing salt-making villages is also reflected in interpersonal communication and neighborhood relations. "Soft rules such as customs and rituals have become important rules for villagers to maintain social relations"[1]. The disintegration of the acquaintance society is an inevitable trend in the process of modernization, but the acquaintance society in the traditional salt-making villages will not collapse completely; instead, it will remain a state of "half-acquaintance, half-stranger" society in the demographic and structural changes. Under such a demographic pattern, it is necessary to guide the healthy development of the social pattern by a new interpersonal communication culture. On the one hand, we should not only give play to the mutual assistance spirit of neighborhood watch on the basis of protecting traditional culture, but also respect the reality of social changes, establish the necessary awareness of interpersonal boundary, and pursue the concept of harmonious coexistence of interpersonal communication in villages, so that the local salt-making villages can become the common spiritual homes of new and old villagers as well as tourists. On the other hand, when inheriting and carrying forward

1 Jia Fenfen. Renaissance of Traditional Culture and Optimization of Village Cultural Function [J]. *Probe*, 2019(2).

the traditional interpersonal communication culture, we should encourage the elites from all walks of life to participate in the construction of local salt-making village culture, reshape the interpersonal communication culture in these villages, give play to the exemplary role of new and old sages here, and cultivate civilized and harmonious rural customs through their fine words and deeds, so as to promote the harmonious, friendly, and healthy development of rural interpersonal communication.

c. Excavating the Institutional Culture: Social Spatial Renewal in Response to Rural Revitalization

For a long period of time, Yanjing salt-making villages have achieved good operation through self-governance, which was mainly reflected in "including informal institutional arrangements that people agreed to or thought were in line with their interests"[1], such as family rules and disciplines, village rules and folk conventions, and patriarchal systems, etc. These informal institutions have the continuity of historical evolution in the interaction of local cultural inheritance and interpersonal communication, thus forming the institutional culture that embodies the local wisdom. The institutional culture in the spatial production logic of Yanjing salt-making villages refers to self-governance as a a kind of institution, which has the consistency between the cultural dimension and the rule dimension[2]. It can be divided into the following three levels: (1) the cornerstone of self-governance formed by the accumulation of traditions, habits, and experiences, namely the patriarchal system culture; (2) the self-governing norm based on the value of benevolence and righteousness and the idea of etiquette, namely the folk regulations; (3) self-governing carriers including social institutions and organizations, etc, namely non-governmental organizations.

First of all, we should excavate the institutional culture with Yanjing characteristics, critically inheriting and developing the patriarchal system culture. In the local traditional patriarchal system culture, there are conservative, backward and even feudal superstition elements, which have constrained the development of Yanjing society for a long period of time. The founding of the People's Republic of

---

1 Yu Keping. *Governance and Good Governance* [M]. Beijing: Social Sciences Academic Press, 2004:10.

2 Zeng Xiaohua. Culture, Institutions, and Institutional Culture [J]. *Journal of Zhejiang Party School of C. P. C*, 2001(2).

China, the deepening of reform and opening-up, and the development of the socialist market economy have gradually done away with the dross and backwardness of the local patriarchal system. The patriarchal system and the husband's authority are no longer the core of the system for the operation of traditional salt-making villages, but the collective consensus of isomorphism of the family and the country, and the sense of social belonging of benevolence and affinity contain the centripetal force of loving the country and neighbors, as well as the cohesion of solidarity , mutual assistance, and hometown revitalization. Therefore, it is necessary for us to actively look for the integrating point between the clan system culture and the modern value of spatial production in Yanjing salt-making villages, explore the internal logic of benign self-governance, and stimulate the endogenous power of comprehensively promoting rural revitalization.

Secondly, we should attach importance to the development and utilization of the value of ethical norms in folk regulations. As an informal system with consensus, folk regulations are a social control method to regulate the behaviors of village community members, adjust rural social relations, provide public services, and realize the villagers' self-governance. With regional differences, they as a whole embody the value function of the fusing Confucian concept of benevolence and righteousness and etiquette system, advocating not only praising virtue and punishing vice through moral education and norms, but also solving interpersonal disputes and public problems by conventional methods. The formulation and improvement of Yanjing folk regulations in the new era should start from moral construction, "shape the villagers' value orientation in an environment of advocating virtue and goodness, valuing faith and righteousness, so as to guide the benign development of rural social atmosphere"[1], and promote the coordinated development of material civilization and spiritual civilization; the concept of co-construction, co-governance, and shared benefits in folk regulations should be refined and sublimated, and the regional characteristic culture be integrated into the self-governing system of these salt-making villages, so as to enhance the subjective role of villagers in their revitalization, and promote their spatial renewal with more contemporary connotation

1 Yao Baosong, Zhou Haowen. The Dilemma of Folk Regulations and its Solution from the Perspective of Rural Revitalization [J]. *Tribune of Study*, 2019(3).

and value of Chinese modernization.

Thirdly, we should integrate interest differentiation by relying on non-governmental organizations, and enhance collective action. Traditional Yanjing salt-making villages take a family as the basic identification and action unit, but as its scope is too small, it is necessary to solve the problem of public goods and welfare supply through a larger range of cooperation. Clan organizations and other village non-governmental organizations have reasons to meet the public needs, and then have created functional organizations above the families. The prevalence of traditional non-governmental organizations in Yanjing salt-making villages is characterized by villagers' self-organizations and other organizations led by the government, enterprises, and third-party social institutions, which enable the patriarchal system and the folk regulations to play the role of self-governance and promote the renewal of social space. The development of commodity economy has not only injected new vitality into the growth of non-governmental organizations in local salt-making villages, but also brought about the differentiation of interests. Therefore, on the basis of facing up to the differentiation of social interests, it is necessary to realize the self-integration of local society through functional organizations, create a new social life community in the traditional salt-making villages, provide basic public services and reconstruct characteristic culture of the villages in collective forms, internalize the functional organization forces into the production and life of villagers, and reshape the new rural social structure and social identity.[1]

3. The Main Approaches to Its Development and Utilization

Effective protection is a prerequisite for rational utilization of cultural heritage, and scientific development is the basic guarantee for its protection and utilization. On the whole, the development and utilization of Yanjing salt cultural heritage has got off to a good start and has achieved remarkable results. However, there are still some problems, for example, it's still in a situation characterized by the single

1 Lin Li, Miao Yiyang. From Capital Logic to Cultural Logic: the Chinese-style Modernization of Spatial Production in Traditional Villages--Based on the Investigation of W Ancient Village in Suzhou [J]. *Tribune of Study*, 2023(4).

development domain, low development intensity, low cultural recognition, and seriously insufficient degree of attention, etc. Only by jumping out of the traditional mode and updating ideas can we achieve the purpose of protection, development, and inheritance. We must explore a more suitable way between development and inheritance, which can fully realize benign interaction and combination.

1) Cultural and Creative Product Development

Cultural and creative products are images or ideas recognized by consumers, which, through the refining and planning of various cultural resources such as architectural art, myths and legends, traditional festivals, natural landscapes, cultural relics and utensils, etc., are transformed from the product symbols that meet the various cultural needs of human beings by innovative thinking and creative means in response to the material and spiritual needs of people's cultural life.[1] According to this, Yanjing cultural and creative products can be understood as new products with both cultural and practical values created through modern science and technology, which rely on rich well salt resources and take their characteristic cultural connotation as the creative inspiration. In recent years, the state has vigorously advocated the development of cultural and creative undertakings as well as cultural and creative industries, leading to an increasing boom in the research and development of cultural and creative products with Chinese characteristics[2], which has also brought new opportunities for the development of such products in Yanjing. What we should do in Yanjing cultural and creative product development are as follows:

a. Identifying the Product Positioning through Market Research

Cultural and creative products, with both commodity attributes and spiritual attributes, are the unity of value and use value. With consumers' higher and higher demands for cultural and creative products, the current market is not lacking in them, but in those that consumers are satisfied with. Under the situation of fierce competition in the domestic cultural market and a surplus of popular cultural

---

1 Yan Sanjiu. *Creativity and Planning of Cultural Industry* [M]. Shanghai: Fudan University Press, 2008.

2 Yang Chun. An Exploration of the Cultural Source of Cultural and Creative Products from the Perspective of Cultural Confidence [J]. *Beauty and Times (Part 1)*, 2021(3).

and creative products, Yanjing cultural and creative products require supply-side structural reforms while adjusting the development direction according to the needs of consumers, in order to meet people's increasing material and spiritual needs.[1]

Market demand research is the front-end link of product development process, which can provide a basis for product decision-making. Only by having a deeper understanding and recognition of the development status of the market, paying attention to the product forms favored by consumers, and knowing the consumption levels and positioning of different consumers, can we design products that are more in line with the market development, improving the competitiveness of our products. So, it is necessary to find the target groups for the salt cultural and creative products bearing Yanjing culture, get to know their social status, age structure, education levels, interests and hobbies, etc., and clarify their psychological needs, which is the preliminary basis for the top-level design and overall planning of the products.

Yanjing cultural and creative products should meet the needs of multi-level and multi-form cultural consumption, and establish their own accurate positioning. Its strategy is to carry out market segmentation on the basis of analyzing and judging the consumption characteristics of the cultural market, establish the design concept of cultural and creative products, and finally design a product positioning scheme that conforms to the development characteristics of a specific cultural industry. Through an analysis of the cultural consumption demand of Yanjing cultural and creative products, it is found that consumers pay special attention to the features of creative and interesting products with practical functions, aesthetic values, modern texture, and cultural connotation, etc. Therefore, for local cultural and creative products, it is necessary to design multi-level and diversified product positioning according to the differences in age group, consumption level, aesthetic taste, lifestyle, and value pursuit, etc. According to the difference in age group, we can design "cute children's fun" series of products for little kids, as well as "fashion tide" series of products for youths; according to the difference in lifestyle, we can design "good life" series products for daily life, as well as "national tide collection" series products for light luxury life; according to the difference in consumption ability, we can

---

1 Zheng Liwei. Research on the Current Situation and Promotion of Regional Cultural and Creative Product Development[J]. *Journal of Shaanxi Youth Vocational College*, 2022(4).

design “laughter-provoking” series of products with affordable prices, as well as “high-end custom” series. Designers may design accurately according to the market demand of cultural tourism consumers in Tibet, especially in eastern Tibet, and meet the personalized spiritual needs and aesthetic pursuits of consumers through the development and design of cultural and creative products with rich types and diverse differences.[1]

Of course, the development of Yanjing cultural and creative products cannot be achieved overnight; instead, it is necessary to continuously promote the updating and iteration of products to meet the needs of consumers. Therefore, it is an important principle of cultural and creative design to clearly define the target audience and conduct hierarchical development.

b. Strengthening Product Characteristics by Excavating the Regional Culture

Regional culture is the representative and symbol of a place, which has accumulated and formed in the long-term historical development. And regional resources mainly include local architecture, special food, religious culture, traditional customs, places of interest, and historic sites, etc. At present, the development of cultural and creative products in many regions still remains in the stage of mechanical imitation, with a single product form and uniform shapes, and even direct replication of cultural symbols, which is generally manifested in product homogeneity. As relevant products are not integrated into the regional culture, tourists have weak desires to buy and aesthetic fatigue, affecting the spread of the local culture. At the same time, with the development of the Internet, some large cultural and creative enterprises have penetrated all over the country. They have absolute advantages in terms of price, development speed, modeling, and other aspects, thus it is difficult for the local cultural and creative industry to compete with them. In this case, only by adhering to the principle of integrating regional cultural elements into the development of cultural and creative products, can Yanjing give full play to the inspiration and creativity in design, so as to activate its economic vitality.

By extracting the essence of Yanjing salt cultural elements and integrating

---

1 Zhang Yibo. Research on the Integration of Regional Cultural Elements into the Development and Design of Cultural and Creative Products of Hebei Museum[J]. *Identification and Appreciation to Cuktural Relics*, 2023(4).

them into cultural and creative products, we can make them uniquely symbolic, which not only helps to enhance the cultural attributes of local well salt cultural and creative products, but also enables them to realize the cognitive function based on local cultural expression. When selecting items for Yanjing cultural and creative product design inspiration, we should first choose cultural relics with high historical and cultural values and high cultural recognition as the creation prototype, such as various horse caravan objects, salt-making tools and other highly recognizable cultural relics, etc., and secondly choose those with thick historical information and rich cultural connotation as the design prototype, such as Yanjing Catholic Church, Gangda Temple, Princess Wencheng Temple and other cultural relics with great stories. By creating aesthetic conception and establishing emotional connection with consumers, we may enter the stories behind the vast historical relics, making dialogues between the past and the present. Lastly, by selecting cultural relics with unique shapes and strong decorations as design prototypes, such as Zagusi stone carvings, sculptures, etc., and extracting the decorative patterns and symbols in murals and paintings, cultural and creative products can meet the decorative and artistic value pursuit of consumers, and realize the unity of practical and aesthetic functions of cultural and creative products.

Of course, in addition to the integration and reorganization of explicit elements such as architecture, patterns, colors, designs, etc., more important is the transformation of implicit elements such as history, customs, social culture, etc. When using symbols of traditional cultural elements, designers should explore new symbols of elements on the basis of paying attention to inheritance, while retaining their unique styles, extending their cultural connotations and images, etc., and achieving scientific innovation of cultural and creative products with modern expressions and innovative design concepts.[1] Through these cultural elements, the uniqueness of Yanjing historical and cultural heritage can be further demonstrated, and more consumers can learn about the local culture and history, which has extraordinary historical and cultural value.

1 Ma Yingying, Jiang Shaohua. Research on the Development of Cultural and Creative Products in Henan Province: A Case Study of the "Reunion Spring Bowl" in the Year of Tiger [J]. *Art Education Research*, 2023(6).

In short, the potential market of Yanjing cultural and creative products being very large, the excavation and utilization of regional cultural characteristics are the key for the development of this industry. At present, there is an urgent need to promote the in-depth integration of cultural creativity and technology, i.e., to design and develop cultural and creative products that highlight local regional cultural characteristics, unique materials and craftsmanship, and strong artistic creativity.

c. Building Well-known Brands by Designing Serialized Products

Serialized design is to establish an internal connection between different cultural and creative products through a unified design element, so that it presents a family style, forming a strong visual impact in the market. This unified element can be one or more of the design elements such as brand name, brand logo, color, graphics, shape, style, etc., but generally no more than three. As a common design method of modern cultural and creative products, serialized design can form large-scale serialized products with simple design elements to meet the needs of different consumers.[1] Yanjing cultural and creative products can be roughly divided into three series:

Firstly, the category of arts and crafts. It is mainly a replication and reproduction of the natural resources as well as the cultural and artistic treasures of Yanjing, which reflects the local humanistic spirit and is an important medium for inheriting the local culture. This category of cultural and creative products, including home furnishing articles, creative models, replicas of cultural relics, etc., is mostly used to decorate and beautify the home environment, thereby adding home beauty, reflecting the owners' taste, and having a high appreciation value. We can design household items such as safety car pendants, festival calendars, dark-red enameled pottery, tea sets, etc., integrating exquisite natural landscapes or production and life scenes into home design by means of printing and carving, etc., and we can also make them into scaled models for sale, so that consumers will not only see the local characteristic attractions, but also take their favorite scenes home.

Secondly, the category of derivative souvenirs. It, which must combine the elements of local history, culture, and residents' emotional imagery with the

---

1 Zheng Liwei. Research on the Current Situation and Promotion of Regional Cultural and Creative Product Development [J]. *Journal of Shaanxi Youth Vocational College*, 2022(4).

functionality of the items and the needs of consumers, is a commodity that integrates the elements such as history and culture, functional needs, aesthetic impressions, commemorative values, and local cognitive recognition, etc.[1] This kind of cultural and creative products, by virtue of creativity, give cultural elements to the inconspicuous ordinary objects in daily life, making them vivid and interesting, such as stationery, packaging, toys, gifts, etc. Due to the low costs, affordable prices, and convenience to carry, their are much favored by consumers.

Thirdly, the category of published works. It is the Yanjing cultural relics, intangible cultural heritage photos, values, research results, etc. published in such forms as special editions, catalogues, multimedia promotional videos, audio and video, etc., so as to give play to their educational and publicizing role. Among them, printed materials often appear in the forms of picture albums, picture books, postcards, creative notepads, etc., which are a relatively intuitive carrier of cultural transmission, allowing consumers to quickly obtain relevant resource information, such as destination overviews and travel advice.[2]

At the same time, in view of the lack of well-known brands in Yanjing cultural and creative products, more attention should be paid to its brand building. The establishment of cultural brand images is the core of designing cultural and creative products, which can awaken their cultural vitality and promote the sustainable development of cultural and creative industry.[3] At present, cross-border co-branding of cultural and creative products with other brands has become a regular marketing operation mode; for example, the co-branded lipstick between the Beijing Palace Museum and Mao Geping is a great innovative product, whose combination of the two can attract the attention of consumers, stimulate their desire to buy, and give

---

1 Chen Taihong, Chen Fang. Discussion on the Development Orientation of Chongqing Salt Historical and Cultural Town and the Ideas of Cultural and Creative Development [J]. *Salt Industry History Research*, 2023(1).

2 Yang Yonghe. Research on the Development Strategy of Tourism Cultural and Creative Products: A Case Study of Lijiang City, Yunnan Province [J]. *Tourism and Photography*, 2023(22).

3 Ma Yingying, Jiang Shaohua. Research on the Development of Cultural and Creative Products in Henan Province: A Case Study of the "Reunion Spring Bowl" in the Year of Tiger [J]. *Art Education Research*, 2023(6).

full play to its maximum value.[1] Therefore, Yanjing can also try to cooperate with designers, artists, institutions, and enterprises in different fields, jointly launching innovative and unique well salt cultural and creative products.

Through brand development, we can improve the product quality, and cultivate loyal fans of the brand, so as to occupy a place in the market. Through serial development of the brand, it will be suitable for the individual needs of consumers of different ages and genders, forming an impact in the cultural and creative product market.

In salt culture research, while strengthening the basic research, it is necessary to strengthen and highlight the application and development research, closely combining salt culture transmission with the development of salt culture industry and local economic construction, and promoting the development and prosperity of salt culture and local economy in the way of "industry-university-research cooperation".

2) Industrialization and Industrial Upgrading

Cultural industry refers to the collection of cultural products, cultural communication services, leisure and entertainment, as well as related production and sales activities, including cultural creativity, film and television production, publication and distribution, printing and reproduction, advertising, performing arts and entertainment, cultural exhibitions, digital content, animation, and other industries. As a new highlight in the current social and economic development, it can greatly boost the social and economic development and growth in Yanjing area. Like many technical intangible cultural heritages, although it has undergone rapid development in recent years, Yanjing salt industry still has not broken away from the workshop manual processing mode, and its industrialization is still in its infancy. The main problems include low industrial efficiency, small industrial scale, incomplete industrial chain, more emphasis on production than on sales, and so on. [2]Therefore, according to the current situation of salt cultural resource endowment and cultural

---

1 Yang Yu. Analysis on the Development Direction of Museum Cultural and Creative Products [J]. *Heihe Journal*, 2023(1).

2 Xin Ru, Zhang Shufen. Industrialization and Industrial Upgrading are New Ideas for the Development and Protection of Intangible Cultural Heritage: A Case Study of Stone Carving and Tourism Development in Quyang, Hebei Province [J]. *Productivity Research*, 2010(5).

industry development in Yanjing area, industrialization and industrial upgrading are one of the important strategies to pass it on in modern society, and the following three aspects can be given priority:

a. To Enrich Product Types and Improve the Quality

The report of the 19th National Congress of the Communist Party of China pointed out that China's economy is now in a period of rapid development and is striving to transform to a stage of high-quality development. However, through an investigation of Yanjing salt industry, it can be seen that the local well salt products are relatively single, rigid in form, neither in line with modern aesthetic needs nor strong in innovation, and the good and bad mixed together, especially lacking high-quality and high-grade cultural products. On the other hand, owing to their poor practicability, some well salt products cannot meet diversified consumption needs, and it's difficult for them to win the favor of consumers. Therefore, it is necessary to strengthen the development of salt product uses and varieties, combinre taditional craftsmanship with modern design concepts, and focus on practical aspects to launch more products suitable for daily life needs.

First of all, we should "increase varieties", i.e., enhance the research and development of salt products at different levels and different directions, rationally allocate resources, and vigorously promote the development speed of new product projects, so as to continuously meet the consumption preferences of various consumers. By subdividing the market needs of consumers inside and outside the region, as well as relying on resource advantages and regional geographical influence, we should actively develop and produce nutritious, healthy, and green functional salt products, enrich medium and high-end products and high value-added products, and strive to form a multi-level, multi-purpose, multi-type, and multi-price series of products to meet the salt needs of residents at different levels inside and outside the region. For example, in the United States, salt products are subdivided according to their use, taste, packaging, specifications, granules, and consumption habits to meet the personalized needs of consumers. In Yanjing area, by learning from this, we may vigorously improve the research and development of salt products, bringing forth the new through the old in all aspects.

Secondly, we should "improve the quality" to promote the effective supply capacity of salt. Product quality is not only the key to the recognition of consumers, but also the guarantee that the production can last for a long time. With the increasing attention of consumers to food safety and health, ecological salt has become an important development direction for the salt industry to "improve the quality". According to the technical and quality standards of well salt products such as raw material requirements, variety specifications, product grades, appearance quality, etc., we should strictly check on the production, strengthen quality inspection, strictly produce according to standards, strengthen management and supervision, accelerate the change of the current situation of serious homogeneity of salt products, and promote the development of high-end salt categories, so as to effectively improve the technical level and product quality of the well salt industry. On the one hand, through the introduction of equipment and technology, the safety risk of salt production will be reduced. On the other hand, through the introduction of the concept of green environmental protection, environmental protection and practicality are paid attention to in the production process and product packaging, and the competitiveness of local well salt products is further improved.

Thirdly, we should "build brands", i.e., on the basis of the existing local salt brands, dig deep into the connotation of the word "salt", enrich the brands of Yanjing salt industry, and establish a good image of Yanjing salt production. Centering on a series of well salt cultural resources such as the origin of well salt culture, ancient salt transportation culture, ancient salt trading culture, salt merchant culture, salt-maker residence culture, and food culture, etc., we may establish the brand of "Yanjing well salt culture", and the brand effect will be used to promote the industrialization of local salt culture. Taking the development of the overall brand strategy as an opportunity, we take the lead in cultivating a group of branch cultural brands with market appeal within the industry, and extend them to related industries to maximize the brand value. For example, by learning the creation and market-oriented operation mode of large-scale live performances such as "Impression Liu Sanjie" and "Impression Lijiang", etc., or dance and stage plays such as "Riverdance" and "Golden Sands", etc., we may create a unique brand symbol belonging to Yanjing well salt culture, gradually building Yanjing salt products into a well-known domestic brand with strong market competitiveness.

b. To Promote the Adjustment of Industrial Structure through Characteristic Cluster Development

From the perspective of the evolution trend of industrial spatial structure, the traditional industrial development mode relying on resource endowment will be the basic carrier of regional economic development in a long period of time[1], but at present, the production of Yanjing well salt is mainly based on small workshops with families as the unit, the production is not large-scale, and the industrial structure can no longer meet the needs of modern social development, which needs to be adjusted and optimized. According to the development status of the salt industry in Yanjing, it has become an effective way to cultivate new economic growth points in the region by building a characteristic industrial cluster of well salt based on the advantages of location and resources.

On the basis of cluster development, the salt chemical industry in Yanjing promotes the optimization and upgrading of the industrial structure, focusing on two aspects: its rationalization and advancement. The rationalization of industrial structure refers to the kind in which various industries, by coordinating with each other, with strong industrial structure conversion ability and good adaptability, can adapt to changes in market demand and bring the best benefits. Specifically, the rationalization of industrial structure based on salt chemical characteristic clusters mainly includes: (1) Rational utilization of salt resources. Improve the efficiency of resource utilization through technological progress, and make full use of human, material, financial, and natural resources inside and outside the system. (2) The play of regional advantages of salt-related resource-rich areas in eastern Tibet. The adjustment of Yanjing industrial structure must be combined with the natural resources conditions and other location advantages of the local area and resource-related areas, and establish an advantageous industrial group that can give full play to its own advantages--salt production, salt chemistry, new material industry chain, so as to form a regional characteristic economy. (3) Self-organizing ability of regional industrial structure. Salt chemical industry has become an important pillar industry

---

1 Zeng Shaolun et al. Development of Regional Distinctive Industry Cluster and Industrial Structure Adjustment: A Case of Salt Chemical Cluster in Zigong City [J]. *Ecological Economy*, 2010(12).

of Yanjing, but whether it can become a leading one, play a guiding role in the regional industrial structure, and drive the development of other industries, its further development will be essential. Moreover, while developing the salt chemical industry, it is necessary to establish related industrial groups around the salt chemical cluster. (4) Harmonious industrial structure ratio. As economic growth is continuously promoted by the dynamic change "equilibrium-disequilibrium-equilibrium" of the ratio between industries, it is necessary to constantly optimize them, so that the appropriate ratio can be maintained. The proportion of Yanjing salt chemical industry in the industrial structure must meet the requirements of resource endowment, economic growth mode transformation, as well as economic and social development. (5) Compatibility between the industrial structure and the demand structure. Under the condition of market economy, the industrial structure should change with the change of demand structure. To this end, for Yanjing salt chemical industry, it is first necessary to grasp the pulse of the market and its dynamic development direction, and adjust the relationship within the industrial cluster accordingly to adapt to the dynamic change in demand structure. (6) In addition, the industrial structure must meet the requirements of macroscopic readjustment and control. Yanjing industrial structure and its industrial structure within the salt chemical cluster should not only conform to the national macroscopic readjustment and control policy, but also accord with the industrial adjustment direction of Yanjing Township, Markam County, and the Tibet Autonomous Region.[1]

The high-grading or advancement of industrial structure, also known as the upgrading of industrial structure, refers to its development from a low level to a higher one according to the history and logical sequence of economic development, including the evolution of the dominant proportion of the primary industry to that of the secondary and tertiary industries in the entire industrial structure, of the dominant proportion of labor-intensive industries to that of capital-intensive and technical knowledge-intensive industries, and of the dominant proportion of industries manufacturing primary products to that of those manufacturing intermediate products and final products. [2]For Yanjing, the development of characteristic industrial clusters,

1 Ibid.

2 Ibid.

extension of industrial chain, and technological innovation have become the main path for the upgrading of local industrial structure.

c. To Integrate the Cultural Industry and Technology

Through an analysis of Yanjing salt cultural heritage cases, it is found that there are still many shortcomings and problems in the current integration and development of salt cultural industry and technology. In order to protect, innovate and develop Yanjing salt cultural heritage and improve the local cultural industrialization process, it is also necessary to explore the scientific mode and way of integrating Yanjing salt cultural industry and technology, so as to better promote the development of salt cultural industry.

To Improve the System and Mechanism for the Integration of Cultural Industry and Technology, and Create a Good Environment for Development

In order to promote the integration of Yanjing salt cultural industry and technological development, the most critical step is to improve the current system and mechanism according to the problems encountered in the process, and establish relevant regulatory departments to supervise the implementation of policies. To cultivate local people's awareness of cultural development, various measures should be taken to increase publicity and improve their comprehensive quality, so that they will realize the importance of the integration of cultural industry and technology, and actively participate in the environment of cultural development and technology integration. In addition, by learning from the development concepts of developed countries and regions, we may give full play to the functions of the government, cultivate the local cultural market in Yanjing, and recruit more scientific and technological talents to contribute to the integrated development of salt culture and technology.

To Construct a New Form of Cultural Industry and Vigorously Develop a New Type of It

Besides the inheritance of Yanjing traditional salt culture, to cultivate new cultural forms is also a new way to promote the integration of local salt culture and technology. The existing salt cultural industry and cultural services might be integrated with science and technology to enhance cultural value and form high value-added cultural and technological products. In addition to the traditional salt culture, the digital culture of Yanjing can also be enriched through the integration of science and technology, so as to develop cultural products of the new era belonging to Yanjing. For example, in recent years, the rise of Chinese cartoon has also made some achievements in the integration of China's cultural industry and technology. Cultural geography and smart tourism, scientific and technological protection of cultural heritage, and the development of TV media can all promote the development of a new salt cultural industry in Yanjing to some extent.

To Build a Platform for the Integration of Salt Culture and Technology, and Improve the Cultural and Technological Innovation System

Many local ethnic cultures in our country have retreated in the development of the times; the reason for this is the lack of a platform for the integration of culture and technology, which leads to the inability of many characteristic cultures to adapt to the development characteristics of the current society, and thus are lost in the colorful cultural ocean. In order to protect Yanjing salt culture as much as possible, a well salt culture demonstration zone should be established in Yanjing Township, which will integrate the cultural industry of Yanjing with science and technology, and it is necessary to focus on the unique culture of the region, fully highlighting the cultural characteristics of Yanjing area; what's more, we should complete cultural service transactions and build a new cultural tourism service platform within the demonstration zone.

To Strengthen Talent Support in the Salt Cultural Industry, and Cultivate and Introduce Technology Composite Talents

In the development process of cultural industry and technology integration, cultural industry talents play an important role. Composite talents with technological knowledge and cultural quality can provide necessary technical support for cultural development. Therefore, it is necessary to increase the cultivation of technology composite talents for Yanjing salt culture. On the one hand, we can increase capital investment in Tibetan universities and colleges and set up relevant key laboratories, so as to help them establish the awareness of local ethnic culture protection in the practice process and provide help for the development of Yanjing salt culture. On the other hand, we can introduce more composite talents from abroad by increasing salaries and other benefits.

In short, in order to better integrate Yanjing salt culture and technology, we must further explore and practice according to the local reality and the characteristics of the cultural industry. While recognizing the shortcomings of the integration of local salt cultural industry and technology, we should establish a "Yanjing mode" of salt culture industrialization led by the government and jointly promoted by cultural industries and enterprises, so as to form a new environment, new business form, and new industry for the development of local people and local culture.[1]

3) Tourism Development of Salt Cultural Heritage

Tourism, as an emerging industry, has become the world's largest industry since its development in the 1960s. However, the "surface" landscape tourism characterized by sensory stimulation already encountered the aesthetic fatigue of tourists, and the development of cultural tourism represented by cultural experience and exploration has become the next direction of world tourism. The salt culture of Yanjing is not only irreproducible because of its originality, ethnicity, religiousness, and freshness, but also has become advantageous tourism resources with great development potential. With the goal of integrating culture and tourism, we should develop new cultural tourism projects based on Yanjing culture while satisfying the development of traditional tourism service functions such as food, accommodation, transportation, traveling, shopping, and entertainment, etc., so as to combine folk customs, ethnic arts, and ecological environment perfectly.

a. To Integrate Tourism Resources and Enhance Competitiveness

Salt culture tourism in Yanjing started late, but the rich and colorful salt cultural resources with complete series and profound connotation may provide good conditions for the multi-level and multi-functional development of tourism. Only by creating high-quality tourism routes, tourism products, and tourism service projects, can we activate the Yanjing salt cultural tourism market and promote the development of local tourism economy.

Firstly, we may develop high-quality salt culture tourism products. The development of salt culture tourism products is an important way to integrate the characteristic brand of Yanjing salt industry with cultural tourism, and its modeling conception, pattern design, material use, structure manufacturing and other aspects

---

1 Chen Liang. Research on the Development Path of the Integration of Intangible Cultural Heritage Industry and Technology [J]. *Science and Technology & Innovation*, 2021(17).

should fully extract the cultural elements and concepts of Yanjing salt, which are manifested as: (1) The ruins, remains, and existing production sites of local well salt should be fully utilized, such as brine wells, ancient salt buildings, etc., which reflect the brilliant evidences of the history of Yanjing salt production, and models of their imitation and replication, as well as various transformed products can be developed. (2) The tools used in well sinking, repairing, brine drawing, sun-drying, etc., such as brine pails, salt scrapers, wooden bats, etc., can be re-innovated for their simple and elegant shapes and beautiful lines, and some daily necessities and stationery series with both aesthetic and use values can be designed. (3) In the use of materials and colors, pseudo-classic design can be carried out to combine classical and modern elements, for example, the color of salt buildings and brine may be applied to go with the bright color of modern materials, returning to nature without losing fashion and magnificence.

Secondly, we may create special tourist attractions of the salt industry. The featured brands of the salt industry can conduct in-depth cooperation with the tourist scenic spots, and demonstrate the history, culture, and craftsmanship of the salt industry through the planning and design, so as to attract tourists to visit for fun, and enhance its popularity and influence. (1) To create a characteristic street of salt culture. In addition to the unified use of salt culture in the name of street roads, culture walls, guideboards, shop signs, etc., as well as the construction or renovation of some theme hotels and guesthouses, we may build a salt culture square and make various sculptures featuring salt-making tools as well as salt production, transportation and marketing,so as to reproduce the prosperity of Yanjing yesterday. (2) To build a tourist attraction for brine bath. We can learn from the Bamboo Salt Well culture of Daying County as well as the type, development mode, operation and management mode of Zigong *hai tang shen quan* brine bath, cooperate with local salt companies to carry out brine bath recuperation projects, and gradually cultivate a diversified of business groups such as health catering, health bathing, and health sports. (3) To build a salt-making technology experience museum. In view of the limited exhibition content of Yanjing Museum, the Yanjing government can invest in the establishment of an exploratory interactive experience project--Yanjing Salt-making Technology Experience Center. Focusing on historical research, technology inheritance, discussion and exchange, research and education, cultural and creative experience, etc., the theme of local salt production technology may be developed in multiple dimensions by using metaverse technology and VR technology, etc.

Thirdly, we may launch theme activities of salt culture tourism. It is possible for characteristic brands of Yanjing salt industry, by cooperating with tourism organizations, to launch theme activities of salt culture tourism to attract tourists' participattion and enhance their own popularity and reputation. (1) To implement salt culture festival activities. Festival activities in Yanjing area are diverse in types and rich in connotation, including music, dance, painting, folk customs, etc., which are the reproduction of specific historical and cultural symbols of the salt industry. For local development, they are not only a folk phenomenon with cultural taste, but also an important economic activity type. (2) To organize salt culture exhibitions. The exhibitions might be divided into several sections, such as ancient salt making, well salt technology, well salt history, well salt literature, well salt craftsmen, well salt legends, etc. By expanding the area, adjusting the space, and adding projects, we may improve the popularity of Yanjing well salt, produce local characteristics, and provide new development opportunities for establishing a unique tourism image. (3) To develop leisure and performing arts venues. It is also a good means of publicity and marketing to choose some restaurants or teahouses for opera performances in Yanjing tourism distributing centers or leisure center areas. Local folk dramas and other folk art forms combined with catering and tea art, tourists, by enjoying the traditional performing arts while tasting wine and food, may get physical and mental pleasures as well as cultural edification, and enhance their feelings about the regional cultural characteristics of Yanjing.

b. To Strengthen Regional Tourism Interaction and Cooperation

Regional tourism cooperation means that tourism economic entities in different regions jointly promote the flow and reasonable allocation of their tourism industry elements under unified principles and systems, with a view to maximizing the economic, social, and ecological benefits of tourism. The ultimate goal of cooperation is to establish barrier-free tourism areas and achieve the integrated development of regional tourism.[1] For the development of Yanjing salt cultural tourism, we should

---

1 Liu Yuting. Research on Tourism Cooperation in Changsha, Zhuzhou, and Xiangtan Region [D]. Changsha: National University of Defense Technology, 2008.

have the concept of large region, large tourism, and large market in mind, and be integrated into the tourism loop of eastern Tibet, Tibet and Sichuan, and even the entire western region, becoming an important node in it. In this way, while achieving resource sharing, mutual benefit and win-win results, it can also introduce tourists and inject new vitality into the local well salt cultural tourism market and economic development.

Due to the wide geographical distribution and diverse forms of ethnic cultural resources in western China, and their different intact degrees, integration can make the development of these resources, including the well salt culture, more dynamic, continuous, and complete. In addition, from the overall point of view, the significance of Yanjing salt cultural tourism under regional tourism cooperation is to promote the integration of tourism economy in eastern Tibet through its comprehensive drive. On the basis of establishing and unifying the basic reference basis--the overall plan for regional tourism development, regional cooperation at multiple levels should be carried out to share resources and markets, and jointly establish a unified regional tourism image; on the basis of integrating the cultural tourism resources of Sichuan, Tibet, and Yunnan, we should launch a high-quality tourism route that not only has the regional characteristics of Yanjing salt culture, but also has the local characteristics of various provinces (autonomous region), so as to avoid the adverse pattern of separate governance and disorderly competition in the development of various parts of the cultural region to the maximum extent, give full play to the advantages of resources to a greater extent, and reduce unnecessary losses.[1]

The important measures to strengthen the regional tourism cooperation and development of Yanjing salt culture are mainly to improve the management mechanism, and conduct unified planning as well as joint publicity. The premise of regional cooperation is the coordination of management; similarly, the development of Yanjing salt cultural resources depends on the cooperation of relevant counties, cities, and departments. Therefore, Yanjing must unite the surrounding counties and cities, as well as departments involved in water conservancy, transportation, urban construction, planning, tourism, natural science, culture, environmental protection,

---

1 Xi Min,Wang Xueping. Development of Huai'an Salt Cultural Tourism Based on Regional Cooperation [J]. *Salt Industry History Research*, 2011(4).

etc., to form a management body--Yanjing Salt Cultural Zone Management Committee, and set up a special office in each place to assume related management and coordination responsibilities and organize the implementation of the project. Under the leadership of the Management Committee, unified planning should be carried out to compile the overall plan for the protection and development of Yanjing salt culture area, as well as the detailed regulatory plan of each section and other plans at different levels, clarifying the scope, period, protection and development principles and priorities, so as to provide planning ground for the implementation of comprehensive development. On the basis of planning, we should study the combination and collocation of tourism products in various counties and cities, formulate a joint promotion plan, shape the overall tourism image of Yanjing salt cultural area, jointly create a high-quality brand of "Yanjing Salt Cultural Tourism", and finally realize the goal of achieving better promotion effects at a lower cost.[1]

The basis for the realization of regional tourism cooperation is that the collaborating parties have complementary or similar tourism resources, and the integration of these resources can make the advantages of all parties apparent, even achieving a magical effect. To this end, Yanjing should take the initiative to strengthen communication and contact with surrounding areas as well as surrounding provinces and cities, establish strategic partnerships, jointly open up new routes and markets, and share potential benefits--even surrender part of the profits.

c. To Alleviate the Investment Pressure of Local Tourism through Multi-channel Financing

As an emerging industry, the sustainable development of tourism requires a large amount of capital investment and financing support. Located in the west, the development Yanjing is relatively backward, and the government cannot put a large amount of funds into the development of salt cultural tourism resources. So we must implement a diversified development strategy, and take the way of multi-channel financing, especially forming the situation that foreign investors, government, departments, collectives, and individuals all pitch into the work by making use of private funds and private capital.

1 Ibid.

To Strive for National, Provincial, and Municipal Support

According to the direction of government investment, the focus will be on public areas that are difficult to obtain direct benefits, including public infrastructure and utilities such as rural roads, river dams, etc., environmental management projects such as forests and water sources, etc., and the protection and maintenance of cultural relics and sites in the region. We should make full use of relevant national inclusive policies and play a good policy combination. We should make full use of various preferential policies to form a policy depression, comprehensively integrate all investment channels to form large-scale investment, and do a good job in the preliminary work of the project according to the investment trend. We should alos actively plan and prepare a number of rural tourism projects that are not only in line with the national, provincial, and municipal investment priorities, but also reflect the characteristics of Yanjing.

To Strengthen Cooperation with Financial Institutions

The funding channels mainly include applying for mortgage or pledge loans from commercial banks (such as Industrial and Commercial Bank of China, China Construction Bank, etc.), applying for subsidized loans from policy banks (such as China Development Bank, Agricultural Bank of China, etc.), seller's credit, financing guarantee through guarantee companies, applying for loans from the World Bank and the Asian Development Bank, etc., and applying for inter-country aid loans (such as FAO and NNDP, etc). The main financing methods are to take the future tickets of local scenic spots or together with other charging rights as a pledge, and the development and operation right of the scenic spots, the right to use the state-owned land or the ownership of related buildings as collateral for loans.

To Attract Foreign Investment through Intensified Efforts

The main methods include recruiting equity financing, private placement financing, and overall project financing, etc. We should make greater efforts to open up investment, attracting funds from inside and outside the region and from abroad, formulate preferential policies, and explore various modes of cooperation. We should also improve the investment environment, establish a concept of "cordial and profitable environment for investors", and prepare the packaging and investment promotion materials for rural tourism projects, including: project proposals, business plans, feasibility study reports for rural tourism projects, planning and programming for rural tourism projects, specific cooperation intentions, various preferential policies, basic legal documents, etc.

To Revitalize Assets through Innovative Modes

On the one hand, we may innovate business modes. By transferring from the development and operation mode of scenic spots, which mainly relies on gate receipts, to a number of businesses such as hotels, residential hotels, catering operations, tourism transportation, tourism souvenir production and sales, et., its financing capacity, including property financing, selling partial property rights, lease financing, etc. will be enhanced. On the other hand, we may create such modes as cultural residence collection, theme residence collection, nostalgic family collection, etc., and innovate resident management methods, such as issuing "green cards" to grant residents the right of permanent residency, etc. By actively introducing middle and high-end groups who are likely to have emotional resonance, rural tourism consumption will be greatly stimulated.

To Guide Local People to Invest

We may guide local people to invest in areas with small capital, quick results, stable returns, and flexible modes of operation, which include tourist catering such as restaurants, bars, teahouses, cafes, etc., residential hotels such as homestays, farmhouses, etc., tourist goods such as tourist souvenirs sales, folk handicrafts production, etc., tourist entertainment such as folk performances, opera performances, special entertainment services, etc, as well as small local travel agencies. The government should provide a fair business environment for rural tourism projects developed by private capital and provide necessary financial support. It may formulate preferential policies and make appropriate policy preferences for rural tourism projects invested and developed by local people. It may also help the people who invest in rural tourism set up non-governmental organizations, so that they can carry out mutual assistance and self-discipline in the industry.[1]

At the same time, in order to facilitate the in-depth integration of salt culture tourism, Yanjing, based on high-quality salt culture tourism projects, should also launch strong tourism publicity and promotion, so as to quickly change the original

---

1 Good Projects Lack No Funds, 7 Channels to Solve Rural Tourism Financing Problems. See https://www.163.com/dy/article/GI3SDFSD0538J4LT.html.

impression in people's minds, establishing a new tourism image of "magic Yanjing". A special salt culture tourism marketing agency can be set up; through careful planning and by expanding the popularity and influence of salt culture tourism online and offline, plus activity promotion, cross-border cooperation, etc., more tourists will be attracted to participate in it. In addition, it is necessary to continue to strengthen the construction of infrastructure supporting facilities and build a perfect tourism transportation network, so that the time for tourists to arrive at Yanjing tourist attractions will be shortened, the mobility of people will be more convenient, and the waste of tourism resources will be avoided to the greatest extent.[1]

1 Song Jie. Development and Utilization of Cultural Tourism Resources with Local Characteristics: A Case Study of Luobei County, Hegang, Heilongjiang Province [J]. *WENCUN YUEKAN*, 2023(7).

CHAPTER NINE

Strategies and Measures for the Internationalization of Yanjing Well Salt Culture

Culture, as a product of a certain society and era, inevitably evolves with societal changes. It continuously absorbs the essence of existing culture and develops forward dynamically, becoming a continuous and ongoing process. Furthermore, with the proliferation and migration of populations, cultural values at the psychological level, such as value concepts, ways of thinking, psychological qualities, ideal personalities, aesthetic tastes, and national characters, are transmitted from one society to another, and from one place to another. Since the psychological level of culture is the core and soul of cultural types, cultural transmission occurs[1].

Under the backdrop of globalization, no cultural system can exist in isolation; going global is an inevitable outcome. The excellent traditional culture of China, especially its local cultural characteristics, while maintaining its individuality, is eager to enter the international arena to seek consensus and intersection, actively absorbing advanced concepts of world culture, especially those cultural values that have gained global consensus, to enrich and develop itself. The thousand-year-old ancient salt fields of Yanjing, Markam County, with a long history of salt production, have a salt culture rooted in local civilization and have become the region's business card and cultural essence. As the degree of western China's opening up to the outside world continues to increase, cultural propaganda products that "start understanding China from the west" will gradually break through. Therefore, how to better

1 Li Lang, Chen Zhuo, Wen Ming-sheng. A Brief Discussion on the Dissemination of Salt Culture [J]. *Salt Industry History Research*, 2012(1):60-65.

communicate Yanjing salt culture to the outside world and internationalize it has become an urgent and practical issue before us.

1. The Significance and Value of Internationalizing Yanjing Well Salt Culture

The Yanjing Well Salt Culture possesses dual qualities of regional specificity and global universality, which determine its potential and inevitability to integrate into the global community. As an integral part of the excellent traditional Chinese culture, it carries a profound historical legacy and embodies a unique industry charm. The internationalization of its dissemination is not only necessary for the development of local cultural industries but also serves as a vital window for Chinese culture to step onto the world stage. Through international communication, it helps the world better understand the unique beauty of China, enhances China's international discourse power, strengthens national soft power, and boosts international competitiveness.

1) Supporting the "Belt and Road" Initiative

In September and October 2013, Chinese President Xi Jinping proposed the cooperative initiatives to build the "New Silk Road Economic Belt" and the "21st Century Maritime Silk Road". This initiative aims to draw on the historical symbol of the ancient Silk Road, to uphold the banner of peaceful development, and to actively develop economic cooperation with countries along the route, jointly creating a community of shared interests, destiny, and responsibility characterized by political mutual trust, economic integration, and cultural inclusiveness. The "Belt and Road" Cultural Development Action Plan (2016-2020) issued by the Ministry of Culture provides policy, financial, and talent support for external cultural exchanges and cultural industry development, aiming to comprehensively improve the level of China's cultural field opening to the outside world. The introduction of this action plan aims to promote countries and regions along the "Belt and Road" to actively participate in cultural exchanges and cooperation, inherit the spirit of the Silk Road, promote mutual learning among civilizations, achieve mutual understanding among peoples, and promote Chinese culture to "go global", expanding the international influence of Chinese culture.

With the in-depth advancement of the "Belt and Road" initiative, various places in China are accelerating opening up and cooperation according to their own advantages and characteristics, achieving foreign economic integration, and exploring the economic development path of "bringing in" and "going out". Tibet, located in the southwestern frontier of China, is a bridge and link connecting Nepal, Bhutan, India, Pakistan, and other South Asian countries. It shares a geo-economic plate with South Asian countries and has close ties with them in terms of ethnicity, religion, and culture. This geographical advantage makes it an irreplaceable and important position for promoting and expanding towards South Asia in the "Belt and Road" construction and an important channel for China to open up to South Asia determined by the Party Central Committee. Under the "Belt and Road" background, advanced technology, ideas, and values are introduced to Tibet through the exchange platform, promoting more frequent exchanges between Tibet and the outside world. Whether it is developing economic and trade activities or carrying out cultural exchanges, Tibet plays an important role and bears a heavy task.

Located in Markam County, Tibet, the Yanjing Township, with its thousand-year-old ancient salt fields and profound well salt culture, has become an important business card and cultural essence of the region. The well salt culture of Yanjing is a concentrated expression of the values and spiritual pursuits of the people of Yanjing, reflecting their unique way of life and wisdom for survival. Due to its unique regional and ethnic characteristics, the well salt culture of Yanjing has significant historical and global value and is an important part of intangible cultural heritage. As one of the regions with the most complete preservation of China's national and traditional forms, the well salt culture resources of Yanjing have a rich historical heritage and cultural accumulation, making the Tibetan cultural form unique and carrying a heavy and far-reaching cultural value. It is the most essential national character and feature that distinguishes the Tibetan nationality from other nationalities and is also a microcosm of the historical culture of the Tibetan people. Therefore, under the "Belt and Road" background, the well salt culture of Yanjing is in line with China's current cultural development direction. It is the specific, practical, and operational implementation of China's cultural "going out" strategy and is a rare local characteristic resource and a strategic resource that should not be missed to promote Chinese culture "going out". It is of great significance to achieving the great rejuvenation of the Chinese nation and enhancing China's cultural soft power.

Yanjing should seize the historical opportunity of the "Belt and Road" construction, integrate into the strategy and practice of Chinese culture "going out", and integrate into the cultural exchange channels of Chinese culture "going out". Excavating, displaying, and spreading the rich, profound, and unique culture of Yanjing, letting the world know, recognize, and love it, and improving the quality of external communication and enhancing its international influence is a sacred mission and an important task for the external communication of Yanjing culture, which can help promote the external communication of Chinese culture.

2) The Need to Enhance Tibet's Soft Power

Soft power is a brand-new concept proposed by Joseph Nye, the dean of Harvard University's Kennedy School of Government, in the late 1980s, emphasizing a country's ability to influence and assimilate other countries through the attraction of its own culture, political thought, and foreign policy. After this concept was proposed and introduced to China, it has had a wide and profound impact on China's intellectual, cultural, media, and political circles. With its widespread dissemination, enhancing "regional soft power" has also entered the agenda of many provinces, municipalities, autonomous regions, and cities in China. For the Tibet region, promoting the well salt culture of Yanjing, which has a profound historical and cultural heritage, to "go global" can better introduce the story of Tibet to countries along the "Belt and Road" and better enhance Tibet's cultural soft power.

a. Enhancing Tibet's International Image

The well salt culture of Yanjing, as an important part of Tibetan culture, helps enhance Tibet's international reputation and reputation. By showcasing the unique charm and profound cultural heritage of Yanjing, it can attract more international tourists to visit and experience, thereby promoting the development of Tibet's tourism industry. (1) Showcasing the uniqueness and diversity of Tibetan culture. The well salt culture of Yanjing is an important part of Tibetan culture, integrating natural and cultural landscapes, and demonstrating the wisdom and skills of ancient people. The overseas dissemination of this culture allows the world to more directly understand the cultural characteristics of Tibet, feel its profound historical heritage, and unique ethnic customs, thereby enhancing Tibet's cultural influence internationally. (2) Enhancing international understanding and cognition of Tibet. The overseas dissemination of the well salt culture of Yanjing allows the international community to more deeply understand the history, culture, customs, and natural environment of Tibet. This understanding helps to eliminate misunderstandings and prejudices about Tibet, enhances international cognition and respect for Tibet, and lays the foundation for Tibet to establish a good image on the international stage. (3) Enhancing Tibet's international reputation and reputation. As a business card of Tibet, the well salt culture of Yanjing demonstrates the unique cultural charm and historical heritage of Tibet. Its overseas dissemination can shape a unique cultural brand of Tibet, enhance Tibet's international reputation and attractiveness, and attract more international tourists to visit and experience.

b. Promoting Cultural Exchange and Mutual Learning between Tibet and the World

The overseas dissemination of the well salt culture of Yanjing is not only a process of showcasing Tibetan culture to the outside world but also a process of cultural exchange and mutual learning with the world. Through cultural exchange activities with other countries and regions, mutual understanding and friendship can be enhanced, and the diversity and prosperity of world culture can be promoted. This cultural exchange helps build a community with a shared future for mankind and promotes the harmonious coexistence of global culture. (1) Enhancing cultural identity. The international dissemination of Yanjing well salt culture demonstrates the unique cultural charm and historical heritage of the Tibet region, helping to enhance the cultural confidence and pride of the Tibetan people. This confidence and pride are important foundations for enhancing cultural identity and remind people to pay more attention to the protection and inheritance of this cultural heritage. (2) Enhancing public awareness of protection. Through international dissemination, more people will understand the importance and uniqueness of the well salt culture of Yanjing, thus paying more attention to and supporting the protection of cultural heritage. This will form a powerful social force, promoting the improvement of protection mechanisms and providing strong support for the long-term survival and development of the well salt culture of Yanjing. (3) Promoting cultural exchange and integration. The overseas dissemination of the well salt culture of Yanjing allows Tibet to absorb and learn from the excellent cultural achievements of other countries and regions, enriching and developing its own cultural connotations, while also providing unique cultural perspectives and insights for other countries and regions, promoting cultural diversity and prosperity.

c. Promoting the Economic and Social Development of Tibet

The dissemination of well salt culture can also drive the development of related industries, providing more employment opportunities and sources of income for local residents. Moreover, as the reputation of Yanjing increases, it can attract more domestic and foreign investors to invest and start businesses, bringing a positive impact on the economic and social development of Tibet. (1) Promoting the development of the tourism industry. With the enhancement of the international reputation of the well salt culture, more and more international tourists are turning their attention to this mysterious and ancient land. This directly promotes the development of the tourism industry in the Tibet region, bringing more tourism revenue to the local area. At the same time, the prosperity of the tourism industry also drives the development of related industries such as catering, accommodation, and transportation, providing more employment opportunities and sources of income for local residents. (2) Promoting the development of the cultural and creative industry. The international dissemination of Yanjing well salt culture also promotes the development of the cultural and creative industry. Tourism souvenirs, handicrafts, and cultural creative products with the theme of salt culture continue to emerge, enriching the cultural product market of the Tibet region. The production and sales of these products not only bring economic benefits to the local area but also promote the inheritance and innovation of culture. (3) Promoting the adjustment of economic structure and industrial upgrading. The overseas dissemination of Yanjing well salt culture helps the vigorous development of the local tourism industry and cultural industry, allowing the Tibet region to gradually transform from traditional agriculture and animal husbandry to a more diversified industrial structure, forming a complete industrial chain including cultural tourism, cultural performance, and cultural creativity. This transformation helps to enhance the economic competitiveness and sustainable development ability of the Tibet region.

In summary, the international dissemination of the well salt culture of Yanjing is of great significance in enhancing the soft power of Tibet. It not only can enhance the cultural identity and self-confidence of the people in Tibet but also promote economic development and social progress, enhance international image and status, and inject new vitality and momentum into the comprehensive development of Tibet.

3) The Need to Sustain the Vitality of the Well Salt Culture

The well salt culture of Yanjing is a vibrant and regionally distinctive culture formed by absorbing and drawing on the Central Plains culture and other foreign cultures on the Tibetan land. It has provided rich nourishment for the development and growth of Tibet, but due to its relatively remote geographical location, its international reputation is limited. To maintain its vitality and development, the well salt culture of Yanjing must "go out" and cannot stick to the old ways.

a. Cultural Inheritance and Protection

International dissemination provides a broader space and opportunities for the inheritance and development of the well salt culture of Yanjing. Through communication and interaction with the international community, it can absorb and draw on the excellent elements of other cultures, promoting the innovation and development of the well salt culture of Yanjing. At the same time, international dissemination also helps to promote the well salt culture of Yanjing to the world stage, allowing more people to understand and appreciate this culture, thus promoting its inheritance and development on a global scale. (1) Preservation of historical memory. The well salt culture of Yanjing carries a long history and rich cultural connotations and is an important part of the cultural heritage of the Tibet region and the Chinese nation. Through overseas dissemination, this unique cultural form can be more widely recognized and inherited, thus avoiding gradual disappearance in the long river of history. (2) Inheritance of skills and knowledge. The well salt making skills are the core of the culture of Yanjing, containing the wisdom and skills passed down from generation to generation. Overseas dissemination can stimulate more people's interest and learning in these traditional skills, ensuring that these precious skills can be inherited and developed. (3) Attracting international funds and technical support. When international dissemination allows this cultural treasure to be widely recognized overseas, it may attract international organizations, foundations, academic institutions, and others to provide financial, technical, and personnel support. This is an important driving force for the protection, restoration, research, and inheritance of the well salt culture of Yanjing.

b. Stimulating Cultural Vitality

International dissemination can significantly enhance the global recognition and influence of the well salt culture of Yanjing. Through various channels such as international media, cultural exchange activities, and academic conferences, the unique charm of the well salt culture of Yanjing can be displayed to all parts of the world, attracting more attention and interest. This widespread recognition not only helps to enhance the understanding and respect of the outside world for the well salt culture of Yanjing but also injects new vitality and vitality into it, promoting the inheritance and development of culture. (1) Deepening and integrating cultural exchanges. Through international dissemination, the well salt culture can engage in dialogue and interaction with the cultural heritage of other countries and regions, jointly exploring the paths of cultural heritage protection, inheritance, and development. This helps to broaden cultural horizons, enhance mutual understanding and respect, and also injects new elements and vitality into the well salt culture of Yanjing, making it more colorful. (2) Promoting cultural innovation. Overseas dissemination is not only a display of traditional culture but also an opportunity for the innovation and development of traditional culture. Through exchanges and collisions with different cultures, the well salt culture of Yanjing can absorb new elements and inspirations, stimulating the vitality of cultural innovation. (3) Prosperity of the cultural industry. The attention and market demand brought by overseas dissemination can promote the development of industries related to the well salt culture of Yanjing, such as cultural tourism, cultural performance, and cultural creative products, creating more economic benefits for the local area and also providing a more solid economic foundation for the inheritance of culture.

c. Meeting the Challenges of Globalization

Against the backdrop of globalization, the overseas dissemination of the well salt culture of Yanjing is an important means to meet the challenges of cultural globalization and maintain cultural characteristics and individuality. International dissemination helps the local well salt culture cope with the challenges brought by globalization, such as cultural homogenization and environmental changes. By international cooperation and exchange, it gathers resources and strength from all parties to form a more powerful protective force.[1] (1) Enhancing influence and competitiveness. By displaying the unique charm and value of the well salt culture on the international stage, it can attract more attention and interest, thereby promoting its integration into the global cultural system. This enhancement of influence helps the well salt culture to occupy a place in the competition of globalization, providing strong support for its long-term survival and development. (2) Strengthening national identity and unity. The international dissemination of the well salt culture can stimulate people's sense of national pride and belonging, promoting exchanges and integration between different ethnic groups. This sense of national identity and unity is an important cornerstone for building a community with a shared future for mankind and helps promote the harmonious and stable development of the global society. (3) Promoting sustainable development. By spreading the ecological concepts and practical experiences of the well salt culture to the international community, it can guide people to pay attention to environmental protection and sustainable development issues, promoting the green development of the global society and providing a sustainable development path for building a community with a shared future for mankind.

In summary, the international dissemination of the well salt culture of Yanjing is not only conducive to inheriting and promoting traditional culture, promoting local economic development, enhancing the sense of national identity and cohesion, and enhancing the national cultural soft power and international influence but also can promote the development of international friendly relations. Therefore, we should actively promote the international dissemination of the well salt culture of Yanjing and contribute to building a community with a shared future for mankind.

2. Issues and Dilemmas in the International Dissemination of Yanjing Well Salt Culture

With the continuous development of Tibet's tourism industry and the national emphasis on the protection of intangible cultural heritage, the salt culture of Yanjing

1 Markam, Tibet: Millennial Ancient Salt Fields Create a New Future, http://www.tibet.cn/cn/news/yc/202404/t20240409_7606033.html.

has ushered in new opportunities for development. By strengthening infrastructure construction and improving the quality of tourism services, the international reputation and influence of Yanjing salt culture can be further enhanced. However, despite the many advantages of the international dissemination of the Yanjing well salt culture, there are still some issues and dilemmas in the actual operation process, mainly including the following aspects:

1) The Dissemination Content is Relatively Singular

The international dissemination of Yanjing salt culture has a problem of relatively singular content and form, which may lead to a one-sided understanding of Yanjing salt culture by the audience and prevent them from fully and deeply experiencing its rich connotations.

a. Lack of Diversity

Diversity is crucial in cultural dissemination, as it not only enriches the content and enhances attractiveness but also promotes the exchange and integration between different cultures. The lack of diversity in the international dissemination of Yanjing well salt culture is indeed a problem that deserves attention. (1) Repetition of history and craftsmanship. In the international dissemination of Yanjing salt culture, there is often an overemphasis on introducing its ancient salt fields, unique salt-making techniques, and rich historical relics, while neglecting the rich connotations of salt culture at multiple levels such as society, art, and religion. Moreover, there is relatively little introduction to the deep connotations of salt culture, its connection with local ethnic culture, and its status in the global history of the salt industry. (2) Lack of storytelling. When introducing Yanjing well salt culture, there is a lack of stories related to the salt industry, such as the life stories of salt workers, the business legends of salt merchants, etc., and it does not fully show the interactive stories with the daily life, festivals, and folklore of local ethnic groups, failing to fully display the humanistic spirit and historical changes of Yanjing well salt culture. Moreover, the singular and empty narration makes the cultural content seem dull and unattractive, making it difficult to elicit emotional resonance due to the lack of appeal. (3) Neglect of modern elements. In external dissemination, the content of Yanjing salt culture overly emphasizes traditional salt-making skills and historical value, while neglecting the application of modern technology in salt-making processes, such as modern salt-making equipment and environmental protection technology. The lack of these modern elements makes the dissemination content lack a sense of the times. At the same time, it also fails to fully show the combination of salt culture with modern lifestyles, such as the application of salt products in modern cooking, health, and beauty fields. The absence of these modern life scenes makes the culture seem disconnected from the lives of modern audiences.

b. Lack of In-depth Exploration

In-depth exploration helps to fully display the rich connotations and unique charm of Yanjing salt culture, enhancing the international audience's cognition and identification with it. However, the lack of this aspect may lead to a one-sided and limited cultural dissemination. (1) Superficial introduction of cultural background. The dissemination content often only stays on the surface introduction of Yanjing well salt culture, with issues such as oversimplification of historical background and lack of geographical and environmental information. For example, it may only mention the age and basic salt-making process of Yanjing salt pans without delving into its historical evolution, relationship with local social structure, and the impact of historical events. In addition, environmental background information that is crucial for understanding the formation and development of salt culture, such as the geographical features, climate conditions, and resource characteristics of Yanjing, is often not fully presented. (2) Neglect of the deep meaning of cultural value. In the dissemination, the value of Yanjing well salt culture is often simplified to the display of salt-making skills and products, without delving into its intrinsic cultural value, such as its spiritual symbolism and social function in local society. Moreover, the dissemination often fails to delve into the inheritance methods, historical evolution, and modern practices of salt culture, and how these inheritances have shaped local cultural identity and social practice. (3) Neglect of in-depth discussion on cultural conflict and integration. The dissemination often overlooks the dialogue and exchange between Yanjing well salt culture and other cultures and discusses its adaptability in the context of globalization less, including how it adapts to the international market and cultural environment, thus failing to show the interaction and influence between different cultures. At the same time, the international dissemination often fails to delve into the construction and expression of cultural identity and how these identities are understood and accepted in cross-cultural communication. (4) Lack of in-depth thinking on cultural protection and sustainable development. The dissemination content often fails to introduce the specific protection measures of Yanjing well salt culture in detail, including the preservation of traditional skills, the maintenance and restoration of cultural sites. At the same time, it does not delve into long-term planning and strategies for local sustainable development, the impact and assessment of salt culture activities on the local environment, and how to implement environmentally friendly cultural practices.

c. Lack of Innovation

Innovation is self-evident in the international dissemination of Yanjing salt culture. It is not only the key to enhancing cultural influence but also the driving force for the continuous development and prosperity of culture. However, the local well salt culture has not been well combined with a modern perspective in the process of international dissemination, showing the development and innovation of Yanjing well salt culture in contemporary society and its connection with modern society. (1) Insufficient application of modern design in cultural elements. In the process of international dissemination, the traditional elements of Yanjing salt culture are often out of touch with modern aesthetics, appearing too old-fashioned and monotonous under modern aesthetic concepts. Moreover, there is a lack of innovative attempts to combine traditional elements with modern design concepts, failing to integrate salt culture with other cultural elements organically to form content with unique charm and wide appeal. (2) Lack of modern design and packaging. The products of Yanjing well salt culture often rely too much on traditional cultural elements and forms, lacking exploration and innovation of new elements and forms. This leads to a lack of vitality and appeal in cultural dissemination, failing to fully integrate with modern aesthetics and market demands, making it difficult to stand out in fierce market competition. (3) Lack of modern marketing strategies. Yanjing well salt culture has a rich history and unique salt-making techniques, but it lacks effective branding and storytelling marketing strategies in international dissemination, making it difficult to enhance its brand value and resulting in a lack of recognition and competitiveness in the international market, limiting its expansion in the international market. (4) Lack of innovative tourism promotion. Although the well salt producing villages have begun to become a tourist village, there is still much room for improvement in how to more deeply integrate salt culture with eco-tourism and how to enhance the international reputation of salt culture through tourism.

2) Ineffective Communication Channels

The term "ineffective communication channels in cultural international communication" refers to the obstacles and restrictions encountered in the process of cross-border cultural dissemination under the background of globalization, which prevent cultural content from being smoothly and effectively conveyed to the international audience. The salt culture of Yanjing also has significant issues in this regard.

a. Limited Coverage of Traditional Communication Channels

The Yanjing well salt culture is mainly promoted through traditional communication channels, such as tourism brochures, cultural exhibitions, and news reports. Although these channels can convey information about salt culture to some extent, their coverage is relatively limited and difficult to reach a global audience. Especially for international audiences far from Yanjing area, they may find it difficult to learn about the unique charm of Yanjing well salt culture through these traditional channels. (1) News Coverage. Although news coverage is an important way to spread culture, it is often limited to domestic or regional media. This results in limited exposure of Yanjing well salt culture in international news reports, making it difficult for international audiences to obtain relevant reports and information through international media, thus limiting the breadth of its international dissemination and preventing Yanjing well salt culture from achieving a high enough level of international fame and influence. (2) Cultural Exchange Activities. Cultural exchange activities for Yanjing well salt culture are relatively few and limited in scale and frequency, usually targeting specific audience groups, such as scholars, artists, and cultural enthusiasts. Although these people have a high interest and understanding of culture, they only represent a part of the audience. This makes it difficult for international audiences to gain an in-depth understanding of Yanjing salt culture through such activities, thus limiting the breadth of its international dissemination. (3) Tourism Brochures. As a traditional medium of promotion, the content presentation of tourism brochures is limited by length and format, making it difficult to fully and deeply showcase the rich connotations and unique charm of Yanjing salt culture to international audiences; moreover, the distribution channels of brochures also limit their dissemination range, making it inconvenient for some international audiences to obtain the brochures of Yanjing salt culture, thus missing the opportunity to learn about this culture. (4) Books and Journals. Due to the relatively few reports and studies on Yanjing well salt culture in relevant fields of books and journals, it is difficult to form a wide influence. Coupled with the relatively long publication cycle of books and journals, which requires a series of cumbersome processes from writing, editing, reviewing to publication, this also affects the timely dissemination of the latest research results and trends of salt culture to international audiences.

b. Insufficient Utilization of Modern Communication Channels

In the digital age, new media channels such as social media and online platforms have become an important force in cultural dissemination. However, Yanjing well salt culture does not fully utilize these modern communication channels in its international dissemination. (1) Incomplete Digital Platform Functions. The existing digital platforms for Yanjing well salt culture may lack interactive functions with the audience, such as online Q&A and comment areas, which prevent the audience from asking questions and sharing feelings in a timely manner, thus reducing the effectiveness of dissemination; at the same time, its digital dissemination mainly relies on official websites or social media platforms, lacking diversified dissemination channels and partners, limiting the widespread dissemination of information. (2) Limited Application of Digital Technology. On the one hand, advanced technologies that can provide more intuitive and immersive experiences, such as Virtual Reality (VR) and Augmented Reality (AR), are to some extent missing in the digital dissemination of Yanjing well salt culture; on the other hand, there may be a lack of collection and analysis of audience behavior and interest data in the local digital dissemination process, leading to a lack of scientific basis for the formulation of dissemination strategies, making it difficult to accurately target audiences and optimize dissemination effects. (3) Insufficient Multimedia Content Creation and Sharing. In the creation process, the cultural differences and audience needs of different countries and regions are not fully considered, making it difficult for Yanjing well salt culture dissemination content to resonate and interest international audiences; in the sharing process, the promotional resources of other propaganda entities such as overseas Chinese, students studying abroad, and foreign institutions are not fully utilized, and there may be a lack of interaction and participation opportunities with international audiences, reducing audience participation and stickiness, affecting the dissemination effect of the content.

c. Limited International Exchange and Cooperation Channels

Yanjing well salt culture has a relatively low international profile and lacks influential international exchange platforms. This makes it difficult for Yanjing salt culture to be widely displayed and exchanged on the international stage, limiting the scope and depth of its international dissemination. (1) Limited Cooperative Institutions. At present, Yanjing well salt culture has relatively few cooperation numbers with domestic and foreign cultural institutions, tourism institutions, educational institutions, etc., and the level of cooperation is relatively low (mainly manifested as superficial exchange activities and project cooperation), with limited cooperative resources (including funds, talents, technology, etc.), and lacks a well-established cooperation mechanism (including the signing of cooperation agreements, the implementation of cooperation projects, the evaluation of cooperation results, etc.), which may make it difficult to form effective international influence. (2) Scarce International Exchange Activities. Yanjing well salt culture has few international exchange activities, lacking regular international exhibitions, forums, seminars, etc., making it difficult to establish close contact and interaction with international audiences, limiting the depth and breadth of its international dissemination; at the same time, its participation in international academic conferences in related fields is low, lacking in-depth exchange and cooperation with the international academic community and research institutions, also making its research and development difficult to keep pace with international standards, making it difficult to gain international recognition and support. (3) Insufficient International Promotion. The international promotion activities of Yanjing well salt culture often lack innovation and interactivity, making it difficult to attract the attention and participation of international tourists. Moreover, due to the lack of effective promotion strategies and methods, there are also deficiencies in the organization and execution of promotion activities, such as unreasonable activity arrangements, poor promotion effects, and other issues, which also make it difficult to attract the interest of international tourists and investors, affecting the international dissemination effect of Yanjing salt culture.

3) Limited International Communication Capacity

The international communication of Yanjing salt culture faces certain challenges. Due to limitations in language, culture, geography, and other factors, it is difficult for Yanjing salt culture to cross borders and spread widely. At the same time, the lack of professional international communication organizations and talents also restricts the international fame and influence of Yanjing salt culture.

a. Insufficient Funding and Resource Investment

The importance of funding and resource investment in cultural dissemination is self-evident; they are key elements in promoting effective communication and sustainable development of Yanjing well salt culture. However, the external communication of the local salt culture may face restrictions in funding and resources, affecting the breadth and depth of dissemination. (1) Lack of publicity funds. Due to limited funds, it is difficult for Yanjing well salt culture to place large-scale advertisements in international media, which restricts the enhancement of its brand awareness and influence. Moreover, this leads to limited production quality of promotional materials, which may not fully showcase the unique charm and value of Yanjing salt culture. (2) Inadequate cultural resource exploration and protection. In-depth research on Yanjing well salt culture requires financial support, but the current lack of investment results in insufficient exploration of the historical background and folk customs of the salt culture. Additionally, a shortage of funds affects the protection of the local salt cultural heritage, potentially leading to damage or loss of valuable cultural relics. (3) Lagging infrastructure construction. Tourism facilities in Yanjing area, such as transportation, accommodation, and catering, may be delayed due to insufficient funds, affecting the visiting experience and satisfaction of tourists. Moreover, a lack of funds also leads to a shortage of dedicated cultural display venues, such as museums and exhibition halls, which limits the display and dissemination of Yanjing well salt culture. (4) Limited international cooperation and exchange. Insufficient funds may reduce the participation of Yanjing well salt culture in international exchange activities, restricting cultural exchanges and cooperation with other countries and regions. Additionally, a lack of funds affects the introduction and training of international talents, limiting the talent support for the international communication of Yanjing well salt culture.

b. Language and Cultural Barriers

The international communication of Yanjing well salt culture also faces language and cultural barriers. Audiences from different countries and regions use different languages, and the lack of multilingual cultural dissemination materials and translation services makes it difficult for international audiences to truly understand and accept Yanjing well salt culture. This is discussed from two aspects: (1) Language barriers. On the one hand, there is a lack of translation. The materials, especially promotional materials, of Yanjing well salt culture often fail to provide multilingual content, limiting the understanding and acceptance of audiences from different linguistic backgrounds. Due to language differences, international tourists may not accurately understand the historical background, production techniques, and folk customs of the salt culture. On the other hand, there are difficulties in communication. In cultural exchange activities, language barriers may make it difficult for both parties to communicate and exchange effectively. This not only limits the spread of Yanjing well salt culture but may also affect the visiting experience and satisfaction of international tourists. (2) Cultural barriers. Firstly, there are cultural differences. Different countries and regions have different cultural backgrounds, values, and aesthetic standards, which may lead to varying degrees of understanding and acceptance of Yanjing well salt culture among international audiences. For example, some international tourists may find it difficult to understand and accept certain traditional customs and beliefs in the salt culture. Secondly, there is cultural sensitivity. In the process of cultural dissemination, it is necessary to consider the sensitivity and taboos of different cultural backgrounds. Some elements of Yanjing well salt culture may touch on the cultural sensitivities of certain international audiences, leading to misunderstandings or conflicts. Thirdly, there is low cultural recognition. Yanjing well salt culture has a relatively low international profile, which makes international audiences lack understanding and cognition of it. Due to low cultural recognition, international tourists may find it difficult to distinguish Yanjing well salt culture from other well-known cultural landscapes or cultural heritages.

c. Lack of Professional Talents

Professional talents can lay a solid foundation for the international communication and long-term development of Yanjing salt culture. However, due to the lack of talent resources in the area, the effectiveness of cultural dissemination is hard to achieve as expected. This is mainly reflected in the following aspects: (1) Lack of foreign language professionals. There is a shortage of multilingual translators and editors proficient in various languages (especially internationally common languages like English), as well as a lack of composite language service talents who are proficient in foreign languages and understand the professional knowledge of the salt industry. There is also a lack of small language professional translation talents serving countries along the "Belt and Road", which may lead to distortion in the translation process or an inability to accurately convey the unique charm of the salt culture. (2) Lack of digital media and new media technology talents, including professionals with skills in digital media production, social media operations, data analysis, etc. This makes the international communication of Yanjing well salt culture lack sufficient interactivity and participation, making it difficult to form a sustained communication effect and influence. (3) Lack of marketing and brand promotion experts. This not only makes Yanjing well salt culture fall into a predicament of blind and ineffective communication in the international market but also makes it monotonous and lacking in novelty, making it difficult to attract the attention and interest of international audiences and lacking competitiveness in the international market. (4) Lack of cultural tourism and event planning experts.[1] This makes it difficult for Yanjing well salt culture to develop tourism products with local characteristics and cultural connotations, making local tourism activities monotonous and lacking in creativity, affecting the attention and participation of international tourists, and ultimately affecting the effective shaping and dissemination of the tourism brand image. (5) Lack of research and academic support talents. This makes the dissemination content of Yanjing well salt culture appear superficial and one-sided, making it difficult to show its profound cultural heritage and unique charm; moreover, its dissemination strategy may lack a scientific basis and targeting, making it difficult for related academic research results to be widely disseminated and recognized.

In summary, the problems and dilemmas faced by the external communication

1 Strengthening International Communication Capacity Building and Cultivating Language Service Professionals, https://www.cssn.cn/skgz/bwyc/202310/t20231025_5692691.shtml.

of Yanjing salt culture involve many aspects, which not only limit the breadth and depth of local cultural dissemination, affect the shaping and promotion of the cultural brand, hinder opportunities for cultural exchange and cooperation, affect the protection and inheritance of cultural heritage, but also restrict the development and innovation of the cultural industry. Therefore, how to effectively solve these problems and escape the dilemma has become an urgent issue in the current context.

3. Strategies and Paths for the International Dissemination of Yanjing Well Salt Culture

To address the issues faced in the international dissemination of Yanjing salt culture, a variety of measures need to be taken, such as enriching the content and forms of dissemination, expanding communication channels, strengthening international communication capabilities, increasing the protection and inheritance efforts, enhancing audience participation, and strengthening communication on cultural differences and misunderstandings. Through the implementation of these measures, the effective dissemination and development of the local salt culture can be promoted, allowing more people to understand and recognize this unique cultural heritage.

1) Brand Building and Promotion

The brand building and promotion of Yanjing salt culture in international communication is a systematic project aimed at enhancing the visibility and influence of the local salt culture, attracting the attention of more domestic and international tourists and scholars. The following is a detailed analysis of the brand building and promotion of this type of salt culture:

a. Brand Building

Brand building not only enhances the recognition of Yanjing well salt culture, promotes cultural inheritance and development, but also strengthens market competitiveness and drives the development of cultural tourism. (1) Core Value Extraction. It is necessary to delve into the historical origins, traditional crafts, and humanistic connotations of Yanjing well salt culture to extract unique and attractive core values; at the same time, emphasize the originality, manual nature, and eco-friendliness of Yanjing well salt culture, highlighting its differences from other salt cultures. (2) Brand Image Design. A recognizable brand image should be designed, including logos, colors, fonts, etc., to form a unified visual image; the image design should reflect the characteristics of Yanjing well salt culture, such as using elements of salt fields and salt drying tools, combining traditional and modern design techniques. (3) Brand Story Dissemination. The legends, historical allusions, etc., of Yanjing well salt culture should be excavated and disseminated through text, images, videos, and other forms, emphasizing the contributions of Yanjing well salt culture to the local economy, society, culture, etc., to enhance the emotional value of the brand.

b. Brand Promotion

Brand promotion not only expands the influence of Yanjing well salt culture and enhances cultural identity but also promotes the integration of cultural tourism, increases product value, and drives cultural inheritance and innovation. (1) Media Promotion. Traditional media such as television, radio, and newspapers can be used for advertising and news reporting to increase brand visibility and exposure; at the same time, well-known media personnel and travel enthusiasts can be invited to visit and experience on-site, writing special reports and travel notes to share the charm of Yanjing well salt culture. (2) Social Media Promotion. Social media platforms such as Weibo, WeChat, and TikTok can be used to publish pictures, videos, and articles about Yanjing well salt culture, interacting with fans; official accounts can also be created to regularly release updated content and hold online activities, such as lotteries and Q&A sessions, to attract more attention and participation. (3) Offline Activities. Salt culture festivals, salt drying experience activities, and cultural exchange meetings can be held, inviting tourists and experts to participate and experience the charm of Yanjing salt culture firsthand; cooperation with cultural and tourism institutions in other regions can also be carried out to jointly hold exhibitions, performances, and other activities to expand the brand's influence. (4) Cooperative Promotion. Cooperative relationships can be established with travel service providers such as travel agencies and hotels to promote Yanjing well salt culture together, providing convenient travel services for tourists; cross-border cooperation with other cultural and tourism brands can also be carried out to jointly create characteristic tourism routes and products. (5) International Communication. Participation in international tourism exhibitions, cultural exhibitions, and other activities can showcase the charm of Yanjing well salt culture and attract the attention of international tourists; strengthening exchanges and cooperation with international tourism and cultural institutions can also promote the international dissemination of Yanjing well salt culture.

c. Matters Needing Attention

Although brand building and promotion are key steps in enhancing the visibility and influence of Yanjing salt culture, the following aspects should be paid attention to during this process: (1) Maintain authenticity. On the one hand, the data and facts quoted in the promotion should be strictly verified to ensure accuracy and avoid exaggeration or fabrication of facts, so as not to damage the brand image and reputation; on the other hand, the true conditions and production process of the well salt can be displayed through on-site shooting and live broadcasting, and the sources of information and references should be clearly marked in the promotional materials to increase transparency. (2) Focus on experience. Rich experience activities and interactive links should be provided for tourists to enhance their sense of participation and satisfaction; measures such as creating immersive experience scenes, focusing on emotional resonance and story dissemination, and optimizing service and experience details can also significantly enhance the tourists' sense of experience and satisfaction. (3) Continue to innovate. New highlights and elements of Yanjing well salt culture should be explored and practiced continuously, starting from product innovation, marketing innovation, cultural innovation, and technological innovation, to maintain the brand's vitality and attractiveness, making it a culturally unique brand.

Through the implementation of the above measures, a unique and attractive brand image of Yanjing salt culture can be shaped, and the brand's visibility and influence can be enhanced through multi-channel and multi-level promotion, attracting more tourists to experience and feel the charm of Yanjing salt culture.

2) Expanding Communication Channels

In the international dissemination of Yanjing salt culture, expanding communication channels is key to enhancing visibility and influence. Here are some specific strategies for expanding communication channels:

a. Utilization of New Media Platforms

Against the backdrop of the new media era, leveraging new media platforms to promote and disseminate Yanjing well salt culture is of great significance for increasing the cultural visibility and economic development of the region. (1) Social Media. Platforms such as YouTube, TikTok, and Twitter can be used to publish texts, videos, and articles about Yanjing well salt culture; official accounts can be created to regularly update content, interact with followers, and host online events like lotteries and Q&A sessions to attract more attention and participation. (2) Video Platforms. Documentaries, promotional videos, and short videos about Yanjing well salt culture can be released on video platforms like Netflix, Hulu, and Triller; collaborations with well-known video creators can also be sought to produce engaging video content, increasing views and dissemination effectiveness. (3) Audio Platforms. Audiobooks, podcasts, and lectures about Yanjing well salt culture can be published on audio platforms such as Spotify and iHeartRadio; interviews and explanations by experts and scholars can also be conducted to enhance the authority and appeal of the content.

b. Cooperation with Traditional Media

To introduce this cultural treasure of Yanjing well salt culture to the world, cooperation with traditional media is equally important for its international communication. (1) Television. Collaborations with television stations can be made to produce documentaries, feature reports, and variety shows about Yanjing well salt culture, or to place advertisements on specific channels like travel and culture channels to increase brand visibility and exposure. (2) Radio. Partnerships with radio stations can be established to create radio programs and advertisements about Yanjing well salt culture, leveraging the broad coverage and diverse audience base of radio for widespread dissemination. (3) Newspapers and Magazines. Articles and advertisements about Yanjing well salt culture can be published in travel and culture newspapers and magazines, or collaborations with well-known writers and journalists can be made to write feature reports and travelogues, enhancing brand visibility and reputation.

c. Integration of Online and Offline Activities

The international communication of Yanjing well salt culture can be more effectively promoted to the world by integrating online and offline methods. (1)

Online Activities. Hosting online photography contests, essay competitions, etc., can encourage visitors and netizens to share their works and stories about Yanjing well salt culture; live streaming events such as salt drying demonstrations, cultural lectures, and folk performances can attract international audience participation; interactive features like online Q&A and message boards can answer questions from international audiences and collect their feedback and suggestions. (2) Offline Activities. Organizing salt culture festivals, salt drying experience activities, and cultural exchange meetings can invite visitors and experts to participate; during these events, exhibitions, performances, and experience sessions can allow visitors to experience the cultural charm of Yanjing firsthand. Collaborations with travel agencies, hotels, and other tourism service providers can also be made to launch themed tourism routes and products featuring Yanjing well salt culture.

d. Cross-Sector Cooperation and Resource Integration

Cross-sector cooperation and resource integration in the international communication of Yanjing well salt culture is a complex and fruitful process involving cooperation across multiple fields and levels, aimed at jointly promoting the international communication and development of Yanjing salt culture. (1) Cross-Sector Cooperation. Collaborations can be made with other cultural and tourism brands to create characteristic tourism routes and products; partnerships can be established with educational institutions to include Yanjing well salt culture in their curriculum, fostering cultural awareness and interest among students; collaborations can also be made with domestic and international cultural institutions, museums, and art schools to jointly conduct research, protection, and dissemination of Yanjing salt culture. (2) Resource Integration. By integrating tourism resources, cultural resources, human resources, and policy resources related to Yanjing salt culture, a complete tourism industry chain can be formed, which is beneficial for creating a brand for Yanjing salt culture tourism and enhancing overall competitiveness and influence.

Through the implementation of the above measures, the communication channels for Yanjing salt culture can be expanded, increasing brand visibility and influence, and attracting more tourists to experience and appreciate the unique charm of Yanjing salt culture. This approach can showcase the cultural charm of Yanjing in an all-around and multi-level manner, attracting the attention of more international tourists and observers.

3) Deepening Cultural Research and Exchange

In the international communication of Yanjing salt culture, deepening cultural research and exchange is a crucial link. This not only enhances the connotation and value of Yanjing salt culture but also promotes understanding and respect between different cultures. Here are some specific strategies and suggestions:

a. Strengthening Cultural Research and Exploration

In the international communication of Yanjing well salt culture, strengthening cultural research and exploration is necessary for inheriting and promoting local excellent traditional culture. It is also an important measure for promoting cultural exchange and mutual learning, driving cultural industry development, enhancing awareness of cultural heritage protection, and providing strong support for the application for world cultural heritage. (1) Historical and Archaeological Research. Deep exploration of the historical origins of Yanjing well salt culture is needed, through archaeological excavation and documentary research, to understand the origins, development, and evolution of the salt fields; it is also necessary to study the historical connections between the salt fields and surrounding areas, discussing the important status and role of salt culture in local history. (2) Anthropological and Sociological Research. Research should be conducted from anthropological and sociological perspectives on the impact of salt fields on local communities, including economic, social, and cultural aspects; it is also necessary to investigate the interactive relationship between salt fields and the lifestyle of local residents, understanding the specific manifestations and inheritance methods of salt culture in residents' lives. (3) Ecological and Environmental Research. Research should be conducted on the relationship between salt fields and the natural environment, exploring the impact of salt fields on the local ecological environment and the impact of the ecological environment on salt fields; at the same time, the sustainability of salt field production methods should be analyzed, and measures and suggestions for protecting the ecological environment should be proposed.

b. Promoting Academic Exchange and Cooperation

By strengthening international academic exchange and cooperation, understanding and recognition of Yanjing well salt culture can be enhanced, its international influence and visibility can be improved, and new impetus can be injected into the inheritance and development of local culture. (1) Organizing Academic Conferences and Seminars. Regular academic conferences and seminars on Yanjing well salt culture can be held, inviting domestic and international experts and scholars for exchange and discussion, sharing research results, promoting academic progress, and enhancing the academic influence of local salt culture. (2) Establishing Academic Research Institutions. Research centers or institutes for Yanjing well salt culture can be established, focusing on the research and inheritance of salt culture, attracting and cultivating professional talents, and promoting in-depth development of salt culture research. (3) Conducting International Cooperation and Exchange. Cooperation can be established with international academic institutions and researchers to jointly carry out research on Yanjing well salt culture, and by participating in international academic conferences and exhibitions, showcase research results and enhance the international visibility of Yanjing well salt culture.

c. Strengthening Cultural Inheritance and Education

By strengthening cultural inheritance and education, the diversity and uniqueness of Yanjing well salt culture can be maintained, national identity and cultural confidence can be enhanced, cultural exchange and mutual learning can be promoted, and professional talents and inheritors can be cultivated. (1) Protection of Intangible Cultural Heritage. Efforts can be made to apply for Yanjing salt culture as intangible cultural heritage, strengthening its protection and inheritance; encourage and support foreign tourists to participate in the inheritance activities of salt culture, such as holding traditional salt drying skills competitions and salt culture festivals. (2) Education and Training. Yanjing well salt culture can be included in the curriculum of overseas Chinese language schools, cultivating foreign students' awareness and interest in salt culture through classroom teaching and practical activities; cultural education and training activities for overseas tourists can also be carried out to enhance their understanding and knowledge of salt culture. (3) Integration of Cultural Tourism. The dissemination of Yanjing well salt culture can be integrated with the tourism industry, developing characteristic cultural tourism products and routes; through the promotion and publicity of cultural tourism, more overseas tourists can be made to understand Yanjing salt culture, playing their role in word-of-mouth dissemination.

In summary, the strategies and paths for the external communication of Yanjing salt culture should comprehensively consider various aspects including brand building, communication channels, cultural research and exchange, cultural tourism product development, and cultural heritage protection and inheritance. By implementing these strategies and paths, the visibility and influence of Yanjing salt culture can be effectively enhanced, promoting its inheritance and development.

4. Translation Model and Strategies for the International Promotion of Yanjing Well Salt Culture

In addition to the strategies and paths discussed above, the translation and interpretation of Yanjing salt culture is a key link in its international dissemination, particularly requiring enhanced quality assessment and the creation of high-quality translations. Therefore, to promote the internationalization of the local salt culture, we need to explore effective translation models and strategies to ensure the accuracy and quality of translation works; by creating high-quality translation works, we can enhance the international reputation and prestige of the salt culture.

1) Construction of Translation Theory Framework

The construction and application of the translation theory framework for the international promotion of Yanjing sell salt culture is a comprehensive process involving multiple aspects. This section aims to discuss this issue in detail, hoping to provide theoretical support and practical guidance for improving translation quality and enhancing the effectiveness of international communication.

As translation studies deepen, there are many theories available for external promotion translation, but given the particularity of Yanjing well salt culture resources, the composition of its theoretical framework is as follows:

a. Functional Equivalence Theory

Functional Equivalence theory emphasizes that translation should enable target language readers to achieve the same response as source language readers, that is, to achieve functional equivalence. On the one hand, this theory considers translation to be a dynamic process, where translators need to adjust according to the culture and context of the target language, making the translated content more vivid and close to the actual experience of the target audience, thereby enhancing the dynamism of the international promotion translation of Yanjing well salt culture. On the other hand, this theory encourages translators to find a balance between fidelity to the original text and adaptation to the target culture, enhancing the flexibility and creativity in the translation process.

b. Cultural Translation Theory

Cultural Translation Theory is a set of theoretical systems that study how to handle cultural differences and cultural elements between the source language and the target language during the translation process. It proposes the principle of "cultural fidelity", focusing on cultural transmission, adaptation, and representation in translation activities, and emphasizes maintaining the uniqueness and integrity of the source language culture. In the translation of Yanjing well salt culture resources, this theory can enhance the adaptability of translation, address cultural difference issues, and deepen cultural understanding, thereby accurately conveying its historical, artistic, and cultural values.

c. Skopos Theory

Skopos Theory emphasizes that translation should serve the purpose of the translation text and the needs of the target language audience, rather than simply language conversion. Therefore, translators should choose translation methods and strategies based on this. It adheres to the principle of purpose, coherence, and fidelity, breaking the traditional translation theory's notion of the supremacy of the original text, making translation more flexible and practical. In the translation of Yanjing well salt Culture resources, this theory can better adapt to the target culture, optimize translation decisions, enhance the targeting of translation, and improve the practicality of translation.

d. Multimodal Translation Theory

Multimodal Translation Theory is a theoretical framework that explores the interaction and joint construction of meaning of different modalities such as language, vision, hearing, and action in translation. This theory not only focuses on traditional language translation but also delves into the unique role and profound impact of non-verbal modalities such as images, sounds, and gestures in translation. In the international promotion translation of Yanjing well salt culture resources, this theory has explanatory, descriptive, and systematic characteristics and advantages, enhancing the accuracy and completeness of information, improving the accessibility of information, and adapting to digital trends.

e. Communicative Translation Theory

Communicative Translation Theory is a translation theory that takes the communicative function of language as its core. It emphasizes that translation activities are not only conversions between languages but also processes of cross-cultural communication and interaction. This theory mainly focuses on how translation achieves effective communication, ensuring the transmission and understanding of information in different linguistic and cultural contexts, including key points such as reader-centeredness, context consideration, information transmission and interpretation, and interactivity. For the international promotion translation of Yanjing well salt culture resources, it is of great significance in enhancing the acceptance of the target audience, improving the communicative efficiency of translation, and enhancing the interactivity of translation.

The above theories can provide important theoretical support and practical guidance for the international promotion translation of Yanjing salt culture, serving as important tools for its journey to the world, helping to achieve more effective cross-cultural communication and international promotion.

2) Principles of Theory Application

The application principles of Yanjing well salt culture resource external promotion translation theory framework cover various aspects such as accuracy, readability, cultural transmission, and purpose orientation, as follows:

a. Fidelity Principle

The fidelity principle in the external promotion translation process of Yanjing well salt culture requires translators to accurately and completely convey the information of the original text while maintaining the writing characteristics of the original text. (1) Content fidelity. The translation should accurately reflect the information in the original text, including the history, technical features, and related folk activities of Yanjing well salt culture, avoiding unnecessary additions, omissions, or alterations, maintaining the authenticity and integrity of the content. (2) Spiritual fidelity. The translation should capture and convey the cultural spirit carried by Yanjing well salt culture (including religious, philosophical, and aesthetic values, etc.), and reproduce the emotional colors and atmosphere of the original text, allowing readers in the target language to experience the emotional depth. (3) Linguistic style equivalence. The translation should also convey the aesthetic value and artistic expression of the original text, including ethnic style, era style, genre style, and the author's language style, etc., allowing readers in the target language to experience the charm of the original work and appreciate the artistic beauty of Yanjing well Salt culture[1].

The fidelity principle can maintain interlingual coherence between the original text and the translation, but it is important to note that it does not mean that the translation must correspond to the original text word for word.

b. Communicative Effectiveness Principle

The communicative effectiveness principle in the external promotion translation of Yanjing well salt culture mainly focuses on how to effectively convey information in the translation process and stimulate the interest and curiosity of the target readers, thereby promoting cultural exchange. (1) Clarity and explicitness. The translation should accurately convey the communicative intent of the original text, avoiding ambiguity; for cultural symbols and metaphors that are difficult to translate directly, appropriate processing should be done to ensure their transparency and clarity. (2) Linguistic fluency. The translation should use fluent, natural, and concise target language, avoiding awkward and obscure expressions, which helps to improve the reader's reading experience and interest. (3) Readability. The translation can be divided into reasonable paragraphs, clear logical structure, and vivid descriptions to resonate with readers and make them feel identified, thus better absorbing and digesting the information.[2]

By following these communicative effectiveness principles, the external promotion translation of Yanjing well salt culture can communicate more effectively with international audiences, achieve the purpose of cross-cultural exchange, and enhance its visibility and influence globally.

---

1 Wang Shijia, Huang Chengxiang. Aesthetic Reproduction in Prose Translation from the Perspective of Translation Aesthetics: A Case Study of Chen Dewen's Translation of "Thousand Songs River Charm" [J]. *Modern Linguistics*, 2024, 12(6).

2 Chen Rui. Report on the Practice of Selected Articles Translation in "Translation Quarterly" [D]. Southwest University of Finance and Economics, 2014.

c. Cultural Adaptability Principle

The cultural adaptability principle in the external promotion translation of Yanjing well salt culture mainly refers to the need to fully consider the characteristics and needs of the target language culture during the translation process, so that the translation can adapt and integrate well into the target language cultural environment. (1) Cultural sensitivity. The translation process should be highly vigilant and have a deep understanding of the differences in values, social customs, religious beliefs, historical traditions, etc., contained in the source and target languages, avoiding content that may cause misunderstanding or offense. (2) Context consideration. Translation should consider the context and usage habits of both the source and target languages, ensuring that while maintaining the contextual meaning of the source language, it is adapted to the target culture, making the translation natural and appropriate in the target culture.[1] (3) Conversion of cultural elements. For certain specific cultural elements that are difficult to translate directly, the elements and expressions of the target culture can be cleverly used for transmission or replacement, allowing the target audience to experience similar cultural feelings.

The core elements of cultural adaptability in external promotion translation, as mentioned above, help ensure that the translation work accurately conveys the essence of the original culture and has a positive impact on the target readers.

d. Reader-Centered Principle

The reader-centered principle in the external promotion translation of the Yanjing well salt culture mainly focuses on the needs, preferences, and cultural backgrounds of the target language readers, ensuring that the translation work can resonate with them. (1) Audience analysis. Before translation, a thorough analysis of the target language readers should be conducted to understand their interests, needs, and cognitive characteristics, in order to better position the translation content and meet their reading expectations and preferences. (2) Targeted translation. During the translation process, based on the needs and background of the target language readers, select relevant and attractive information and make appropriate adjustments and optimizations to the translation content, making it closer to their reading

1 Intercultural Translation: Cultural Adaptation and Transformation, https://theory.gmw.cn/2024-04/13/content_37261249.htm.

habits. (3) Interaction and feedback. Encourage the target audience to participate in interactions and communications through social media platforms, and establish feedback channels to collect feedback from the target audience on the translation work, which can be used to improve future translation practices[1].

By following the reader-centered principle, the external promotion translation of Yanjing well salt culture can touch the emotions of the target audience and enhance their interest and identification with Yanjing well salt culture.

In summary, the external promotion translation of Yanjing well salt culture requires translators to have a solid language foundation, rich cultural background knowledge, and good aesthetic literacy. These require translators to continue learning, accumulating, and practicing in order to better contribute to the international dissemination of Yanjing well salt culture.

3) Methods of Translation Practice

The practice of external publicity translation for Yanjing salt culture is a complex process involving cultural dissemination and language conversion. To ensure the accuracy, fluency, and cultural adaptability of the translation, here are some suggested translation methods:

a. Thorough Preparation and In-Depth Research

In the practice of international promotion translation of Yanjing well salt culture, thorough preparation and in-depth research are key steps to ensure the quality and effectiveness of the translation.

The former mainly includes: (1) Understanding the background knowledge of Yanjing well salt culture, such as its history, culture, technical style, and its influence domestically and internationally, in order to better comprehend the original text and grasp its cultural connotations and artistic features. (2) Mastering professional terminology and unique expressions to ensure the accuracy and professionalism of the translation. (3) Clarifying the purpose of translation and the needs of the audience, which helps in selecting appropriate translation strategies and methods to meet the audience's needs and expectations. (4) Preparing translation tools and

---

1 New Explorations in Translation and Social Media Studies--A Review of "Translation and Social Media: Theory, Teaching, and Practice", http://www.fbxslw.com/wslw/29672.html.

resources, such as dictionaries, translation software, and parallel texts, which aid in more efficiently completing translation tasks.

The latter mainly includes: (1) Analyzing the context and style of the original text, which helps to better understand the meaning and expression of the original text, thereby choosing more suitable translation strategies and methods. (2) Studying the expressive habits of the target language, such as grammatical structure, vocabulary selection, and sentence characteristics, which helps to improve the fluency and readability of the translation. (3) Comparing the differences between Yanjing well salt culture and other salt cultures can highlight its uniqueness, helping to enhance the appeal and persuasiveness of the translation. (4) Exploring the cultural connotations and artistic features and giving them full display, which helps to promote the cultural value and artistic taste of the translation. (5) Referring to relevant research results and materials, such as academic papers, monographs, and exhibition materials, can provide beneficial references, helping to more accurately understand the original text and translation content.

b. Integration of Theory and Practice

In the practice of international promotion translation of Yanjing well salt culture, integrating theory with practice is crucial. This not only helps to enhance the accuracy of the translation but also ensures that the cultural connotations of the original text are properly conveyed. Here are some specific strategies and methods: (1) Constructing a dual-subject mechanism. In the practice of international promotion translation of Yanjing well salt culture, constructing a dual-subject mechanism between translation theory and practice requires translators to place both on an equal footing and fully recognize the dialectical relationship between them. It is important to value both the study and application of translation theory and the importance of translation practice. (2) The guiding role of translation theory. It can help translators clarify the direction and objectives of translation, preventing them from losing their way during the translation process. It also provides translators with ideas and methods for solving problems, helping them find solutions when encountering difficulties. (3) Feedback from practice to theory. Through practice, translators can continuously accumulate experience and discover the insufficiencies and limitations of translation theory in practical application. This helps translators continuously improve their translation methods and strategies, enhancing the quality of translation.

c. Quality Control and Review

To ensure the accuracy and communication effectiveness of the international promotion translation content of Yanjing well salt culture, it is necessary to establish a strict quality control process and review process, which mainly includes: (1) Preliminary review. Completed by professional translators, it checks the accuracy, fluency, and cultural adaptability of the translation, focusing on its basic quality, ensuring there are no obvious errors or omissions. (2) Expert review. Experts or scholars in the field of well salt research are invited to review and provide professional opinions and suggestions to ensure that the translation meets the standards and requirements of the field. (3) Language review. Completed by senior language experts, it checks the accuracy and fluency of the translation to ensure it conforms to the expressive habits and grammatical rules of the target language. (4) Final review. Completed by the project leader or translation team leader, it ensures that the translation meets the project requirements and the needs of the target audience. In the final review stage, attention should be focused on the overall effect and communication effect of the translation content, enabling it to accurately convey the value and charm of Yanjing well salt culture.

d. Continuous Learning and Improvement

In the practice of international promotion translation of Yanjing well salt culture, continuous learning and improvement allow translation staff to adapt to changes and enhance their professionalism and industry competitiveness. (1) Content of continuous learning includes language skills such as listening, speaking, reading, and writing in the target language, professional knowledge such as the latest developments and research results in the field of Yanjing well salt culture, translation skills such as interpretive theory and functional equivalence theory, and cross-cultural communication abilities for appropriate cultural adjustments. (2) Ways of continuous learning include participating in professional translation training courses and training courses in the field of Yanjing well salt culture research, reading professional literature and materials in the field of Yanjing well salt culture research, actively participating in translation practice projects and communicating and discussing with peers, and using online resources for self-study and online communication. (3) Strategies for continuous improvement include setting clear learning goals and plans based on one's actual situation and needs, maintaining enthusiasm and interest in learning through a positive attitude and values, continuously reflecting and summarizing experiences and lessons in the learning and practice process, and actively seeking feedback and guidance from peers or experts to understand one's own deficiencies and areas for improvement.

In summary, the practice of international promotion translation of Yanjing well salt culture needs to follow certain principles and methods. Through scientific translation practice, the unique charm of Yanjing well salt culture can be accurately conveyed, attracting more visitors to experience and promoting the prosperity of local cultural tourism.

Made in United States
Orlando, FL
04 November 2026

90200394R00260